A CULTURAL HISTORY OF PLANTS

VOLUME 2

A Cultural History of Plants
General Editors: Annette Giesecke and David J. Mabberley

Volume 1
A Cultural History of Plants in Antiquity
Edited by Annette Giesecke

Volume 2
A Cultural History of Plants in the Post-Classical Era
Edited by Alain Touwaide

Volume 3
A Cultural History of Plants in the Early Modern Era
Edited by Andrew Dalby and Annette Giesecke

Volume 4
A Cultural History of Plants in the Seventeenth and Eighteenth Centuries
Edited by Jennifer Milam

Volume 5
A Cultural History of Plants in the Nineteenth Century
Edited by David J. Mabberley

Volume 6
A Cultural History of Plants in the Modern Era
Edited by Stephen Forbes

A CULTURAL HISTORY OF PLANTS

IN THE POST-CLASSICAL ERA

VOLUME 2

Edited by Alain Touwaide

BLOOMSBURY ACADEMIC
LONDON • NEW YORK • OXFORD • NEW DELHI • SYDNEY

BLOOMSBURY ACADEMIC
Bloomsbury Publishing Plc
50 Bedford Square, London, WC1B 3DP, UK
1385 Broadway, New York, NY 10018, USA
29 Earlsfort Terrace, Dublin 2, Ireland

BLOOMSBURY, BLOOMSBURY ACADEMIC and the Diana logo
are trademarks of Bloomsbury Publishing Plc

First published in Great Britain 2022
Paperback edition published in 2025

Series design: Raven Design
Cover image © Bibliotheque Nationale, Paris, France,
De Agostini Picture Library/J. E. Bulloz/Bridgeman Images

A catalogue record for this book is available from the British Library.

Library of Congress Control Number: 2021932844

ISBN: PB: 978-1-3505-5059-9
Set: 978-1-3505-5064-3
ePUB: 978-1-3502-5929-4
ePDF: 978-1-3502-5928-7

Series: The Cultural History Series

Typeset by Integra Software Services Pvt. Ltd.
Printed and bound in Great Britain

CONTENTS

ILLUSTRATIONS

SERIES PREFACE

The connectedness of humans to plants is the most fundamental of human relationships. Plants are, and historically have been, sources of food, shelter, bedding, tools, medicine, and, most importantly, the very air we breathe. Plants have inspired awe, a sense of well-being, religious fervor, and acquisitiveness alike. They have been collected, propagated, and mutated, as well as endangered or driven into extinction by human impacts such as global warming, deforestation, fire suppression, and over-grazing. *A Cultural History of Plants* traces the global dependence of human life and civilization on plants from antiquity to the twenty-first century and comprises contributions by experts and scholars in a wide range of fields, including anthropology, archaeology, art history, botany, classics, garden history, history, literature, and environmental studies more broadly. The series consists of six illustrated volumes, each devoted to an examination of plants as grounded in, and shaping, the cultural experiences of a particular historical period. Each of the six volumes, in turn, is structured in the same way, beginning with an introductory chapter that offers a sweeping view of the cultural history of plants in the period in question, followed by chapters on plants as staple foods, plants as luxury foods, trade and exploration, plant technology and science, plants and medicine, plants in (popular) culture, plants as natural ornaments, and the representation of plants. This cohesive structure offers readers the opportunity both to explore a meaningful cross-section of humans' uses of plants in a given period and to trace a particular use—as in medicine, for example—through time from volume to volume. The six volumes comprising *A Cultural History of Plants* are as follows:

Volume 1: *A Cultural History of Plants in Antiquity* (*c.* 10,000 BCE–500 CE)
Volume 2: *A Cultural History of Plants in the Post-Classical Era* (500–1400)
Volume 3: *A Cultural History of Plants in the Early Modern Era* (1400–1650)
Volume 4: *A Cultural History of Plants in the Seventeenth and Eighteenth Centuries* (1650–1800)
Volume 5: *A Cultural History of Plants in the Nineteenth Century* (1800–1920)
Volume 6: *A Cultural History of Plants in the Modern Era* (1920–present).

By way of guidance to our readers, it should be noted that the plant names used in these volumes accord with those in the fourth edition of *Mabberley's Plant-book* (Cambridge University Press, 2017). When they are discussed, individual plants are identified using their common names and, at their first mention in each chapter, with their scientific names: e.g. bay laurel (*Laurus nobilis*). As is recommended for general works such as this, the authorities to whom the scientific names are attributed (e.g. *Laurus nobilis* L., where L. identifies Linnaeus as the identifying authority) have been omitted.

Annette Giesecke and David J. Mabberley,
General Editors

Introduction

ALAIN TOUWAIDE

The present volume covers the period 500–1400 CE, which is traditionally identified as the Middle Ages, and each chapter focuses on a particular region and cultural group (or groups) during this period. Both this span of years and this conventional designation call for an explanation. Neither 500 nor 1400 correspond to landmarks in European or World history, and the very name *Middle Ages* implies a *transitional* nature between two poles, negating the possibility of specificity during said period.

DEFINING THE TIME PERIOD

Nevertheless, it is the case that the years around the year 500 CE issued in some of the trends that would shape the general character of the period under consideration here: King Clovis (*c.* 466–511) united all Frankish groups and founded the Merovingian Dynasty; Danish populations moved to Jutland; the Ostrogoth king Theodoric (*c.* 454–526), who was established in Ravenna, visited Rome; Hun tribes occupied Penjab and the area of the Indus; Buddhism spread to southern China, and Buddhist art expanded to China, Korea, and Japan. As for the years *c.* 1400, these hint at the end of an epoch and a deep transformation of existing geo-political and cultural environments: the Byzantine scholar Manuel Chrysoloras (*c.* 1350–1415) was teaching Greek in Florence in a reversal of the relations between East and West, and the Byzantine Emperor Manuel II (1350–1425) was in the West, seeking assistance against the Turks; in Asia Minor, the Turco-Mongol khan Timur (1336–1405), best known as Tamerlane, took the city of Sivas from the Ottomans, and raided Syria before entering into an agreement with Byzantium; and, to mention just a few such projects, the construction of Pavia Chartreuse and Anvers Cathedral started.

Assuming that the period considered here can be defined with precision, one would think that its ending year should be 1492 rather than 1400, that is, the arrival of Christopher Columbus (1451–1506) to what became since the *Americas*. Columbus' exploration did indeed reveal a world to the West that *ipso facto* became the New World, not only having populations that differed greatly from Old World peoples and their cultures, but also evidencing a different aspect of nature, with "new," previously unknown plants and animals, and even new skies. This enlargement of the world and, by way of consequence, of the European experience was confirmed by the subsequent voyages of Christopher Columbus and further pursued by the Portuguese Vasco de Gama (*c.* 1460–1524), who reached the land of spices (India) in 1499 after two years of navigation.

At the same time, however, this expansion went together with a reduction that is less frequently viewed as a historical milestone in European history: Granada, the capital of the Nasrid kingdom and the last Arabo-Islamic bastion in Andalusia (Islamic southern Spain), was conquered by the Christians, and the Jews were expelled from the kingdoms of Castile and Aragon by the Catholic Monarchs Isabella I of Castile (1451–1504) and Ferdinand II of Aragon (1452–1516). These two events took place in the same year (1492), in a simultaneity that was not coincidental. The Fall of Granada put an end to the presence of the Arabo-Muslim culture in Europe, which had produced some of the most achieved accomplishments of the Arab world; suffice to think of the Alhambra. The expulsion of the Jews from Spain meant the end of the so-called *convivencia*, that is, the productive co-existence of the three communities living in the Iberian Peninsula defined by their religion: the Christians, the Muslims, and the Jews. These two facts resulted in a contraction of Europe, whereas the discoveries opened Europe onto new horizons.

Epoch-making and dramatic events such as these were not unprecedented in the fifteenth century, however. Less than fifty years earlier the Byzantine Empire fell. On May 29, 1453, the troops of the Ottoman sultan Mehmed II (1432–81), which had already unsuccessfully besieged Constantinople, eventually took the city. Mehmed II, identified as "the Conqueror," entered the Agia Sofia cathedral on his horse, putting an end to Byzantium. It was the result of centuries of shrinking of the Byzantine world and the beginning of an emigration of Greeks, who arrived first to Italy, sometimes passing through Crete, and often carried on through Europe, mostly France and Spain, spreading Greek culture.

This conjunction of events meant the end of an epoch, which might seem at first glance to have re-centered the Mediterranean world on itself, quite in opposition to the implication of New World discoveries. These events evoke similar ones, that were equally ominous: the Fall of the Roman Empire in 476. That year, the Roman Emperor Romulus Augustulus (d. after 476 CE) was overthrown by the leader of a group of Germanic troops, Odoacer (435–93 CE). Possibly of Hunnic origin, Odoacer had been crowned by his troops and was ruling over Italy until he was murdered in 493 by Theodoric. This cascade of events put an end to the Roman Empire, which was born as a unique, universal entity from the transformation of the Roman Republic in 27 BCE. In the fourth century, however, it was divided into the Western and the Eastern Empires. With the fall of the Western Empire, the Empire of the previous centuries was reduced to the Eastern Empire, that is, Byzantium.

The year 476 and the decades 1453–92 might better define the beginning and the end of the period under consideration here, as they were characterized by two moves of the center of gravity of the Old World: first a shift of the old Roman Empire from the western Mediterranean world (Rome) to the eastern Mediterranean (Constantinople) with the fall of the Western Empire, and then a return of the pendulum from the eastern Mediterranean to the West, with the fall of Constantinople and the Eastern Empire. This double movement "consecrated" the rise of what had become Europe in the meantime. Furthermore, each of these two moves was accompanied by an affirmation of Christianity, which had first been embraced in the fourth century by the founder of the Byzantine Empire, Constantine I (*c.* 272–337 CE), and became the official religion of the new state; in the fifteenth century, Christianity was explicitly reinforced by the fall of the Muslim kingdom of Granada and, at the very least by default, by the expulsion of the Jews from Spain.

It thus seems that the period covered by this volume would be best delimited by the major transformations announced by the year 476 and the decades 1453–92, rather than by exact dates not necessarily as eventful and epoch-making as a segmentation of history might suggest. In any case, years and decades should not be taken as milestones, but as grey zones of transition announcing deep mutations with all their implications, including contradictory trends, uncertainties, and challenges.

A MIRROR OF HISTORY

Interestingly enough, the periods identified by the years and decades above correspond to epoch-making transformations of the book. At some point between the third to fifth centuries CE, the material used to make books, papyrus (*Cyperus papyrus*), was replaced by parchment. Being more flexible than papyrus, parchment limited damage that resulted from reading books on papyrus. Papyrus rolls had to be unrolled with the right hand and rolled back up again with the left hand. After a roll had been read, it needed to be unrolled and rolled back again, but in the opposite direction, so as to have the beginning of the text at the beginning of the roll. These manipulations led not only to abrasion of the surface, but also to possible damage of the papyrus fibers, which sometimes broke and resulted in losses of text. Parchment compensated for these problems thanks to its flexibility. Furthermore, parchment could be covered with writing on both sides, unlike papyrus. This two-side use of the medium (in the way of the paged book) resulted in a better economy of book production (since the same quantity of medium could contain twice the quantity of information of a papyrus roll). This two-side use also transformed the technique of book-making with an epochal transformation from roll to codex, that is, from reel to the book in the shape that we still use.

This major modification of the book at some point in the beginning of our period was echoed by another during the decades 1453–92: printing with movable type. The German goldsmith Johannes Gutenberg (1398–1468), living and working in Mainz (Germany), developed the technique of movable metal type pieces, thus becoming the Western world's first printer. He was not the first in human history, however, as printing with movable type (though of a different kind) had already been developed in China centuries before. Returning to Gutenberg, in 1454 or 1455, he started producing the first printed copies of the Bible, the few remaining copies of which are currently identified as the *Gutenberg Bible*. Until then, the reproduction of texts was accomplished by hand on the part of copyists, who, contrary to stereotype, were not always monks but professionals specializing in this activity. However trained and efficient they might have been, they worked slowly and were always at risk of unavoidable mistakes inadvertently made even by the most attentive and trained professionals. In addition to generating copies that were all identical, printing reduced the time necessary to reproduce any text once the small individual metal letters (type pieces) were set and several lines were placed in a wooden frame (galley). Beyond accelerating production and, consequently, increasing diffusion as it has been claimed, printing transformed reading and generated a new culture which benefited science, including plant knowledge.

AN INTER-CONNECTED WORLD

Between these two major transformations that contribute to delimiting the period considered here, there was another important modification of the book that is relevant for the reconstruction of the history of botanical knowledge and culture. At some point during the eleventh century, parchment started to be replaced by a medium that was totally new: paper. The significance of the introduction of this new material for books is not so much—from our viewpoint here—its nature, the material it was made of, or its production process (both of which were radically different from those of parchment and actually did revolutionize book production), but its origin. Paper reached the Mediterranean through the Arab world, where it had already been used for some time. The Arab world's source of paper, in turn, was China, where it had been produced as early as the first century CE. Originally made of recycled plant fiber, paper was further developed and was available in different forms and types. Under the Tang Dynasty (618–907 CE), for example, Chinese paper-makers were able to produce different qualities, some of the highest quality, including one that was snowy white. This was made of mulberry (*Morus alba*) grown in a rich soil and prepared with pure water. From China, paper-making reached the Arab world. As the legend goes, the Arabs defeated Chinese troops in Central Asia. Some among the captives they took were paper-makers who started producing paper in conquered lands, specifically in Samarkand. The facts were more complex. Paper was known in the region of Samarkand as early as the eighth century, where it might have been brought from China by Buddhist monks. From there, it was diffused to the eastern Arab world. It then reached the western Arab world, that is, Andalusia. Its production expanded from there to southern France and, further, to Italy, which became the major paper provider of the Middle Ages.

The interest of the story is that paper works as an indicator of contacts and exchange between different worlds during the period covered by the present volume. Whereas our delimitation of the period (from 476 to 1453–92) implied a focus on the Mediterranean—even though the pendulum oscillated first toward the eastern Mediterranean and then returned to the West—the world of that time was by no means confined to the Mediterranean. It extended from the Atlantic and the Iberian Peninsula to China and the Pacific Ocean. As the history of paper indicates, many of the different societies between these two poles were interconnected. Although they did not necessarily know each other, these societies had access to products from one another. Silk was another such product. Already known in the Roman Empire in the first century CE—and even so much that an edict had to restrict its acquisition to stop the financial hemorrhage its purchase created—it was produced in Byzantine territory as early as the sixth century CE thanks to a Byzantine mission to China. This connected world covered a vast geographic area. The land road from the easternmost point of China (Fuyuan County in the current Heilongjiang Province) and Lisboa totals over 7,800 miles (12,500 km). It was a mosaic of juxtaposed—and sometimes opposed—empires, the history of which is of interest here.

RISE AND FALL OF EMPIRES

This history was that of changing empires, with lifecycles characterized by rises and declines, sometimes phases of revival, and also an exceptionally short life in some cases. The Byzantine and the subsequent Ottoman Empire were truly exceptional: Byzantium existed for over one millennium (330–1453), and the Ottoman Empire for more than six hundred years (*c.* 1299–1922), even though both went through several phases of crisis, with ups and downs that sometimes threatened their very existence. Going beyond longevity and lifecycle, the most salient characteristics of the world from China to the Iberian Peninsula were, beside interconnectedness, repeated gains and losses of territories, constant attempts to reconstruct the totality of the past, and a search for universalism.

Although the history of all the societies included in the vast area from the Pacific westward to the Atlantic during the period from 476 to 1453–92 deserves attention, we shall focus on the Mediterranean world and the Near East, doing so without losing sight of the backdrop of Eurasia and Eastern Asia. This period was marked by a recurrent, if not omnipresent ideology of universalism coined by the Roman Empire at its maximal extension from Britain to deep into Syria, and from Germany to northern Africa. After the division of the Roman Empire into the Western and the Eastern Empires and the foundation of Constantinople, the Visigoths raided Italy and sacked Rome in 410; they then invaded Spain in 416–18 and settled in in southern France before returning to Spain. Under the leading of Attila (*c.* 406–53), the Huns moved westwards through Germany, and invaded Gaul. They were defeated by a coalition of German and Roman troops in 451 and then turned to Italy, which they raided in the north. Going south they arrived roughly 60 miles (100 km) north of Rome, where they surprisingly stopped in 452, without conquering Rome. Further south, Italy was in the hands of the Ostrogoths, before being invaded by the Lombards in 568, who remained settled in the Peninsula until the eleventh century.

While the West was going through these waves of upheaval, the Byzantine Emperor Justinian (482–565) recovered parts of Spain, Italy, and northern Africa, that is, many of the Mediterranean territories of the Roman Empire at the time of its greatest expanse. This partial reunification of the Western and Eastern Empires was shortly thereafter followed by territorial losses. In the East, Byzantium fought the neighboring Sassanian Empire, which it defeated in 628. Immediately afterwards, it was attacked by the Arabs who launched a vast and astounding campaign of conquest of the Near East, taking Damascus as early as 635 and Alexandria in 641. Twenty years later (661), they established the capital of the newly emerged Arab Empire in Damascus, with the Umayyad Dynasty. They pursued their expansion of the West, capturing Cordoba in 711 and Toledo, the capital of the Visigothic kingdom, the same year. From Spain, they went north, up to France, where they were stopped in 732. At the end of the century, Charlemagne (742–814) was consecrated Emperor of the Romans by the pope at a particularly significant date: Christmas of 800. He was the first emperor ruling over the West after the division of Roman Empire and the overthrow of Romulus Augustulus. He was not followed by any other until Charles the Fifth (1500–58), who reunited most of Europe through his parents.

In the Arab world, the dynasty of the Umayyad was overthrown by the Abbasids in 750. In Spain, however, a member of the Umayyad Dynasty who escaped massacre by the Abbasids seized power and installed the dynasty in Cordoba. This Emirate took its independence in 929, when the Emir proclaimed himself Caliph. In the East, the Abbasids created a new capital for the Empire in Baghdad *c.* 762. This was the beginning of the Golden Age of the Arab world. Byzantium, also, was at the eve of a period of renewed splendor even though it was losing ground in Italy. From the very end of the tenth or the beginning of the eleventh century the Peninsula was invaded by the Normans who successfully established a kingdom on the continent, and also took Sicily from the Arabs in 1091.

As often in history, decadence followed shortly thereafter. In Spain, the Caliphate of Cordoba disintegrated in 1031. It was fragmented into several smaller, independent kingdoms, among them that of Granada, established in 1230. In the East, the Seljuk Turks were attacking the Byzantine Empire. They were not stopped by the Byzantine army and won a significant victory in 1071 at the so-called Battle of Manzikert (in Eastern

Anatolia, now Eastern Turkey). From there they rapidly invaded Asia Minor and created the Sultanate of Rum with its capital in Nicaea, on the southern side of the Bosphorus, across from Constantinople, which they threatened.

The thirteenth century was a period of profound transformations. In 1204 the Western Crusaders on their way to Jerusalem stopped in Constantinople and seized it. They established the Latin Kingdom of Constantinople, which was short-lived, however: in 1261 the Byzantines reconquered their capital. The Arab world, too, was the victim of attacks from outside and, in 1258, its capital Baghdad was conquered by the Mongols. Under the leadership of Gengis Khan (1162–1227), the Mongols would, in fact, invade most of the space from the Pacific Ocean to the Mediterranean, thus creating a vast empire that was essentially monolithic, while at the same time being divided into smaller khanates.

Nicaea, lying opposite Constantinople, was taken from the Seljuks in 1331 by the Turkish Ottomans who expanded into the Balkans, increasing the pressure on Constantinople. In Iran, Timur (1336–1405), known also as Tamerlane, reunified the several minor polities born from the fragmentation of the area and reigned on the territory from Eastern Turkestan and northern India to Ottoman Anatolia. After his death, Persia was briefly occupied by the Turkomans, and the Ottomans captured Constantinople in 1453, putting an end to the Byzantine Empire.

Fratricidal conflicts, infighting, overthrow of rulers, invasions, and raids, in addition to wars, occupations, military coups, and other major geopolitical reversals should not overshadow the interconnectedness of the several powers mentioned above. Interconnections resulted from the hiring of foreign mercenaries; embassies between states; political marriages between reigning houses; bi-, if not multilingualism (not only among the populations at the borders of states, but also across regions and states); permeability of frontiers; trade exchanges; missions of evangelization and conversions (voluntary or forced); inter-faith dialogue; Romanization or Arabization of the state and its administration; *linguae francae*; emigration; and border-crossing processes of all kinds (including foreign artists taking advantage of patronage and offering their services to reigning dynasties). All such processes might have been less visible, but were nonetheless real. Botany was characterized by multidirectional exchanges across time and space.

OLD WORLD BOTANY IN THE NEW WORLD

The case of the New World is significant from the vantage point of such multidirectional exchange. Although the relevant documentation is later than the time period under consideration here, it provides interesting information about approaches to the plant world in the West during our period. The first report of New World flora was by the Spaniard Gonzalo Fernández de Oviedo y Valdés (1478–1557). Entitled *De la natural hystoria de las Indias* (On the Natural History of the Indies) and published in Toledo (Spain) in 1526, it included descriptions of plants (fols xxxiiii recto, chapter lxii–xliii verso, chapter lxxx), together with their uses, frequently making comparisons with the natural world of Spain.

The so-called *Codex Badianus* had different origins. It is a manuscript that was at the Vatican Library, classified as MS Barberini *Latinus* 241, until it was returned to Mexico in 1990 by Pope John Paul II. It is now at the Biblioteca Nacional de Antropologia y Historia (National Library of Anthropology and History) in Mexico City. The *Codex* contains what the European scientists of that time called a herbal, that is, a book on medicinal plants. It resulted from the collaboration of two native Nahua individuals: one of them was a healer (who spoke the local native language, Nahuatl) and the other a translator (from Nahuatl into Latin). Both were identified with Spanish names: the healer was Martin de la Cruz (whose years of birth and death are unknown) and the translator was Juan Badiano (1484–after 1552), after whom the manuscript (codex) referred to here (*Codex Badianus*) is traditionally identified. Their work was commissioned by Francisco de Mendoza (1508–66), the son of the Viceroy of New Spain, and it was intended to be offered to the Spanish royal family. The title of the manuscript (in Latin) is notable. The English translations here reflects the original language:

> Short treatise on the medicinal plants of the Indians, that some Indian, physician of the Holy Cross College, compiled, a doctor without theoretical learning, taught solely by experience. In the year 1552 of Salvation.

This is not what we would now call ethnobotanical fieldwork, conducted by an observer collecting information from local practitioners. It is a transcription by natives of their own knowledge. The text is illustrated with color representations of plants that are not naturalistic, but nevertheless allow for an identification. Even though Martin de la Cruz was not a doctor according to the definition of that time, he proceeded methodically, unless it is the case that he was instructed by the translator how he needed to list the plants. Plants are grouped in a way that is reminiscent of the ancient and medieval practice, namely by "affected" organs or major categories of medical conditions, arranged almost from head to toe (*a capite ad calcem*), followed by more general and specific conditions: head affections; eyes; ears; nose; mouth and teeth; chest and stomach; bladder, urination, gout, feet; blood, fever, and general afflictions; epilepsy and other afflictions; obstetrics and gynecology; neonatology; signs of death. Although this might have been a taxonomy deriving from the local practice, it might also hint at what we could call a "contamination" of local information by the Western system of classification of medical conditions and, on this basis, of medicinal plants. It is probably significant that Juan Badianus de la Cruz frequented the Colegio de Santa Cruz de Tlatelolco (The Holy Cross College of Tlatelolco) in Mexico City. This contamination might have been imposed onto local medicine as a result of the conviction that local culture was inferior to that of the Conquistadores as the following lines in the text of the *Codex Badianus* might suggest (f. 1v):

> ... you will recollect that we, poor unhappy Indians, are inferior to all mortals, and, for that reason, the poverty and insignificance that is in us by nature, merit your indulgence.

This possible process of contamination recurred in subsequent works. Here, however, it took a different form: instead of resulting from the imposition of a medieval taxonomic system onto local nature, it was projected by the Spaniards themselves onto that nature. And it reflected the Spaniards' botanical knowledge. The first of these contaminated works is the *Historia general de las cosas de la Nueva España* (General History of the Things of New Spain) by Bernardino de Sahagún. The work is known through a manuscript traditionally identified as the *Codex Florentinus* (Florentine Manuscript) consisting of three large volumes preserved at the Biblioteca Medicea Laurenziana in Florence (MSS. Palatino 218–20) (see Figure 0.1).

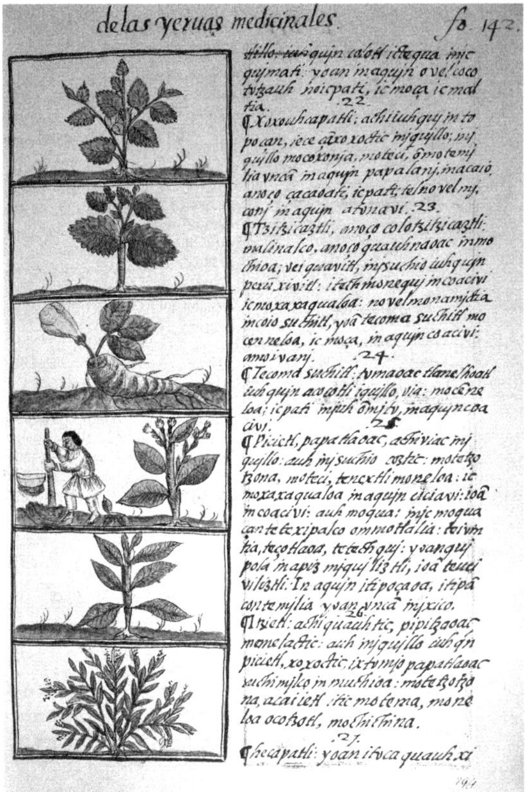

FIGURE 0.1 Medicinal plants, text in Nahuatl, from the manuscript *General History of the Things of New Spain*, also known as the Florentine Codex, f. 142r, by Bernardino de Sahagún. Photo by DeAgostini. Courtesy of Getty Images.

Bernardino de Sahagún (1550 [or before]–90) was a Franciscan friar who went as a missionary to New Spain (that is, present Mexico) in 1529. Besides evangelizing the natives, he conducted true botanical fieldwork from 1540 to 1577. He learned Nahuatl and translated the Psalms, the Gospels, and a catechism into Nahuatl. In his *Historia general*, written in Nahuatl and Spanish in two facing columns, he did not limit himself to an account of the conquest of the New World, but he also included a vast natural-history report as he himself stated in the introduction (Florence, Biblioteca Medicea Laurenziana, MS Palatino 218, f. 1r):

> I, friar Bernardino de Sahagún … did write [these] books on the matters of this New Spain, divine, … human, and natural …

The work had a strange story: after it reached Madrid sometime around 1580, it was condemned by the Inquisition because it described pagan rites. Its manuscript arrived in the hands of Ferdinando de' Medici (1549–1609), the son of the Grand Duke of Tuscany, Cosimo I de' Medici (1519–74), and Eleanor of Toledo (1522–62). With him it arrived in Florence in 1588. Since then, it has been known as the *Codex Florentinus* (Florentine Manuscript).

The *Historia general* is composed of twelve books, the eleventh of which is devoted to natural history. Divided into thirteen chapters, the eleventh book opens with the animals, before progressing to the plant world. At this point it proceeds by successive divisions, with the trees, including timber, arboriculture, and fruits; edible roots, followed by herbs, with several further subdivisions defined by the plants' effects, nature, or uses: psychotropic plants, mushrooms, plants comestible when boiled, herbs eaten raw, medicinal plants (not only herbs, but also trees, with 142 species according to the numbering in the *Codex Florentinus*), including some minerals, scented plants, plants neither edible nor medicinal, flowers and floral arrangements.

Interestingly enough, Bernardino de Sahagún, who might be identified as a precursor of modern ethnobotanists, identified his sources, that is, the natives from whom he collected the information compiled in the *Historia general*. They can be compared to modern informants. All were identified by their names and the church of their neighborhood (Firenze, Biblioteca Medicea Laurenziana, MS Palatino 220, ff. 180v–181r).

Through the taxonomy he used, Bernardino de Sahagún revealed his botanical education. He first structured the natural world in three parts (animal, plant, mineral) and he then divided the plant world on the basis of different classificatory parameters in combination: dimension, nature, and uses (the latter including preparation). Both classifications were not original: they date back to Classical antiquity, specifically the Greek Theophrastus of Eresos (*c.* 371–*c.* 287 BCE), identified in Western scholarship as the Father of Botany, and Dioscorides (first century CE), who can be granted the title of Father of Pharmaco-therapeutics. In *Historia plantarum* (Enquiry into Plants), Theophrastus divided the world of plants into four major categories defined by the size of the plants, from tall trees to small herbs. In *De materia medica* (On the Natural Substances to be Used for the Preparation of Medicines), Dioscorides divided the natural world in three major categories (plant, animal, and mineral), although he did not present them in the same order as Bernardino de Sahagún. Rather, he divided the plant kingdom into two parts, with the colored and scented plants followed by the trees and fruit trees on the one hand, and all the other vegetation on the other hand. Between the two groups he inserted the animals. The presence of these two taxonomic systems in the *General History of the Things of New Spain* attests to Bernardino de Sahagún's classificatory intervention in the arrangement of the material he collected. It also reveals his botanical and natural-history education, which appears to be typical of the teaching of sciences in medieval universities.

The case of the material collected in the New World by Francisco Hernández de Toledo (*c.* 1514/17–87) is more difficult to interpret because Hernández's work was repeatedly edited. Hernández was a physician with a doctoral degree earned at the University of Alcalá (Alcalá de Henares, close to Madrid). In 1567 he became *protomedico* (Chief Physician) of the king of Spain, Philip II (1527–98). Three years later, the king appointed him chief medical officer of the Indies, with a very specific mission: studying the plants, herbs, and medicinal seeds of the Indies; consulting doctors, medicine men, herbalists, and any other person who had understanding and knowledge of the matter; reporting the uses of plants in practice, their powers, and the quantities administered, in addition to their cultivation, habitat, and varieties; and also personally experimenting and testing all this material. In the Indies, Hernández collected a vast quantity of material, which he brought back to Spain in 1577.

Like his predecessors, Hernández was familiar with the scientific and botanical literature of Classical antiquity. As early as 1560, he started working on an edition and Spanish translation of the *Naturalis Historia* (Natural History) by the Roman encyclopedist Pliny

(23/24–79 CE). At that time, this work was the major reference for the knowledge and study of natural sciences broadly understood. Hernández was still working on it when he left Seville for the New World, and he completed his edition and translation during his stay there. There he also collected a vast quantity of data about the natural resources of Mexico, which he needed to properly organize. None of the published versions of his work indicates how he proceeded because they all resulted from heavy editing (including drastic cuts). However, the unpublished manuscript of the full work (which was recently discovered) provides useful information. A close analysis of this manuscript indicates that Hernández combined the medieval and native systems. Whereas the medieval system consisted in listing plants according to the alphabetical order of their most common name, the native system classified them on the basis of botany. In Hernández's work, the entries were not on individual plants (as in medieval scientific literature), but the genera of these plants. These genera were listed in alphabetical order of their Nahuatl name. Within these genera, the plants were grouped on the basis of their "virtue" according to the Nahuatl botanical system, which had also a medical dimension. However complex it might seem—and it was, among others, because of the difficulty of transcribing Nahuatl names—this taxonomy was not totally unprecedented. It relied on that of *De materia medica* by Dioscorides. In this work, Dioscorides assembled plants in coherent groups defined by botanical structures, medicinal properties, or any other common trait. The resulting groups were located on a *scala naturae* (natural ladder, or Ladder of Being), that is, they were listed from higher to lower groups in a continuum of gradual loss of quality. The difference between *De materia medica* and Hernández was that the latter did not arrange the groups according to their qualities on the *scala naturae*, but according to the alphabetical order of their name (according to medieval practice). In so doing, Hernández went beyond Pliny and merged the most authentic Dioscoridean taxonomy with the medieval classificatory principle, alphabetical in nature.

On his way back to Spain, Hernández brought along two copies of his work. One of them was offered to the king and was stored at his library at San Lorenzo de El Escorial, not far from Madrid. The other copy remained in the hands of Hernández, who was still editing it, adding new information, deleting some, and revising the whole. In 1580, the king ordered one of his physicians, the Italian Nardo Antonio Recchi (*c.* 1540–94), to make a selection from Hernández's full work as it was so voluminous. Recchi did not limit his interventions to extracting significant information, but rearranged Hernández's material according to a system he was familiar with (that is, the medieval system). Although Hernández was not satisfied by Recchi's work, copies of this short, rearranged, and edited version were made. These were used in the publication of Hernández's work, instead of the full original compilation. These publications illustrate the sometimes tortuous itineraries taken by ancient manuscripts. One copy of Recchi's selection reached Mexico, where its text was further edited, translated into Spanish, and published in 1615. Another copy of Recchi's manuscript was obtained by the Italian Federico Cesi (1585–1630), who had founded the Accademia dei Lincei in Rome in 1603. He wished to publish it as a major accomplishment of the newly founded academy. Academics soon started their work on Recchi's edited version, taking several decades to compile their commentaries. This explains why their work was not published until 1651 in a 950-folio volume known as the *Mexican Treasury*. However, the history of Hernández's research was not over. Twenty years after the publication of the *Thesaurus* (that is, in 1671) the original manuscript that Hernández had offered to the king and deposited at the Escorial disappeared in the fire that ravaged the library. Hernández's own copy, comprising six massive volumes, was lost for quite some time and reappeared only recently, in several libraries in Spain. Although Recchi's work distorted Hernández's taxonomy, the history of the latter's text is significant for understanding the intellectual climate of the period under discussion here. With a classical education gained through his university training and his long and intense study of Pliny's *Natural History*, Hernández first used a classical (ancient and medieval) taxonomic system in his *Index Medicamentorum*; then, he abandoned such a system, shifting to a local one, which combined the medieval tradition with the knowledge and practice of native populations. From our viewpoint, Hernández's shift represents the end of the scientific model used during the Middle Ages, even though the two published versions extracted from his work seem to be aligned with the medieval taxonomic model.

THE LEGACY OF ANTIQUITY

Of the three classical works that have been used to supply taxonomic models—Theophrastus, *Historia plantarum*; Dioscorides, *De materia medica*; and Pliny, *Naturalis Historia*—the most influential during the period covered by the present volume was without doubt *De materia medica*. It attested to both the internationalization of botany across the different societies to whose history we have alluded, and the daily practice of physicians treating patients. Its compilation is beyond the chronological range covered by this volume, but its subsequent history is relevant.

Although we do not have the exact original text of *De materia medica*, we might state with some plausibility that it was divided into five papyrus rolls (that were further identified as *books*, that is, units defined by a specific content). It was a vast collection of data totaling over a thousand entries, each of which was devoted to a *materia medica,* plant, animal, or mineral in nature. It might be the case that the work was originally illustrated with color representations of plants. Very soon it circulated through the whole Mediterranean world. Both its information and its comprehensiveness destined it to an intensive practical use that, in turn, favored the introduction of personal annotations, improvements, and modifications, in addition to possible mistakes in transmission. This process of expansion through the introduction of heterogeneous material might have gone alongside another process, also generated by the vastness of the work, but acting in the opposite direction: selection and fragmentation. *De materia medica* was, indeed, an exceptional synthesis of material that resulted in a cumbersome opus probably difficult to handle on a daily basis.

Significantly enough, *De materia medica* was often coupled with another work with a similar object: the treatise *De simplicium medicamentorum temperamentis et facultatibus* (On the Mixtures and Properties of Simple Medicines) by the Greek physician Galen (129–after ?216 CE). Galen's treatise was as idiosyncratic as Dioscorides' *De materia medica* was comprehensive. In its first part (Books I–V), Galen defined his own theory of the properties and degrees of action of therapeutic substances, and, in the second (VI–XI), he applied this theory to each and every *materia medica*. He divided the whole field of *materia medica* by natural kingdoms, with the plants, the minerals, and the animals, their parts and by-products. Galen's monumental collection of data radically differed from Dioscorides' *De materia medica* in its way of ordering the several materia medica it listed. Each materia is dealt with in a single chapter as in Dioscorides. However, whereas Dioscorides' chapters were ordered according to a *scala naturae*, those of Galen were listed according to the alphabetical sequence of their common Greek names.

It has been hypothesized that Dioscorides' text was soon alphabetized, possibly as early as the end of the third or the beginning of the fourth century, something that would have facilitated the integration of Dioscorides and Galen's treatises. Although there are several alphabetized versions of *De materia medica*, none of them corresponds. Notwithstanding, as early as the sixth century, an alphabetized version is attested in the so-called *Herbal*, lavishly illustrated and most commonly known through the copy now preserved at the National Library of Austria, the so-called *Vindobonensis* manuscript (*medicus graecus* 1, dated to *c.* 512 CE) (see Figure 0.2). We do not know when and where this *Herbal* was created, by whom, in what context, and for what purpose, even though the manuscript has traditionally been considered as having been created in the early 6th century for the princess Anicia Juliana, to whom it was dedicated. At first glance, it seems to result from a multi-step process of reorganization of Dioscorides' chapters consisting of distinguishing the plants and the other *materia medica;* among the plants, selecting only the herbs; and organizing the selected plants according to the alphabetical order of their common name. Though seemingly logical, the selection of the plants contained in the *Herbal* might have been more complex, and it might have resulted from a combination of availability and therapeutic applications. A closer analysis of the illustrations of the *Herbal* in the Vienna manuscript allows pushing the analysis further; it suggests that the *Herbal* might have assembled two series of plants selected from the full text of *De materia medica*. Each of these two series might have made it possible to treat the same range of medical conditions.

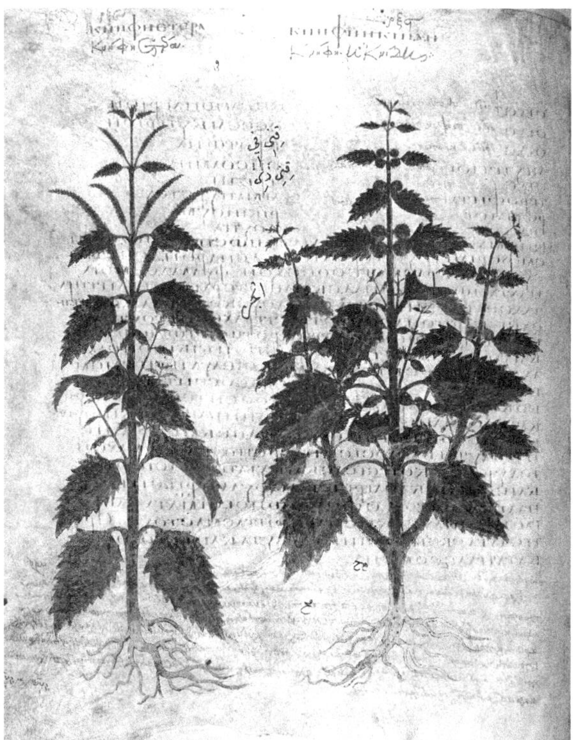

FIGURE 0.2 Two species of Nettle (f. 171v, Dioscorides, *De materia medica*, manuscript *Vindobonensis medicus graecus* 1, traditionally dated to the early sixth century, probably Constantinople). Photo by SSPL. Courtesy of Getty Images.

Presently, we cannot determine with any plausibility where such a phased process of creation of the *Herbal* took place. Was it in Constantinople, where the *Vindobonensis* was present? Or in Italy, where another copy of the *Herbal* might have been produced? Or was it at some unidentified place in the eastern Mediterranean, where manuscripts of *De materia medica* with plant illustrations similar to those of the *Herbal* were known? It might be that the whole process, with its different phases, did not happen in one place and at one time, but resulted from successive accretions across the Mediterranean world.

The appearance of short specialized treatises from the third century onward might be indicative of a process of fragmentation of vast encyclopedias like *De materia medica*. One such smaller work was *Medicinae lex holeribus et pomis* (Medicines from Vegetables and Fruits) by Gargilius Martialis (d. 260 CE). Another similar work was the co-called *Medicina Plinii* (fourth century) and its later abbreviated version identified as *Physica Plinii* (sixth or seventh century). A similar work was *De simplici medicina* that might have been inspired by Galen, *De simplicium medicamentorum temperamentis et facultatibus*, and circulated under the name of Theodorus Priscianus (fifth century). Several other short treatises on the medicinal uses of plants were produced in that time: the *Herbarius* attributed to the second century philosopher and writer Apuleius (*c.* 124–*c.* 170), but possibly compiled in the fourth century; *De medicina* (On Medicine) by Cassius Felix (fifth century); *Euporista* (Medicines Easy to Procure) by Theodorus Priscianus; *Ex herbis feminis* (From Feminine Herbs) (fifth century) with seventy-one plants, and the anonymous work identified by its title as *Alphabetum Galieni* (Galen's Alphabetical [list]) (sixth or seventh century), which analyzes three hundred *materia medica* (plant, mineral, animal). Some such treatises were even devoted to only one plant, as *De peonia* (sixth or seventh century) about the peony (*Paeonia officinalis*).

ACROSS CULTURES

The sixth century was a turning point. In the West, it was the time of a new translation of *De materia medica* into Latin that is known through a manuscript in a writing typical of the Lombard milieu and identified for this reason as the *Dioscorides Longobardus* in Latin (Lombard Dioscorides). It was the last Late Antique Latin translation of Dioscorides, which concluded centuries of assimilation and re-arrangement of the ancient legacy. It was followed by a transition period that lasted until the affirmation of the Carolingians.

In the East, this century marked the start of a vast and long-lasting process of exchanges of knowledge that was much larger than botany alone and had deep consequences. Greek philosophical and scientific texts were translated into Syriac and, later, into Arabic. As the case of *De materia medica* indicates, the versions of the texts used for translation into Arabic were editions made in the previous centuries in Byzantium. Editing was necessary because of the transformation of the texts generated by their practical use and the introduction of personal notes, additions, and other interventions by readers and users mentioned above. Editing aimed at possibly restoring the pristine, allegedly original state of the texts. In the specific case of *De materia medica*, this editing work preceded its Arabic translations and might be expected to have been made at the latest in the sixth century. As for where this took place, it might have been the eastern Mediterranean as Greek, Syriac, and Arabic versions of this edition were known and produced in that region broadly defined. But it could very well have been Constantinople, as this edition can be traced in the capital in the ninth century. Whatever the exact time and place, the "place" can be defined as eastern Mediterranean, reflecting the move of the center of gravity of scientific and cultural activity from the West to the East.

With the eighth century, we reach the firm ground of preserved Greek manuscripts and, toward the end of the century, possibly also the first Arabic translations in Baghdad, specifically the Arabic rendering of *De materia medica*, shortly followed by that of *De simplicium medicamentorum temperamentis et facultatibus* (see Figure 0.3).

At this point the evolution of the Greek and Arab worlds diverged. In the Arab world, the body of botanical data resulting from translations generated an abundant production and intense speculation along three major lines of research. Curiously enough, such activity started with speculation on the activity of *materia medica* similar to that of Galen by al-Kindi (d. *c.* 865), who not only pursued Galen's research, but pushed it to its limits by formulating a mathematical algorithm aimed at determining the action of compound medicines. Galenism was reproduced also by Ibn Sina (980–1037), best known as Avicenna, in his *al-Qanun fi al-Tibb* (Canon of Medicine). Dioscoridean encyclopedism, together with alphabetization, were soon echoed by al Biruni (973–1050) in the eastern part of the Arab Empire.

FIGURE 0.3 A page from a manuscript copy of the Arabic translation of Dioscorides, *De materia medica* (f. 22r, MS Leiden Or. 289 of the Universiteitsbibliotheek, at Leiden, dated to 1083 CE). Photo by Werner Forman, Universal Images Group. Courtesy of Getty Images.

Both encyclopedism and alphabetization were later brought to their most accomplished achievement in the western part of the Arab Empire, where a thriving school of botany had been established. In the eighth century, when the Umayyad Abd al-Rahman moved from Syria to Cordoba, he introduced to Andalusia plants native to his own environment, and he promoted a program of acclimatization and propagation of these plants throughout Andalusia. In the tenth century, the ninth-century Arabic translation of *De materia medica* made in Baghdad was known to the scientists in Cordoba. As the story goes, a Byzantine embassy was sent to the Caliphe. Among the gifts presented to the Caliphe was an illustrated copy of *De materia medica* in Greek. As none of the scientists in Cordoba was able to read Greek, a request was sent to Constantinople to send a scholar who could help local scientists to read the text. A monk named Nikolaos was dispatched; he is supposed to have worked in collaboration with the Cordoban botanists, allegedly contributing to the development of the local school of botany. Whatever the veracity of the story and the accuracy of the facts, later on al-Ghafiqi (twelfth century), possibly from a town close to Cordoba, and Ibn al-Baytar (1197–1248), from Malaga, compiled the largest, Dioscoridean-like botanical encyclopedias ever produced in the Arab world.

This process of assimilation of ancient botany with all its forms was completed at the same time by an enterprise of Arabization of *De materia medica* in the Eastern Arab Empire. Activity proceeded in two different geographical regions of the empire (East and West) and at different phases of its history: to the East, in the late tenth century, a certain al-Natili, who might have been a teacher of Ibn Sina, revised the text in Samarkand; to the West, in the twelfth century, among the small polities derived from the fragmentation of the earlier Abbasid Empire in Upper Mesopotamia, two different linguistic revisions were made.

TOWARD A PAN-MEDITERRANEAN CANON

Activity in the eastern Arab Empire was abruptly interrupted in 1258 by the conquest of Baghdad by the Mongols, which had long-term consequences as we shall see. Returning to the tenth century, Constantinople was living a period of renewed scientific, artistic, and cultural activity. Scholars were engaged in a program of reconstructing the full text of *De materia medica*. They recovered and assembled all the segments of the work that circulated independently as if they were treatises in their own right, and they treated each such piece as if it were a "book" (discreet section) of *De materia medica* as follows: herbs (that is, the *Herbal*, which made Book I), animals (Book II), oils (Book III), minerals (Book IV), and trees and their fruits (Book V). As a result, the new text was made of five books ostensibly comprising the full original text, even though these new books did not correspond at all to those of Dioscorides' version. In each such book, the different chapters appeared according to the alphabetical order of the name of the substances they were devoted to.

The compilation method was also used to combine *De materia medica* and Galen, *De simplicium medicamentorum temperamentis et facultatibus*. This was a more complex endeavor since the two works did not have exactly the same content and, in any case, they did not organize it in the same way. In Galen's treatise, the substances were listed according to the alphabetical order of their most common name, whereas in Dioscorides they were organized according to their ranking on the *scala naturae*, that is, the hierarchy of the natural world. Assembling the two works required first reducing *De materia medica* to a series of individual entries, and then ordering these entries in alphabetical order. The enterprise is currently known through a unique manuscript, something that hints at a limited impact.

The most significant characteristic of this epoch was the introduction of data related to *materia medica*—and not living plants—coming from the Arab world in a reversal of the flux of information between the Byzantine and the Arab worlds. This injection of data from outside the Byzantine Empire might have started with a mission from the Arab world to the Byzantine Emperor during the tenth century, exactly as had been the case at almost the same time from Constantinople to Cordoba. Exchanges increased during the eleventh century. They are best represented by a small treatise traditionally identified as *De alimentorum facultatibus* (On the Properties of Foodstuffs) by the eleventh-century author Symeon Seth, possibly from Antioch. Being bilingual in Greek and Arabic, Seth also translated an Arabic novel. *De alimentorum facultatibus* is not only about foodstuffs, but also about *materia medica,* including plants. This process of introduction of data on plants used in Arab medicine, which were probably known in the form of dry drugs and not as living plants, also took place in Sicily under the Normans in the eleventh century.

In the West, this time marks the beginning of a new enterprise of translation. Here, too, the direction of the flux was inverted, not only geographically from East to West, but also linguistically from Arabic into Latin. A certain Constantine from Kairouan (d. before 1098/9), traditionally identified as Constantine the African, was struck by the poor knowledge of medicine in the Latin world. After he moved to Italy, he became a monk at Montecassino monastery, and he launched a vast program of translation of Arabic medical literature into Latin. The facts were probably different. In the eleventh century, the city of Salerno, south of Montecassino abbey, had been for a certain time in contact with Byzantium. Its bishop Alfano (1015/20–1085) had sojourned in Constantinople, from where he brought Greek books. And he translated some texts from Greek into Latin. The Constantine story, however factual it might have been, is a prosopopea of a reversal in the transmission of information that resulted in the introduction of ancient medicine from its Arabic form into the Western world. Renewed analysis of Constantine the African's translations has revealed that his interest was not so much in Arab medicine in and of itself, but rather in the source of Arab medicine, that is, Greek medicine (see Figure 0.4).

Whatever its origin and rationale, the translation enterprise launched in Italy was paralleled in Spain, particularly in Toledo, where a group of scholars actively worked on the assimilation of Arab medicine, botany, and science more generally into Latin. One of these translators in Toledo was Gerard of Cremona (1114–87), who rendered Ibn Sina's *Qanun* into Latin. This translation provided the basis of the medical teaching in European universities for centuries, until almost the mid-fifteenth century at the University of Montpellier. Among the many relevant texts translated during this period, *De plantis* (On Plants) attributed to Aristotle (384–322 BCE) is of particular interest here. It had a long story of adaptation, disappearance in the Byzantine world, and translation into Syriac, Arabic, and Hebrew in the East. In the late twelfth or early thirteenth century, its Arabic form was translated into Latin. This version was known to Albertus Magnus (*c.* 1193–1280), who paraphrased it in his *De vegetabilibus*. Later on, it was translated into Greek in an apparently contradictory reversal of cultural exchanges across the Mediterranean.

At this point, the translation movement in the West shifted. Instead of searching for Greek medicine through its Arabized form, Western scholars went directly to the source of Arab medicine, that is, Greek. This interest explains how, in the twelfth century, the

FIGURE 0.4 Nutmeg in the so-called *Tacuinum sanitatis* (Tables of Health) (Vienna, National Library, MS Series Nova, 2644, f. 14r, dated to *c.* 1400). Photo by Prisma, UIG. Courtesy of Getty Images.

Pisan judge Burgundio (*c.* 1100–93), who sojourned in Constantinople, translated into Latin several Galenic treatises in addition to other scientific works. Western scholars were also active in the Byzantine Empire. The most famous among them might be the Belgian Dominican, Wilhelm of Moerbeke (between 1215 and 1235–before 1286). Moerbeke had access to Greek manuscripts that offered more accurate versions of the texts they contained. The Fourth Crusade and the seizing of Constantinople by the Latins in 1204 interrupted this scientific activity. Ironically, the destruction and the looting that accompanied the sack of the city opened the doors of libraries to the Crusaders. This is probably how a manuscript containing representations of plants reproducing those of the so-called *Vienna Dioscorides* that appeared later in the West was produced. Whereas the Vienna manuscript had been preserved in the imperial library and was not accessible until then, it probably now became available and was reproduced by some Westerner in Constantinople.

In 1261, the Byzantines reconquered their capital. Constantinople was then a cosmopolitan city frequented by Genovese, Pisan, Amalfitan, and Venetian merchants trading local and Oriental products to the West, while, at the same time, scholars and other individuals were coming from the Far East, like the Nestorian monks who arrived from China to Byzantium in 1278. More than any others, Arab scientists were present in the city. The reconquest of Constantinople had been preceded shortly before by the capture of Baghdad by the Mongols in 1258. From the second half of the thirteenth century onward, Arab medicine was widely known in Byzantium. It is tempting to hypothesize that Arab scientists left Baghdad and emigrated to Constantinople, offering their services. This might have been so because Arab medicine was then the most advanced in that part of the world, and Byzantium needed to rebuild its sanitary system. Arab physicians probably brought with them the drugs—we do not know whether in the form of dry products or living plants—that they had been prescribing in their practice in the Abbasid Empire. Byzantine manuscripts of the fourteenth century contain bilingual lexica of plant names that list the Arabic names of these drugs and their Greek equivalents. Arab physicians were interacting with their Greek colleagues in order to make it possible for the latter to integrate the Arab pharmacopoeia, especially the plants, into their practice.

While this Arab-Byzantine collaboration expanded the range of plants used in medicine, the process of transfer of the Greek plant knowledge to the West that had started in the pre-crusade period developed anew. The best representative of this new movement was arguably Pietro d'Abano (*c.* 1250–1315 or 1316), a teacher at the University of Padua. He traveled to Constantinople and acquired texts by Galen and Dioscorides that he further translated into Latin. Of *De materia medica* he knew two different versions, which he compared when he commented on the work in his teaching back in Padua. According to the medieval practice that we have seen, he followed the alphabetical version which does not convey any taxonomical information contrary to the five-book original version.

THE END OF A PERIOD

The reconstruction of Constantinople after 1261 was accompanied by the recovery of the city's library and its heritage, its restoration when necessary, and also its reorganization. Many of the manuscripts of *De materia medica* produced in earlier periods and preserved in the capital were collected and grouped in one library. This operation was not just an intellectual exercise of tracing manuscript origins and descent; rather, it provided actual medical practice with a reference collection. The text of some manuscripts was reproduced and used for an editorial project that resulted in producing an updated version of *De materia medica*. Since the manuscripts brought together contained different versions of *De materia medica*, the enterprise did not simply consist in comparing the several copies; it required the compilation of an index of each version and then the establishment of a table of concordance that allowed one to find the text of each entry in all versions. The resulting work was a new edition of *De materia medica* that returned to the full text of *De materia medica* (instead of any of its segmented derivatives), with its five-book structure (instead of the alphabetical order of the common name of the plants). It thus reinstated and reaffirmed the original project of Dioscorides, with its taxonomy reflecting the natural order of nature. This work on *De materia medica* went further and included a concordance of the treatise with Galen, *De simplicium medicamentorum temperamentis et facultatibus*, as well as the transfer of the plant illustrations from a manuscript of Dioscorides' text to the tenth-century manuscript that merged *De materia medica* and Galen's treatise and followed Galen's alphabetical order. The illustrations of the Vienna copy of *De materia medica* also were reproduced. However, several of these new copies did not include the text, being true botanical albums. Through the lexica of plant names, the reproductions of the plants in these albums could be connected with the Arabo-Persian knowledge.

If this activity achieved the search for universalism and completeness that animated Byzantine scientists through the centuries, it did it in a form that differed greatly from previous attempts and possibly also from expectations. It did not result in a unique, comprehensive, and all-encompassing body of knowledge, but rather in a collection of discrete and particular elements that were assembled in a coherent structure. The Fall of Constantinople in 1453 put an end to the scientific activity in the city. Byzantine scientists emigrated to the West. They brought with them copies of the edition of *De materia medica* prepared in Constantinople in the fourteenth century, and one such copy became the model for a printed version of the treatise produced as early as 1499. This volume provided Western scientists with the full text of *De materia medica*, its description of plants, and its taxonomy, all of which contributed to new developments after they had been assimilated. Illustrated copies of the treatise made their way into library collections only later, at a time when botany had made such illustrations obsolete, transforming these manuscripts into works of art and their acquisition into antiquarianism. As for lexica of plant names, those were forgotten, being rediscovered only recently as sources for inquiry into the history of the ancient botanical lexicon.

With this renewed transfer of Dioscorides' *De materia medica* from Byzantium, the West received the most complete collection of botanical knowledge that had been produced in antiquity and was constantly transmitted through the centuries, though in different forms. Its Greek text was printed shortly after Columbus' arrival in the New World (actually in 1499), providing the West with a reference for botanical studies and an instrument for the exploration of the flora of the New World. The starting and ending points of the period dealt with in this volume were thereby connected.

CHAPTER ONE

Plants as Staple Foods

MELITTA WEISS ADAMSON

The recipes in medieval cookbook manuscripts from across western Europe, nearly all of them written between 1300 and 1500, may give modern readers the impression that almonds, rice, and sugar were important staple foods and yet nothing could be further from the truth. They were all luxury goods that reflected a cuisine enjoyed by the rich and aspired to by the less affluent. If one is to define staple foods as those foodstuffs that provide a large part of the food energy intake of a given population, then a handful of cereals, legumes, roots, tubers, fruits, and nuts fed much of medieval Europe. What these staples had in common was that in raw or processed form they could be stored for longer periods and used in times of scarcity. Over-reliance on one staple food can have disastrous consequences, as a more recent example, the Irish Potato Famine (1845–9), caused by the plant disease commonly known as potato blight, has shown. While the medieval diet was more diversified than that of nineteenth-century Ireland, especially in the first and last two centuries of the era, famines brought on by bad weather, wars, or epidemics were nevertheless a frequent occurrence.

The period under investigation in this chapter will be the European Middle Ages broadly defined as the millennium from *c.* 500 to 1500 for which the following subdivision is frequently used by medievalists: early Middle Ages (*c.* 500–1000), High Middle Ages (*c.* 1000–1300), and late Middle Ages (*c.* 1300–1500). Geographically, the focus will be on western Europe with some consideration also being given to Byzantium (330–1453). The centuries between the Fall of Rome (476) and the beginning of the Renaissance in the fifteenth century brought many innovations in agriculture in Europe, which in turn led to an increase in population and the growth of towns and cities. A new social system, technical advances, and the introduction of new foodstuffs were accompanied by a warmer climate in the middle of the period, from *c.* 800 to 1300 CE. And although the Middle Ages ended as they began, with an extended cold period, today often referred to as the Late Antique Little Ice Age (LALIA), and the Little Ice Age (LIA) respectively, and with the so-called Plague of Justinian in the sixth century in the East, and the Black Death in the West in 1348, the continent had in the intervening centuries been irrevocably transformed. Not only did it recover relatively quickly from the onslaughts of plague epidemics, but it did so with a more varied, protein-rich diet for all segments of society.

AGRICULTURAL TRENDS

The staples of the Roman world during the Republic (509–27 BCE) and the Empire (27 BCE–476 CE) were the plant-based products of culture, bread, wine, and olive oil. These were in stark contrast to the staples of the Germanic and Celtic world at the time, in which meat, milk, and butter figured prominently (Flandrin and Montanari 1999: 165–77). Although the Romans also consumed meat, and the Germanic tribes used grains to brew a type of ale, the Roman diet centered on cultivated fields while the Germanic diet centered on forests and pastures. With the end of the Roman Empire, the cultural center in the West moved from the Mediterranean north to the Carolingian world and Charlemagne's capital Aachen (in modern Germany) in the eighth century, which in turn led to a fusion of the two food systems. Meat and dairy products were increasingly combined with grains in the form of gruel and bread, as well as wine in western Europe which resulted in a relatively balanced diet for the population in general. Western Europe's conversion to Christianity was another factor that elevated the status of bread and wine as symbols of the body and blood of Christ. And with animal fats forbidden during Lent, olive oil became an important substitute, at least for those who could afford it.

The expansion of agriculture across Europe made improvements to the ard, the Roman plow which was essentially a simple scratch-plow, necessary. While it worked well in the light soil of the Mediterranean, it was not well suited for the heavy, wet, but fertile soils further north. By around 600 CE the Slavs in the East began to use a much-improved plow pulled by a team of oxen. It featured a coulter or vertical knife blade to cut deep into the soil, and a moldboard that moved the sod to one side (Trager 1995: 39). By the eighth century this new type of plow was equipped with wheels, and horses gradually replaced oxen as draft animals (Gies 1994: 45f.). The introduction of the medieval heavy plow changed the layout of fields from the earlier square plots to long strips which reduced the number of times the draft animals had to turn around (see Figure 1.1). Furthermore, the resulting ridge-and-furrow pattern was better suited for the wet climate north of the Alps (Gies 1994: 112).

The increasing importance of horses in Europe had much wider implications than agriculture, however. By the ninth century a new social system, feudalism, had evolved that clearly separated farmers from warriors. As peasants, the former were primarily in charge of food production, and as knights, the latter became the ruling elite engaged in hunting and fighting on horseback. Several innovations, all coming from the east, were needed to make the horse an effective animal for peasant and warrior alike. A padded horse collar, which had been used in Asia for thousands of years before it reached Europe in the eighth century CE, prevented the horse from choking and made it a better draft animal, surpassing the ox in speed and endurance. The horseshoe benefited both peasants and knights by protecting the hooves from the wet northern soil, and from stony ground. The stirrup became the preferred way for medieval knights to mount their horses, to which were later added spurs and the curb bit (Gies 1994: 55, 57). Swords, lances, and axes were the tools of the warrior class, plowshares, harrows, scythes, spades, and pitchforks those of the peasants. And both groups were equally served by the all-important blacksmith (Gies 1994: 45, 126).

In addition to better farm equipment, a new crop-rotation system was developed in the eighth century and promoted by Charlemagne (742–814) in his realm. The traditional two-field system was now being replaced by a three-field system leaving only a third rather than half of the farmland fallow in a given year, thereby increasing the yield. In the new system, one field was planted in the fall with wheat or rye, and one in the spring with barley, oats, and legumes such as broad beans, peas, chickpeas, or lentils, while the third could be used for grazing (Gies 1994: 110–12; Trager 1995: 46). This had other advantages as well: the legumes provided valuable nutrients for humans, their

FIGURE 1.1 March: Plowing. From *Très Riches Heures du Duc de Berry* (MS 65 in the Musée Condé, Chantilly, France, fifteenth century). Photo by Leemage, Corbis. Courtesy of Getty Images.

stalks nitrogen for the soil, and as the use of horses increased so did the need for oats as animal fodder. The manure of farm animals, including horses, was used as fertilizer. Another method to enrich the soil was to spread marl, a clay that contained carbonate of lime (Gies 1974: 151).

Major new impulses in western European agriculture in the early Middle Ages also came from the Arabs. Crossing the Strait of Gibraltar, they arrived in Spain in 711 and by the ninth century had conquered much of the Iberian Peninsula as well as Sicily and the south of the Italian peninsula. They introduced (or reintroduced) many plants to Europe, some of which in time became staple foods. Among the new foodstuffs were rice, sorghum, durum wheat (*Triticum turgidum* Durum Group), sugarcane, watermelon, eggplants, spinach, artichokes (*Cynara carduncellus*), bitter oranges, lemons, and limes (Flandrin and Montanari 1999: 210; Gies 1994: 102). To cultivate many of those plants, crop-rotation, fertilizers, and intensive watering were the methods they used. Experts in irrigation, the Arabs built (or restored) dams, canals, and tunnels for draining water, and in doing so also increased the amount of land being cultivated. As they did in Spain, where some dams had survived from Roman times and many new ones were added, the Arabs in Sicily made use of and substantially expanded the existing irrigation system with knowhow developed in Asia. They built water towers and reservoirs, some of which can still be found today in places like Palermo (Adamson 2002: 114f.). The Arab innovations in agriculture, ranging from the types of foodstuffs to their cultivation, left their mark not just on the landscape but on the Spanish, Italian, and other European languages. The names of foodstuffs such as lemons, limes, oranges, apricots, artichokes, saffron, sugar, tamarind, and carob came to Europe via the Arabs.

The reception of the plants introduced in the West by the Arabs can be gauged by the catalog of foodstuffs contained in the manual known in the late Middle Ages as the *Tacuinum Sanitatis* (Tables of Health) (Cogliati Arano 1976; Mendelsohn 2013). The *Tacuinum* is the translation into Latin of a work originally written in Arabic by the eleventh-century physician from Baghdad Ibn Butlan (d. 1066) under the title *Taqwim al-sihha* (Tables of Health). This *Regimen sanitatis* or regimen of health, focused on food and drink, but also touched on the other lifestyle aspects usually treated in the regimen literature, namely air, exercise and rest, sleeping and waking, repletion and excretion, and emotional well-being. Of the 280 entries in the Arabic original, 75 percent dealt with foodstuffs ranging from staples to luxury foods, as well as some forty prepared dishes. The Arabic text was translated into Latin in the thirteenth century, presumably in Palermo at the court of King Manfred (1232–66). A comparison of the Arabic and Latin texts shows that the original content was modified in the Latin translation. The Arab dishes were removed, and a variety of foodstuffs, many of which by the thirteenth century had become an integral part of European food culture, were added. They included the fruits medlars, cherries, and strawberries (*Fragaria vesca*), the grains spelt (*Triticum aestivum* Spelta Group), rye, oats, millets including foxtail millet (*Setaria italica*), a wide range of garden herbs, meats from a number of wild and domesticated mammals and birds, curdled milk and ricotta cheese, lampreys and eels, the nuts sweet almonds, pine nuts, and chestnuts, as well as turnips (*Brassica rapa* Rapifera Group) and rutabagas (*B. napus* Napobrassica Group).

Another driver of the agricultural revolution of the Middle Ages, aside from the peasants, were the religious orders, first and foremost the Benedictine Order (est. 529) that included manual labor in its motto *ora et labora* ("pray and work"), and the Cistercian Order (est. 1098). They not only revived some of the agricultural practices of the ancient world such as grafting (first mentioned in a Greek medical text, the Pseudo-Hippocratic treatise *On the Nature of the Child*, in the fifth century BCE, and in a Roman agricultural text, Cato's *De agri cultura*, in the second BCE) (Mudge et al. 2009: 452, 455), which were preserved in written sources in the monastic libraries of western Europe established between the sixth and thirteenth centuries, but they were also enthusiastic early adopters of new technology, such as water-powered machines for a number of different purposes besides milling. Furthermore, monastic orders were what one historian has called the "shock troops" of forest clearance (Fagan 2008: 37).

After centuries of subsistence farming in western Europe, the agricultural revolution in combination with a milder climate between the eleventh and thirteenth centuries resulted for the first time in food surpluses, which led to a doubling and in some places tripling of the population and the rapid rise of towns. With more and more mouths to feed in town and country, the need for arable land became ever more urgent. At the same time, as the population of Europe boomed, the number of monastic houses of some orders grew dramatically as well. The Cistercian Order, for instance, founded some seven hundred communities between 1098 and 1351, each engaged in converting forests into farmland (Fagan 2008: 33–8). By the time the plague arrived in the middle of the fourteenth century, Europe had lost nearly half of its forests and marginal land had been turned into fields at much higher altitudes and much further north than ever before. While forest clearing and swamp draining were at times encouraged by lords in France, Germany, and elsewhere during the Medieval Warm Period (*c.* 800–1300), the nobility ultimately emerged as the main protectors of the forest by making it the playground for their favorite pastime, the hunt (Gies 1994: 290f.). The fourteenth century finally brought a reversal of the trend towards a constant expansion of farmland, when a harsher climate, coupled with the plague, and wars led to a radical drop in the population. Less need for food, and a severe labor shortage to cultivate crops resulted in more arable land falling fallow, being used for grazing, or reverting to new-growth forest.

PRODUCTION AND CONSUMPTION OF STAPLE FOODS

As Europe moved from a diversified diet to one based more and more on grains as early as the first centuries of the Middle Ages, plants became the staple food for the vast majority of the population, the peasants. Another group subscribing to a largely vegetarian diet by choice rather than by necessity were the religious orders. The staple plants that sustained much of medieval Europe which shall be discussed here were wheat, barley, rye, oats, millet, peas, beans, chickpeas, cabbage, turnips, parsnips, beets, chestnuts, olives, apples, and grapes. With the exception of wheat, none of these foodstuffs enjoyed a particularly high prestige, and made relatively few appearances in the cookbooks of the fourteenth and fifteenth centuries. Although most of them were indigenous to other regions of the world, mostly Asia and the Middle East, they had been cultivated in Europe since Greco-Roman times or even earlier, in the case of barley as early as 5000 BCE. As such they were not rare and exotic new imports. Another reason why they were not regarded very highly may have been their rank in the hierarchy of the world created by God in the "Great Chain of Being." There the path led from inanimate objects via the four elements, earth, water, air, and fire, upwards to God. Of the food coming from the earth, the lowest of the four elements, bulbs were at the bottom, followed by roots, and the fruits of herbaceous plants, shrubs, and trees (Flandrin and Montanari 1999: 308).

Cereals, unless they were used whole to make porridge, had to be ground into flour, and since Neolithic times (10,000 BCE) people the world over had used hand querns (mills) for this purpose. In antiquity (third–second century BCE), water power was first applied to drive the mechanical process of milling. The invention of the horizontal water wheel in Byzantium (the city, not the Byzantine Empire) (early third century BCE) was soon followed by that of the vertical water wheel in Alexandria (*c.* 240 BCE). As Europe's dependence on grain as the main staple grew (eighth–eleventh century CE), so did the number of water mills which initially featured the less expensive but inferior horizontal water wheel. The *Domesday Book* of 1086, compiled for William the Conqueror (*c.* 1028–87), lists 5,624 mills in England which only a century earlier counted fewer than one hundred (Gies 1994: 113–17). France saw a similar explosion in the number of water mills in the High Middle Ages (eleventh–thirteenth centuries). Lords soon began to assert their power by outlawing the use of hand querns, thereby forcing peasants to have their grain ground at their mills. The price for the service was a thirteenth of the amount of grain or flour ground.

Evidence is now coming to light of the existence of some independent water mills operated by free or even feudal tenants (Gies 1994: 116). Mill races, that is the channels whose swift currents drove the mill wheels, sometimes extended over kilometers and were laboriously dug out of soil and rock. The power created by ocean tides was also utilized quite early in places like Ireland, to operate so-called tidal mills. It was in eastern

Persia between 500 and 900 that the first horizontal wind mills were invented which may or may not have inspired the vertical wind mills that started appearing in Europe around 1180, in particular in eastern England, northern France, and the Low Countries. The mills used in Byzantium were primarily hand mills, animal mills powered by donkeys or horses, and horizontal water mills. Aside from grain mills, oil mills used to produce olive oil, a Mediterranean staple, were common in Byzantium (Stathakopoulos 2007).

In addition to technical innovations in milling, baker's ovens, too, went through several stages of development. In the Middle Ages full-sized ovens were heated from inside, and once the inner temperature was sufficiently high, the firewood was removed, the food to be baked put inside and the oven sealed. Small portable ovens buried in the coals were used in some places, notably the South of France at the time of the Avignon papacy (1309–76) (Adamson 2004: 56f.). Bread was either prepared in the home and then baked in a communal oven, or it was baked and sold by professional bakers. As was the case with mills, bakers' ovens, wine, and oil presses, too, became subject to the lord's power, known as *ban*, in the Holy Roman Empire from the eleventh century on.

Cereals

Pure wheat bread of the highest quality was of light color, and was made of bolted or sieved wholemeal flour without the wheat germ removed. More bran content resulted in darker, lower quality bread. With wheat or the means to afford it in short supply, people combined it with cheaper grains. Wheat and rye were a popular blend, and in times of crisis, barley, oats, legumes, vetches (*Vicia* spp.), or chestnuts were also ground into flour and mixed in (see Figure 1.2).

Medieval Europeans grew several types of wheat, from bread wheat (*Triticum aestivum*), and spelt to emmer (*T. turgidum* Dicoccon Group), and *einkorn* (*T. monococcum*) (Adamson 2004: 1–3; Davidson 2006: 847f.). Durum wheat, introduced to Europe by the Arabs during their rule in Sicily and southern Italy (831–1091), proved especially suited for pasta making. Wheat was used in bread-making, and in the form of flour or bread played a role in medieval cookery. It was used as a thickener in soups, potages, sausages, and stuffings, as wheat mush known as frumenty, and as dough for fritters, cakes, pies, and tarts as evidenced in the cookbooks of the fourteenth and fifteenth centuries. Leavened bread was an invention of the ancient Egyptians (third millennium BCE) which was adopted by the Greeks (*c.* 800 BCE). The Romans (*c.* 200 BCE) used brewer's yeast borrowed from the Germanic tribes to make the dough rise.

Barley, which was cultivated even earlier than wheat (sixth century BCE), was in antiquity used to prepare unleavened bread, but also porridge and beer (Adamson 2004: 4, 48f.; Davidson 2006: 61f.). Leavened barley bread is darker, denser, and coarser than wheat bread, which made it a staple food of the poor in Europe throughout the Middle Ages. The European elite used it as a plate substitute, known in England as a "trencher." Barley water was administered to sick people as illustrated by its entry in the *Tacuinum sanitatis* (Mendelsohn 2013: 84). Barley production was much higher, even exceeding that of wheat, in the regions discussed below where ale and beer were popular drinks.

Ale, often produced in small batches for domestic consumption in England by housewives or in larger quantities by so-called alewives and sold in the streets or in alehouses in the fourteenth and fifteenth centuries, used barley malt, yeast, and water as its principal ingredients. Drunk pure, or with spices, milk, or later brandy added, the popular beverage was consumed daily by the masses. The practice of adding hops to the mix, thereby substantially extending the shelf life but also adding a bitter taste to the

FIGURE 1.2 July: Harvest Time. From *Très Riches Heures du Duc de Berry* (MS 65 in the Musée Condé, Chantilly, France, fifteenth century). Photo by Christophel Fine Art, Universal Images Group. Courtesy of Getty Images.

sweet ale, was introduced by Carolingian monks (early ninth century), and slowly spread from Germany to Bohemia (thirteenth century), Holland (fourteenth century), Flanders, Brabant, and England (fifteenth century). The drink, now called beer, a word whose root may be the Latin verb for drinking, *bibere,* became especially popular in the Low Countries, which by 1500 produced it on an industrial scale and were major exporters of the beverage (Davidson 2006: 71f. ; Unger 2007). In England, however, beer met with stiff resistance from ale brewers and the authorities, and so it was not until the sixteenth century that beer-brewing became an integral part of English life.

Rye, which grows well in a wet and cold climate such as that of northern and eastern Europe, where it was the favorite bread cereal in the Middle Ages, was introduced to Britain by the Anglo-Saxons (fifth to seventh centuries) (Adamson 2004: 4; Davidson 2006: 678). Its flour, like barley flour, resulted in a darker and denser bread. When mixed with wheat flour, it produced a higher-grade bread, with barley flour a lower-grade bread.

Oats, a cereal crop that also thrived in a wet and cold climate, and fostered in Britain by the Romans, was a staple food for the masses in medieval Europe, from Russia and northern Germany to Scandinavia and parts of the British Isles, notably Scotland and Wales where the cereal was eaten as porridge or oatmeal (Adamson 2004: 4; Davidson 2006: 550). For bread-making it was mainly used in times of famine. Much of the oats planted in medieval Europe served as food for animals, in particular horses, especially as they gained in importance for peasants and knights alike from the eighth century on.

Millets, which flourish in a somewhat warmer and drier climate than barley, rye, and oats, were also a poor man's grains throughout the Middle Ages, added to soups or eaten as porridge, if not used as animal feed.

Legumes

Cultivated in Europe for thousands of years, in the case of lentils in Greece as early as 11,000 BCE, legumes continued to be an important staple in the medieval diet (Adamson 2004: 5f.). Beans (*Vicia faba*), chickpeas, and peas, in particular, provided much needed protein for the population. Fava beans or broad beans were eaten green or dry by the poor and by monks in the Holy Roman Empire as well as in Byzantium (Davidson 2006: 107f.). During Lent, beans were a convenient substitute for meat for the less affluent, and in times of famine they were ground into bean meal that was then added to bread dough. In the culinary literature of the Middle Ages, nearly all of it written in the fourteenth and fifteenth centuries, which by and large reflected upper-class cuisine, beans are rarely mentioned, primarily because of their association with poverty, but perhaps also on account of the flatulence they cause. There is no mention of beans in the oldest German cookbook from around 1350, for instance. While fava beans were eaten across medieval Europe, chickpeas were more common in the Mediterranean basin. They were eaten by the Romans and played an important role in Arab cookery and the Arab-influenced cuisines along the northern Mediterranean coast, from Spain to southern Italy and Sicily (Davidson 2006: 168f.). Peas, another age-old staple for Europeans, were also eaten green or dry. Like the other legumes, they could be stored and consumed in the winter, on fast days, or in times of famine (Davidson 2006: 588). But unlike fava beans, peas were not as closely associated with the poor. They do appear quite frequently in late-medieval cookbooks, as an ingredient in soups and potages, or combined with bacon. A German upper-class dish, Recipe 45 in *Daz buoch von guoter spise* (The Book of Good Food), was made of peas and eggs mashed together, the mixture then fried, roasted on a spit, and basted with eggs and herbs (Adamson 2000: 76, 102).

Vegetables

The vegetables eaten in large quantities in the Middle Ages were indigenous to Europe and, with the exception of cabbages, they were all root vegetables. Kale, which has recently become fashionable again, was a headless form of cabbage popular with the Scots, while headed cabbages were a staple in Germany and the Low Countries (Adamson 2004: 8; Davidson 2006: 428f., 121f.). Further south, in Italy, the cauliflower and broccoli, other selected cultivars of wild cabbage (*Brassica oleracea*), were more common. Headed cabbage had the advantage that it could be eaten fresh, raw, or cooked, and that it could be fermented and turned into sauerkraut, which would keep over the winter. As with so many of the staple foods in the Middle Ages, cabbage had the reputation as food for the poor, and is hardly mentioned in the cookbooks of the time. This is also the case with the root vegetables, turnips, and parsnips (Adamson 2004: 8f.; Davidson 2006: 579f., 816). Eaten by the peasants and craftspeople, turnips were boiled or pickled, and the greens were added to potages. Parsnips had the added advantage that they were sweet and starchy and could be used as a cheap substitute for honey, the traditional sweetener, and sugar (*Saccharum officinarum*), the fancy new arrival on the culinary scene. Peasants stored parsnips for the winter by either leaving them in the ground until needed, or keeping them in a cool place. When it came to beets (*Beta vulgaris*), which could range in color from yellow to red, both the roots and leaves were eaten by the lower classes, while in upper-class cuisine they played a minor role, except for red beet (beetroot) which was a popular coloring agent at a time when food coloring was all the rage in aristocratic circles (Adamson 2004: 9f.; Davidson 2006: 72f.).

Fruits

The most important fruits in the medieval diet were olives, apples, and grapes. In the ancient world, olives, the fruits of the olive tree which was first domesticated in the eastern Mediterranean some six thousand to eight thousand years ago, were a staple eaten raw and salted or cured in brine (Adamson 2004: 29f.; Davidson 2006: 554f.). The fruit, when fully ripe, was used to extract olive oil. Besides Italy, the southern part of Spain which the Arabs called Andalusia was an important producer of olive oil in the Middle Ages. North of the Alps, olives and olive oil were expensive imports, used sparingly by the Church, the medical community, and affluent consumers.

Apples, on the other hand, which were a luxury good in Greco-Roman times, during which the art of grafting was developed, were grown in most of medieval Europe and came in many different cultivars (Adamson 2004: 19f.; Davidson 2006: 26–30). Ubiquitous and incredibly versatile, they were eaten fresh, cooked, were stored in cool and dry places whole, or cut up and dried. In France and England, they were pressed and turned into cider, an age-old tradition the Romans encountered over two thousand years ago when they landed in Britain, and which subsequently spread across the Empire. In medieval cookbooks of the fourteenth and fifteenth centuries apples are used in fillings of various kinds, in pies and fritters, or were turned into apple sauce. Cheap, readily available, durable, and suitable for fast days, apples played an important role in the medieval diet for all segments of society.

The grapevine was introduced to Greece by the Phoenicians shortly after 1000 BCE, gradually spreading throughout the Mediterranean and, with the expansion of the Roman Empire, to much of Europe (Adamson 2004: 22f.; Davidson 2006: 26–30). The uses of grapes were manifold in the Middle Ages. When not consumed fresh, they were dried and turned into raisins; unripe, sour grapes were added to the tart food additive known as verjuice; grapes were pressed into grape juice, must, or fermented into wine; and if the wine went sour, the resulting vinegar was the universal seasoning of the masses who could not afford expensive spices. The grape harvest was one of the most important events in the medieval calendar, and was often carefully recorded in chronicles. In fact, it was the dates of grape harvests that modern scholars have used as one indicator of medieval climate fluctuations. In Burgundy, for instance, record keeping for the harvest of the Pinot Noir grape began in 1370 and was continued for six hundred years. More than ale and beer which were consumed primarily in the central and northern parts of Europe, wine was the drink of choice for people across the continent. With wine as a symbol of the blood of Christ, the Church and religious orders not only contributed to the further proliferation of vineyards in Europe, but also to an improvement in the quality of the alcoholic drink (Adamson 2004: 49–51). To extract as much of the juice as possible from the grapes, they were pressed several times, with the first pressing yielding the best wine. Red wine, which also contained the skin of the grapes, was more expensive and more durable. The amphorae of Classical antiquity had by the beginning of the Middle Ages in Europe around 500 CE been replaced with wooden barrels, but corked wine bottles of glass only came into use in the seventeenth century.

In southern Europe, from Greece to the Iberian Peninsula, wine was drunk by all classes of society, in central and northern Europe more ale and beer were drunk by the masses, with wine, especially quality wine, reserved for the rich or for special occasions. Like ale, wine was sometimes infused with spices. Best known among those mulled or spiced wines were *hippocras* and *claret* which contained blends of powdered spices ranging from ginger, grains of paradise (*Aframomum melegueta*), pepper (*Piper nigrum*), and galingale (*Cyperus longus*) to cinnamon, cloves, spikenard (*Nardostachys jatamansi*), nutmeg, and mace (both *Myristica fragrans*) (Scully 1995: 150; Adamson 2004: 50). Wine, ale, or beer were consumed at all times of the day in great quantity, their alcohol content presumably lighter than today. Wine was not just drunk, and used in cooking for soups, sauces, and the like, but was also the liquid that bread was dunked into for breakfast. Famous wine regions included Beaune and Bordeaux, Alsace, the banks of the Rhine and Moselle rivers, Rivoglio in Italy, Sicily, as well as Spain and Greece which produced sought-after sweet wines.

Nuts

Often overlooked as a staple for the poor in medieval Europe was the chestnut. In Classical antiquity, the Greeks were familiar with the starchy nut, and between the first and fifth centuries CE, the Romans introduced it to France and Britain (Adamson 2004: 26; Davidson 2006: 166f.). In the High Middle Ages, when Europe was engaged in major land clearance brought on by massive population growth, not all forests were cut down in order to be turned into grain fields (Montanari 2015: 107–16). In the mountainous regions from Italy and the Balkans all the way to the Iberian Peninsula, many oak forests were replaced with groves of chestnut trees whose cultivation was carefully managed. Called "mountain bread," chestnuts were used as a substitute for bread and consumed together with other poor man's food such as beans and millet. Different cultivars allowed for successive harvests spread out over several months. Chestnuts were either gathered from the ground or the trees were beaten with poles to harvest the green chestnuts in their burr. When not eaten fresh, most chestnuts were dried in the sun or over a fire, and later ground to flour. In some places this was done on a large scale in chestnut mills. The consumption of chestnuts could be as high as four pounds (2 kg) per day per person for half of the year, as has been estimated for some parts of France (Montanari 2015: 111). While for the poor the chestnut was clearly a means to survive, the nut was not completely absent from the tables of the nobility or urban elite, where it was used to stuff chickens, ducks, and geese, or consumed in candied form. In years of surplus production, Italian chestnuts were also exported to places as far away as Byzantium, Egypt, and Paris (Montanari 2015: 113).

Feed for Livestock

As has been indicated in the discussion of some staple foods eaten by the poor, they were also used as animal feed, which made a balancing of the nutritional needs of humans and domesticated animals necessary. With the disappearance of many oak trees in Europe in the period between 1000 and 1250, acorns, the traditional food to fatten pigs in the fall, as depicted in the illumination for the month of November in *Très Riches Heures* of the Duc de Berry, were harder to come by. Beechnuts were another traditional forage option for pigs, where available (Fagan 2008: 5). In some places, such as Sambuca in the Appennine region in Italy, laws regulated the access to or passage through chestnut groves and oak stands by pig farmers during harvest season (Montanari 2015: 110). In areas of the Mediterranean with a lack of forage in the High and late Middle Ages, animals were fed some of the same lesser grains eaten by humans of lower social status or in times of need: barley, spelt, sorghum, and millet (Flandrin and Montanari 1999: 269). Overall, the massive deforestation of Europe in the twelfth century and beyond led to an increase in animals feeding on pasture land, or in fields that had been opened up for grazing. After the Black Death epidemic of 1348, with the population in Europe dramatically reduced and farm labor in high demand, there was an even bigger shift from crop farming to grazing of livestock, in England notably of sheep (Gies 1994: 172). This in turn led to an increased consumption of meat among the lower classes as well. Hay was the most important animal fodder during the cold season, and the hay harvest in June and July was a crucial event in the medieval calendar year (Fagan 2008: 6) (see Figure 1.3). Bad weather during the hay harvest often spelled disaster since the hay would rot in the barn and even develop methane gas if it was not completely dry (Fagan 2000: 40). Oats, the feed of choice in the Middle Ages for horses, was one of the crops for which animals had to compete with humans in times of famine. Many animals were slaughtered in the fall to provide food for the coming months and avoid feeding the livestock through the winter which was costly.

FIGURE 1.3 June: Haymaking. From *Très Riches Heures du Duc de Berry* (MS 65 in the Musée Condé, Chantilly, France, fifteenth century). Photo by Buyenlarge. Courtesy of Getty Images.

SUPPLY OF STAPLE FOODS, CRISES OF SUBSISTENCE, AND EPIDEMICS

In the early Middle Ages (500–1000), subsistence farming was the norm in Europe with most of the food being consumed at the place of production. As the population boomed from the eleventh to the thirteen centuries, and towns grew in number and size, consumers and producers were no longer one and the same. Surplus food from the countryside had to be transported to the urban markets, which required means of transportation, an improved road system, the building of bridges and canals, better navigation techniques, and a gradual shift from self-sufficiency and barter to a money economy. To secure the grain supply of London, the largest city in England, the hinterland from which it drew extended to four thousand square miles (about 6,400 km), and in some decades could amount to nearly a fifth of England's arable land during the Medieval Warm Period (800–1300) (Fagan 2008: 32). Bad harvests, whether the result of natural or manmade disasters, increasingly led to food shortages that affected the urban population as well as the peasants. It was at the time of increased urbanization in the twelfth and thirteen centuries that some famous food markets had their modest beginnings. Les Halles, the central food market in Paris, started as a fish market in 1110. In London in 1123, the Smithfield meat market was established, and in 1265 Covent Garden market, which was initially a stand where the monks of St. Peter's Abbey sold their fruit and vegetable surplus (Trager 1995: 56, 61).

With bread having emerged as the main staple across Europe by 1000, on account of its cost-effectiveness compared with other foods which were more expensive and provided fewer calories, legislation was passed in many jurisdictions addressing aspects of its production and sale. Following an earlier law from 1202, the Assize of Bread was instituted in England in 1266, for instance, which regulated the weight, price, and quality of the bread sold and the activities of bakers (Trager 1995: 59, 61f.).

In response to an influx of paupers, towns also began to provide social assistance which, aside from clothing and firewood, often included such plant-based staples as bread, peas, oil, wine, or beer, in fifteenth-century Belgium for instance (Adamson 2004: 172). Famine and war could drive the number of poor and indigent from an estimated 20 to 30 percent of a town's population up to as high as 80 percent. The ability to pay taxes is one measure used by historians to gauge the number of paupers in a medieval town. In Carcassonne (France) in 1304, for instance, 33 percent of the population were classified as too poor to pay taxes. If one considers that grain prices were the food prices that fluctuated the most, and that craftspeople in medieval towns spent approximately two-thirds of their income on food, soaring grain prices in times of crisis such as the 1437/8 famine would have quickly forced this group to join the ranks of the needy (Adamson 2002: 158).

In addition to towns and cities being supplied with food from the surrounding countryside, there was also maritime trade in grain and other staples in the Middle Ages. Since bread had been so central to the Roman diet in antiquity, long-established trade routes existed across the Mediterranean that the southern part of the Holy Roman Empire (800–1806) and the Byzantine Empire (330–1453) could build on. In the north where the strong dependence on grain was more recent, a new trading alliance began to emerge in the middle of the twelfth century that was centered in Lübeck (in modern Germany) and later came to be known as the Hanseatic League or Hansa. At the height of its power, this alliance of merchant guilds had a trading monopoly extending from the Baltic to the North Sea. Goods from the east, from as far away as Poland and Russia, were transported west to consumers in Flanders and England, while mostly manufactured goods from the west were transported east. The plant staples traded by the Hanseatic League were primarily cheap wheat and rye from the Baltic, the new granary of Europe, as well as beer (Nash 1995; Trager 1995: 70, 72).

Despite the increase in local and long-distance trade in the course of the Middle Ages, disruptions in the supply of staple foods leading to crises of subsistence were common. The causes for famines in the early centuries of the medieval West were by and large the same as those that have been identified for the late Roman and early Byzantine Empire (330–750). With regard to climate, they were drought, rain, floods, winds, and extreme cold. Also induced by nature were food shortages brought on by pests such as locusts, rats, and mice. Sieges and warfare were the typical human-induced crises often leading to famine (Stathakopoulos 2004: 35–48). For sixth-century Byzantium there is evidence of thirty-seven famines, a number which dropped to nineteen in the following century (Stathakopoulos 2004: 23). In western Europe the fact that urban life had all but disappeared in the sixth and seventh centuries and that people in the countryside, as has been pointed out earlier, enjoyed a diversified diet, lessened the severity and frequency of famines to some degree (Trager 1995: 47). In the High Middle Ages, famines brought on by crop failures occurred sporadically in regions across the continent. Eastern Europe, by the thirteenth century a major grain producer, was particularly hard hit in 1215 when unusually cold weather in Poland and Russia reduced families to sell their children and eat the bark of trees (Fagan 2000: 28). However, none of these famines compared to the Great Famine that descended on most of Europe in 1315, lasted for years, and brought the population boom to an abrupt end (Fagan 2000: 23–44).

Cold weather and heavy rain in the spring and summer of 1315 prevented grain from ripening, the planting of wheat and rye in the fall also failed, and with hay not properly cured, there was a severe shortage of animal fodder (Fagan 2000: 32). The rain continued the following year, negatively affecting anything from the spring sowing of oats, barley, and spelt, and the harvest of grains and legumes, to the production of wine and of sea salt (Fagan 2000: 40). For cereals, 1316 is now regarded as the worst year of the whole Middle Ages (Fagan 2000: 39). The wet weather continued into the summer of 1317, by which time the population in Europe was severely starved, and succumbing to disease, with deaths at record numbers. Plants and animals, too, were attacked by disease, from mildew on grapevines, to rinderpest in cattle, and liver fluke in sheep and goats (Fagan 2000: 40–1). Although 1315 to 1317 are often given as the dates for the Great Famine, harvests took until 1322 to reach more normal levels, and the overall food supply several more years.

Man-made crises of subsistence were caused primarily by wars and sieges that occurred frequently in medieval Europe. The former could range in length from brief skirmishes to off-and-on conflicts lasting years, decades and, in the case of the Hundred Years War, between England and France, over a century, from 1337 to 1453. But even the history of a mid-sized episcopal town like Würzburg (in modern Germany) in this period is dotted with wars, either with other towns, as was the case in 1303 and 1444, or more frequently between its citizens and the bishop of Würzburg, as was the case in 1354, 1373, 1374, 1380–1418, 1428, 1432, and 1435 (Adamson 2002: 158). Chronicles of the time occasionally provide a good picture of how a medieval town under siege prepared for and weathered such a crisis. From 1449 to 1450, Nuremberg (Germany), a town of some thirty thousand people, was at war with the Margrave of Brandenburg, but had sufficient time to prepare for the conflict by stockpiling grain and meat and making provisions for their distribution (Adamson 2002: 159f.). The amount of grain available per person was impressive: approximately 180 kg (400 lbs), plus a significant quantity of flour the town council also stored. When it came to the distribution of the food, the town made the rich share their provisions with churches and Jews, and later with all those in need. As in so many other times of crisis, conflicts arose between the millers and bakers and the townspeople and their municipal government. In other towns a decade earlier, accusations ranged from bakers producing bread of smaller size in Lübeck, to bakers adding bran and stones to the flour in Konstanz. Bakers in Köln (Cologne) at the time responded to threats by town councils by refusing to sell bread of the same weight and price as before the crisis, and millers in Basel in 1438 by shutting down their mills (Adamson 2004: 173). In the case of Nuremberg, the bakers in 1449/50 refused to sell their bread at the prices set by the town council, with the result that town council took over the production of bread. By taking that measure, Nuremberg followed other towns such as Frankfurt, Erfurt, and Zürich in the fifteenth century, that hired bakers, be they locals or outsiders, to produce the food staple and sell it to those in need at a cheap price. When the siege of Nuremberg ended in 1450, the town not only had meat left over, it also managed to keep the price of bread much lower than during the 1437 and 1438 famines (Adamson 2002: 160). The story of the siege of Nuremberg outlined above is an example of successful crisis management in the late Middle Ages, and yet it could have ended quite differently, had the conflict not been anticipated, had it lasted much longer, or the population already been weakened by years of famine as a result of bad harvests (Abel 1980: 96–105).

When it comes to epidemics, the Middle Ages in Eurasia began and ended with the deadliest of all, the Plague, and in each case the outbreak was preceded by famine. With malnourished bodies more susceptible to disease, the fact that the sixth century counted twice as many famines as the seventh, may help explain the high number of epidemics, more than twice as many as in the seventh century in Byzantium (Stathakopoulos 2004: 34.) The most devastating epidemic of the period was the so-called Plague of Justinian, a pandemic whose first outbreak lasted a year, from 541 to 542, and affected Constantinople and the rest of the Byzantine Empire during the reign (527–65) of Emperor Justinian, as well as the Persian Empire (Stathakopoulos 2004: 110–54). Up until the middle of the eighth century, there were seventeen more outbreaks, leaving millions of dead in their wake (Stathakopoulos 2004: 123). The bacterium that caused the Black Death in the middle of the fourteenth century, *Yersinia pestis*, was also responsible for the earlier Plague of Justinian, and in both cases it was transmitted by fleas on rats. Since the first report of the disease in 541 came from Pelusium near Suez, the rats and fleas presumably reached Constantinople on grain ships from Egypt, a major supplier of grain to Byzantium at the time.

The plague that reached Europe in the fourteenth century had its beginning in Central Asia in 1338/9 and from there spread to China and India (Fagan 2000: 81f.). By 1347 it had arrived in Caffa (modern Feodosia), a port city in the Crimea, that was then besieged by the Mongols. Genoese merchants fleeing Caffa brought rats and fleas carrying the disease on their ships to Constantinople, several Italian ports, and also to Marseille. By 1348, the disease had spread northwest to France, the Iberian Peninsula, and England, in the following two years to Germany and northern Europe, and in 1351 it had reached Russia. At the same time the epidemic also traveled eastward through much of the Middle East (Dols 1977; Herlihy 1997). Of the three types of plague, bubonic, pneumonic, and septicemic, the one most often described in the sources is the bubonic plague with its characteristic buboes or boils in the groin, armpits, and neck (Stathakopoulos 2004: 125–34). The death toll of the plague, now commonly referred to as the Black Death, was enormous, and higher than that of the Plague of Justinian. Although death rates varied in the different regions, on average about half the population or more may have succumbed to the disease. That Europe still suffered the long-term effects of the Great Famine earlier in the century probably contributed to the staggering loss of life. Like the Plague of Justinian, the Black Death returned countless times after the initial onslaught far into the late Middle Ages and the Early Modern Period, albeit in a more localized and less virulent form with each iteration.

Less deadly than the plague, but directly connected with the consumption of staple crops was another disease in the Middle Ages, ergotism, also known as Saint Anthony's Fire. It is caused by the fungus *Claviceps purpurea* (ergot), which produces reddish bodies slightly larger than cereal grains, infects rye, and occasionally also wheat, after a cold winter and wet spring (Matossian 1989: 7–14; Stathakopoulos 2004: 106–8). Since little rye was eaten in the Mediterranean, ergotism was more common in the northern parts of Europe. The ergot bodies, when ground up along with the grain and used in the production of food, would release alkaloids that can lead to two types of poisoning: convulsive and gangrenous. The first affects the central nervous system with symptoms ranging from seizures, spasms, itching, headache, confusion, and panic, to hallucinations, and even psychosis. The second reduces the blood supply of fingers and toes by constricting the blood vessels which ultimately leads to their loss. One of the first reports of an outbreak goes back to 857 in the Rhine Valley (Trager 1995: 48). In 943 four thousand people reportedly died from eating rye bread in Limoges, followed by two further outbreaks in France in 1039 and 1089 (Trager 1995: 53, 55). Ergotism has, in the late twentieth century, also been connected with accusations of witchcraft in history, most notably the Salem Witchcraft Affair (Matossian 1989: 113–22). The frequent depiction of St. Anthony, and fantastic images of demons, chimeras, types of punishment, and hell in the art of Hieronymus Bosch (*c.* 1450–1516), above all "The Garden of Earthly Delights," have led to speculation among modern scientists that Bosch himself may have suffered from convulsive ergotism and survived the disease. Examples of gangrenous ergotism can also be found in his art.

CHANGES IN CLIMATE AND RESULTANT CHANGES IN GROWING PRACTICES: CONCLUDING REMARKS

Until recently, climate has rarely been mentioned as a major factor in the food choices people made and the social and political upheaval that could result from climate change. One of the reasons for this lacuna may be the lack of reliable data from the pre-instrumental period when climate information was sketchy and included in such diverse narrative texts as histories, chronicles, church histories, or saints' lives, or had to be derived from proxy evidence in the form of harvest dates for grain, grapes, hay, and the like (Telelis 2005: 41f.).

Much climate research has been done in recent years on the Byzantine Empire, especially up to the eighth century. The written sources have now been well studied with regard to climate and climate fluctuations, and resulting incidents of famine and disease (Stathakopoulos 2004; Telelis 2005). Drought was the biggest concern in the sources from late Rome and early Byzantium (330–750 CE), and, connected with it, lack of rain for longer durations. Of the floods mentioned, the Nile flood was of particular importance for agriculture and the grain supply of Byzantium. Extreme cold was rare in the Mediterranean, but a harsh winter in Italy in 604–5 caused shortages of the staples grapes and grains. The presence or absence of wind was commented on in the sources especially because of the effect it had on navigation, and the shipment of grain from Alexandria to Constantinople (Stathakopoulos 2004: 36–40).

When the meteorological data from the Byzantine sources are mapped on to the Mediterranean, temperate semi-arid, and desert climatic regions, some of the trends that emerge for Byzantium in the early Middle Ages are: a higher frequency of hot episodes in the Mediterranean regions from 500 to 540; a higher frequency of dry episodes in the temperate semi-arid regions from 530 to 580, and the Mediterranean regions from 560 to 590; a higher frequency of cold episodes in the temperate semi-arid regions from 580–690; a higher frequency of wet episodes in the desert region from 540 to 580; and the temperate semi-arid regions from 660 to 700 (Telelis 2005: 46f.). In other words, the sources seem to indicate a cooling of the temperate semi-arid regions in the late sixth and seventh centuries, as well as more rain in the desert region in the sixth, and the temperate semi-arid regions in the seventh.

Paleoclimatology based on written sources alone in pre-instrumental times has many drawbacks, but science has in recent years developed a number of indirect methods to study climate change going back thousands of years by analyzing proxies such as ice cores, deep-sea and lake cores, coral records, and tree rings (Fagan 2008: 8–9). It was dendrochronology, the study of tree rings, that in 2016 provided proof of a severe cold period in Europe and Asia from 536 to 660, which has been given the name Late Antique Little Ice Age (LALIA) (Büntgen 2016). Factors that contributed to the cooling were three large volcanic eruptions in 536, 540, and 547, ocean and sea-ice feedback, and a solar minimum. The volcanoes in question are still under debate. One volcano that erupted around 540 and has been named as a possible candidate is the Rabaul caldera on New Britain (Papua New Guinea) in the western Pacific, another the Ilopango in El Salvador. The socio-political changes this cooler climate helped bring about were far-reaching and, as the scientists who carried out the study suggest, included famine followed by the Plague of Justinian (541–2), which weakened the Eastern Roman Empire, the migration of Slavic peoples to the eastern part of Europe, and of peoples from Central Asia to China in search of pastureland, the collapse of the Sassanian Empire in Persia and its conquest in the 640s by the Arabs, whose rapid expansion at the time is thought to have benefited from a wetter climate and more vegetation, which in turn increased the camel population vital for the Arab military campaigns.

If the LALIA with its unusually cold temperatures was a contributing factor in the subsistence crises, epidemics, and social upheavals that gripped Eurasia in the sixth and seventh centuries, then the milder climate that followed had a decidedly more positive effect. By around 800 the southward drift of glaciers in Europe was reversed, and with less frost, warmer summers, less precipitation, and generally more settled weather, the conditions were perfect for a major expansion in agriculture in the following centuries (Gies 1994: 43; Fagan 2008: 6–7). The Medieval Warm Period (MWP), as this meteorological change is called today, is considered to have lasted from 800 to 1300 CE. With a rise in average summer temperatures of approximately two degrees Fahrenheit (1 degree Celsius) in western Europe, and slightly more in central Europe, the effects were a raising of the tree lines, and a melting of the ice caps which in turn raised the sea level in the North Sea by approximately 60 to 80 cm (25–30 inches), and altered the coast line of eastern England and the Low Countries (Fagan 2000: 17; 2008: 15). When combined with severe storms, the floods in low-lying areas could create an inland sea, as happened in The Netherlands with the Zuider Zee in the thirteenth century (Fagan 2008: 15). The warming extended beyond continental Europe and the British Isles to the far North where Arctic pack-ice receded thereby allowing Norse voyagers to sail west and establish settlements in Iceland, Greenland, and one even as far away as Labrador (Fagan 2008: 45).

Preceded by some Irish monks who had reached Iceland by 790, the Norsemen arrived soon after and by 874 had settled permanently on the island. They grew barley in Iceland, harvested enough animal fodder for the winter to support dairying, and were otherwise engaged in seal-hunting and cod-fishing (Fagan 2000: 7–10). Around 985, the Icelandic Norseman Erik the Red (Erik Thorvaldssen, *c.* 950–1003) landed in Greenland, where two colonies were eventually established on the southwest coast that offered good grazing land for the animals (Fagan 2008: 90). About 1000, Leiv Eriksson (*c.* 970–1020), the son of Erik the Red, reached Newfoundland and founded a small settlement on the northern tip of the island, which is today called L'Anse aux Meadows. He also sailed further south to the mouth of the St. Lawrence River and beyond, giving the region the name Vinland perhaps in reference to the wild grapes (*Vitis labrusca*) found there (Fagan 2008: 90–2). It was in these far north settlements that the meteorological change to cooler temperatures was felt first. By the twelfth century, cereal could no longer be grown in Iceland, by 1250 fewer ships set out from Iceland to the Norse colonies in Greenland, and by the middle of the fourteenth century the settlements in western Greenland were abandoned altogether (Fagan 2000: 62, 66f.; 2008: 90).

During the MWP (*c.* 800–1300), wheat was grown as far north as Trondheim in Norway, and, in southwest England, crops were grown as high up as around 350 meters (1,200 ft) above sea level (Fagan 2000: 17; 2008: 20f.). Vineyards, too, thrived some three to five hundred kilometers (roughly 200 to 300 mi) further north than they did in the twentieth century (Fagan 2000: 17). Around 1100, the monk William of Malmesbury (*c.* 1095–1143) enthused about English vineyards and the sweetness of the grapes, calling them "little inferior to the French in sweetness" (Fagan 2008: 12). In fact, English wine was exported to France in large quantities in the twelfth and thirteenth centuries. Other unlikely wine-growing regions from a modern perspective were eastern Prussia, southern Norway, and the Black Forest at altitudes as high as five hundred and sixty meters (about 1,800 ft) above sea level (Fagan 2008: 12). The expansion of land under cultivation, thanks to favorable weather conditions and deforestation in combination with an agricultural revolution, led to a dramatic population increase from an estimated twenty-seven million in 700 to sixty million–seventy million in 1200 in continental Europe (Gies 1994: 109). The numbers continued to rise, with the result that by 1300 food production could no longer keep up with the number of mouths to feed. With 10 or more percent of the population now living in crammed quarters in the hundreds of new towns that sprang up in the High Middle Ages, the conditions were ripe for an increase in violence (Fagan 2008: 30–2).

As discussed above, a cooling trend in the thirteenth century signaling the end of the MWP manifested itself first in the far north, in particular in Greenland and Iceland, but also in eastern Europe. Glaciers in Greenland and the Alps began to advance again during that time (Fagan 2000: 28, 49). Given the gradual cooling that eventually extended across Europe and much of the world, dates for this particular period now generally referred to as the Little Ice Age, vary somewhat, with 1300 and 1850 stated by many scientists as the approximate beginning and end dates (Fagan 2000: 49). Possible causes for the climate change suggested by modern science include some already encountered in connection with the LALIA, namely a solar minimum, and volcanic eruptions, in particular one major eruption in Indonesia in 1257 followed by three smaller ones in the next three decades, and another major eruption in Vanuatu in the middle of the fifteenth century (Fagan 2000: iii; Stothers 2000). Ocean circulation affected by the fresh water from melting icecaps during the MWP, natural reforestation after the famines and epidemics of the thirteenth century leading to increased carbon dioxide uptake in the atmosphere, and deforestation at higher latitudes resulting in an increased reflection of solar rays from snow-covered ground, as well as spontaneous fluctuations in the earth's climate are some of the others cited (Ruddiman 2003).

By 1310, the climate in Europe became more extreme, unpredictable, cooler, and stormier, and it remained that way for over half a millennium (Fagan 2000: 47). Starvation, disease, and war, which were the hallmarks of the fourteenth century, continued into the fifteenth century and led to a dramatic depopulation of the countryside. In France, three thousand villages were abandoned, arable land was no longer cultivated due to a shortage of labor, armed conflict, and bad weather. Food crises were a regular occurrence in cities like Paris throughout the first half of the fifteenth century. Time and again, prolonged frost, storms, and cold and wet summers caused severe damage to staple crops resulting in bad harvests, and famines, the worst being the continent-wide famine which lasted from 1433 to 1438. By 1440 the sweet wine from Britain, too, once praised by William of Malmesbury, was a thing of the past on account of the sour, unripe grapes that the vineyards now yielded (Fagan 2000: 82–4). The new, harsher climate is perhaps best reflected in the illuminations of the *Labors of the Months*, such as those contained in the *Très Riches Heures* of John, Duc de Berry (1340–1416), from the early fifteenth century. With the countryside covered in snow, the peasants in the month of February are depicted trying to stay warm by bundling up, huddling by the fire, and venturing out into the forest to make firewood (see Figure 1.4).

FIGURE 1.4 February: The Severity of Winter. From *Très Riches Heures du Duc de Berry* (MS 65 in the Musée Condé, Chantilly, France, fifteenth century). Photo by Leemage, Corbis. Courtesy of Getty Images.

CHAPTER TWO

Plants as Luxury Foods

ALAIN TOUWAIDE

A study of plants as luxury foods, whatever the period and place of focus, raises the question of the essence of *luxe*. Is it a quality inherent in the object considered as luxury? Is it the "luxury object's" rarity, foreign origin, or high financial value? Or does the quality reside in the perception of the object considered as luxury by both those who have access to and enjoy it and those who do not and envy the former? Literature ranging from the comedies of the ancient Athenian playwright Aristophanes (*c.* 460–*c.* 380 BCE) to *Cyrano de Bergerac*, by French playwright Edmond Rostand (1868–1918), makes comparisons that shed light on this question. In Classical antiquity, the Athenians who woke up before dawn and walked a long way on dusty roads to participate in the meetings of the legislative assemblies were all too happy to receive a piece of bread, two onions, and three olives (Aristophanes, *Eccleziasusae* [*Assemblywomen* or *Women in Parliament*], vv 307–8). In the seventeenth century, Cyrano (1619–55) also was delighted to accept only one grape, half a macaron, and a glass of water (Rostand, *Cyrano de Bergerac*, Act I, scene 4) after his duel with Valvert which he won for his ability to handle both the sword and to compose a rhyme ("A la fin de l'envoi, je touche" [For when the refrain ends, I strike!]). As Aristophanes elsewhere reveals, the meal offered to the attendees of the meetings was the same as that for recruits in the army (Aristophanes, Fragm. 275). It was, after all, life in democratic Athens, that Aristophanes was describing. Conversely, Cyrano's modest meal contrasted sharply with the magnificence of the French royal Court under King Louis XIII and probably also of practices in Cyrano's regiment (the *Cadets de Gascogne* or Captains of Gascony), which recruited heirs of the old aristocratic families of Gascony. Would this indicate that *luxe* resides in the enjoyment rather than in its object? And, in the case of foods, in the experience that its consumption generates, enhancing the taste of foodstuffs? Whatever its interpretation, luxury is a concept that pervades the world's many cultures. This chapter focuses on a large swath of territory, specifically the Mediterranean region, in the course of the long period between the years 500 and 1400 CE.

THE CLASSICAL HERITAGE

At some point at the beginning of the fifth century, a short work entitled *De re coquinaria* (On Culinary Matters) was circulating. It is a collection of recipes divided into ten books, each of which is devoted to a specific type of food. Although only the third and the fifth books are devoted to vegetables, all the others mention several kinds of plant ingredients.

The most striking characteristic of the collection is its display of an unprecedented taste for culinary extravagance as, for example, in the recipe for a cumin (*Cuminum cyminum*) sauce for oysters and shellfish:

> Pepper, lovage, parsley, dried mint, spikenard, cinnamon, plenty of cumin, honey, vinegar and liquamen (fish sauce).
>
> (*De re coquinaria* 1.15.1)

If lovage (*Levisticum officinale*), parsley, and mint were common in Italy, black pepper, spikenard (*Nardostachys jatamansi*), cinnamon, and cumin were not. They were not native to Italy and were imported at high prices. Other recipes required a broader range of ingredients, sometimes to accompany simple vegetables and transform them into exquisite preparations. A case in point is the following recipe for a sauce for lettuce:

> Lettuce. Serve with oxyporium, with vinegar, and a little liquamen (fish sauce) for digestion and to ease wind and to prevent the lettuce from doing harm: 2 oz. cumin, 1 oz. ginger, green rue, 12 scruples date flesh, 1 oz. pepper, 9 oz. honey, Ethiopian, Syrian, or Libyan cumin. Pound the cumin after you have steeped it in vinegar. When it has dried mix all the ingredients with the honey. When required mix 1/2 teaspoon with the vinegar and a little liquamen or take 1/2 teaspoon after dinner.
>
> (*De re coquinaria* 3.18.3)

This recipe was relatively complex: one of its ingredients (*oxyporium*) was itself a multi-ingredient sauce or "dressing," the details of its preparation also being provided in *De re coquinaria* (1.18):

> 2 oz. cumin, 1 oz. ginger, 1 oz. fresh rue, 6 scruples soda, 12 scruples date flesh, 1 oz. pepper, 9 oz. honey. Soak Ethiopian or Syrian or Libyan cumin in vinegar, dry and grind it. Afterwards you mix everything with the honey. When necessary you use it with *oxygarum*.

In its various details, the preparation of the aforementioned salad dressing would appear to be a near-infinite process. What would be involved in preparing the final ingredient, *oxygarum*? Fortunately, the *oxygarum* recipe was a simple one: just the famous garum (fermented fish sauce) of antiquity mixed with vinegar in equal quantity, with the resulting mixture probably cooked. At the same time, *De re coquinaria* also offered multiple common-sense recipes and simple culinary tips such as the following:

> How to make all kinds of leaf vegetables emerald green. Every kind of leaf vegetable may become emerald green if it is cooked in soda.
>
> (*De re coquinaria* 3.1)

In the Renaissance this anonymous recipe book was attributed to Caelius Apicius, probably referring to the wealthy first-century CE Roman gourmet Marcus Gavius Apicius. According to tradition, M. Gavius Apicius entertained the upper-class of Roman society with magnificent dinners at which he served extraordinary dishes masterfully prepared with great culinary talent, depleting his immense wealth with these ostentatious celebrations. When he realized that he no longer had the means to pursue this luxurious lifestyle, he elected to commit suicide rather than adjust to a simpler, more modest, and common way of life. The story of Apicius, regardless of its veracity, designated him as the perfect author of the collection. *De re coquinaria* has been considered his work ever since.

Research over the past century has established that, even though Apicius might have excelled in the preparation of refined meals, *De re coquinaria* was not his own cookbook.

It collected recipes that were circulating in the early fifth century and might not even all have been of Roman origin, as many recipes in the collection are characterized by aspects more closely associated with Greek cuisine, including parts of their titles. Nevertheless, Apicius did compile recipe books: one for complex dishes and another for sauces, entitled *De condituris* (On Sauces). The full text of neither has been preserved. Some excerpts can be found, however, in a short collection assembled by an otherwise unknown compiler named Vinidarius. The first extract establishes the work's tone:

> A summary of the flavorings which ought to be in the home so that nothing is missing from the seasonings: saffron, pepper, ginger, laser, leaf of spikenard, myrtle berries, costus, clove, Indian nard, adden[d]a, cardamom, spikenard.
>
> (Vinidarius, *Apici Excerpta*, 1)

The significant point is that, apart from myrtle berries, all the spices were plants from the east, imported to Rome at a high cost. And all of them were deemed indispensable for the good management of the kitchen of the upper class.

As mentioned above, the compiler of *De re coquinaria* in the early fifth century is not known. It might be that he was neither a chef nor an expert in the culinary arts, but rather a scholar who, in a useful way, assembled a series of recipes scattered throughout several books. The resulting collection was more than an exercise in antiquarianism however; it bore witness to a projection backward in time of an imaginary golden age characterized by a luxurious lifestyle and its attendant cuisine that was probably unthinkable in the early fifth century. This period was marked by the division of the Roman Empire into its eastern and western parts, the rise of the Eastern Empire, and the decline of the Western Empire, which was raided by groups coming from outside. These invading foreign populations introduced a new way of eating characterized by, among other things, an increased consumption of meat, including lard. In his short treatise on foodstuffs in the form of a letter dated to *c.* 511 and addressed to the Ostrogothic king Theodoric (*c.* 454–526), the Greek physician Anthimus (fourth/fifth century) reported both the abundant use of lard by the population, which he identified as Frankish, and the many medicinal properties that lard was credited with (*De observatione ciborum* 14).

Framed in this context of deep transformation of the range of food and culinary tastes, *De re coquinaria* invites interpretation of *luxe* as a measure of the difference between habits of consumption. This was probably already the case during Apicius' lifetime. It was a period of moral restoration fostered by the Emperor Augustus (63 BCE–14 CE), who prided himself on living a simple life in the most authentic Roman way. Apicius' lavish lifestyle implied a sense of transgression and extraordinariness that contrasted sharply with "ordinary" Roman citizens' way of life.

The text of *De re coquinaria* was uninterruptedly transmitted in manuscript form until the Renaissance, before being printed in Venice, as early as 1498, in an anonymous incunabulum edition (early book printed with movable type). Interestingly enough, the oldest copy preserved, now in the collection of the New York Academy of Medicine, dates to *c.* 830. It was copied at the monastery of Fulda, in Germany, during the so-called Renaissance of the Carolingians. At that time, the Holy Roman Emperor Charlemagne (742–814) had promulgated an edict in which he listed the vegetables to be cultivated in the imperial cities, all of which vegetables were ordinary staples, and not exotic imports. In this context, the copy of *De re coquinaria* attests to a fascination with extraordinary foods that, consequently, constituted *luxe*.

MOVING EASTWARD

A similar phenomenon can be identified in Byzantium in the late twelfth century. Before that century, Byzantine agriculture had developed techniques for producing refined luxury foodstuffs. Byzantium inherited a long tradition of agricultural literature from Classical antiquity dating as far back as Hesiod (eighth century BCE), *Opera et dies* (Works and Days), and Xenophon (*c.* 430–*c.* 354 BCE), *Oeconomicus*. With the exception of these two works, most of this literature is now lost, however. Its existence is known through lists of authors' names in the Latin agronomic corpus, which in turn was derived from Greek agronomic literature. Nevertheless, fragments of ancient Greek treatises had been included in later compilations, particularly the corpus of agricultural literature that circulated in Byzantium in several forms, the most recent of which is the so-called *Geoponica* (Agricultural Matters or, according to a modern adaptation of the title, Farm Work). Although the work makes reference to over thirty authors, very few of the excerpts that make up its text actually derive from them. Taking their lead from the Greeks, Roman authors had themselves produced agronomic works; these works ranged from the work of the austere Cato (234–149 BCE) who, in his *De agricultura* (On Agriculture), promoted the exclusive use of cabbage, to that of Palladius (late fourth/early fifth century CE) and his *Opus agriculturae* (Treatise on Agriculture). The high point of agronomic writing in the Roman world was undoubtedly that of Vergil (70–19 BCE), who transformed his technical subject matter into poetry in his *Georgics*, and Columella (first century CE), author of the comprehensive treatise *De re rustica* (On Agriculture).

Agricultural techniques in the Byzantine Empire were collected in the *Geoponica* mentioned above. This textbook, in the form that we know it today, had a long history. It went back to a compilation of earlier agricultural material collected by the otherwise unknown Vindonius Anatolius, often identified, but without supporting evidence, with the lawyer Anatolius of Berytos (d. 360 CE). It seems more probable that Vindonius Anatolius lived in the fourth or the fifth century, that is, in the early days of the Byzantine Empire. His treatise identified as *Collection of Agricultural Material*, according to an English translation of its Greek title, was exploited by the Latin agronomist Palladius in *De agricultura*. Its original text is now lost; it is known only through translations and its tenth-century re-arrangement discussed here. However, a textual archaeology of this tenth-century version has made it possible to identify various editions of the original compilation.

In the fifth or early sixth century, a Byzantine scholar named Cassianus Bassus expanded Vindonius' initial collection of material. He had access to a good source of relevant texts (a library?) of which he himself had a solid knowledge. The resulting work, entitled *Collection on Agriculture*, is no longer preserved in its initial integral form, but is known through translations, in addition to the tenth-century *Geoponica* which used it as a source. Bassus' version, in turn, underwent a further process of expansion at the hands of an unknown author living at an unknown time. It seems, however, that this revision was made shortly after Cassianus Bassus' death, and could therefore be identified as Late Antique. It is characterized by a certain knowledge of the topic, not as good as that of Cassianus Bassus, however, but certainly superior to that of the later editors of the work. Up to this point, the history of the *Geoponica* is marked by a gradual process of expansion through the introduction of new, relevant material from original sources into the initial text over a short period of time. In the second half of the ninth century, this collection of heterogeneous material, which may not have been stylistically standardized, was edited by an unknown scholar, who does not seem to have been acquainted with agricultural literature, and appears rather as an editor *sensu proprio*. This individual preserved the name of Cassianus Bassus and the title of the latter's compilation, even though he extensively modified the text, including the introduction of an agricultural calendar for the region of Constantinople and also episodes from the classical Greek mythology related to various plants. In the tenth century, this new, enlarged edition, which substantially transformed the original material, went through a further process of revision during the reign of the Byzantine Emperor Constantine VII Porphyrogenitus (905–59), who fostered a program of recovery of earlier scientific literature. The resulting edition is the text of the *Geoponica* that we currently know.

In this "final" form, the *Geoponica* offers a unique collection of material on plants together with knowledge about their uses in Byzantium. The work is composed of twenty books that cover a great variety of topics related to farming, agriculture, nutrition, and therapeutics: astronomy, the calendar, and the weather (Book I and also III, with the activities month by month); the raising of domestic animals, including poultry and cattle, and fish farming (Books XIV–XX); the sowing, manuring, weeding, and harvesting of plants (Book II); wheat, bread, and pulses (Book II.27–41); planting and tending of grape vines (Books IV–V) as well as wine-making and drinking (Books VI and VII, respectively); olive trees and oil (Book IX); fruit trees and other trees (Books X and XI.1–14, respectively); ornamental and edible plants, together with the therapeutic applications of the latter (Books XI.15–30 and XII, respectively); and plant pests (Book XIII).

No book, or part, of the twenty books is specifically devoted to luxury foods. However, several of them include chapters dealing with the improvement of some plant species aiming at transforming them into delicacies that were certainly not common in Byzantium and might have been luxury products reserved for a small, wealthy elite. These include seedless fruits such as grapes (IV.9), gourds and melons (XII.19), pomegranate (X.30), stoneless peaches (X.16), and also rindless walnuts (X.66). Fruits received particular attention, especially explaining how to produce fruit of superior quality: date (X.4–5), citron (X.7–8), peach (X.13), apple (X.18–21), pear (X.22–5), quince (X.26), pomegranate (X.29–38), plum (X.39–40), cherry (X.41–2), jujube (X.43–4), and medlar (X.71). Berries were represented by mulberries (X.69) alone, whereas the nuts discussed were pistachio (X.12), almond (X.58–9), chestnut (X.63), and walnut (X.64–7). Some of the outlined cultivation methods were alleged to produce exotic results. Citrons were to be made red by grafting them onto mulberry or pomegranate stocks (X.7), actually a technical impossibility, as were peaches, their reddening achieved by planting roses under the fruit trees or by dyeing stones with mineral dye (cinnabar) and burying them (X.15), all similarly fantastic. The color of apples could also be worked on to become redder (X.19) and mulberries, instead, were made white (X.69). Figs were to become bi-colored, white on one side and black on the other (X.53). Besides these fruits of a more striking appearance than normal (if they ever existed), certainly reserved for a fortunate few, fruits could also be transformed into works of art representing birds and other animal species, and even human faces: examples were citrus, pear, apple, pomegranate (X.9), and quince (X.27). The shape to be represented was made with a mould of gypsum or clay, and the half-grown fruit was tied into it so as to take the shape. Such fruits might have been served in an elegantly arranged basket in luxury meals. If the human face was that of a beloved woman, the fruits shaped in that way might have played a role in a "game" of seduction in an intimate meal—or the game might have been a subtler one: fruits and nuts conveying "secret" messages that were perceptible only to the object of desire might have been written on figs and almonds (X.47 and 60, respectively). Some vegetables, too, could be made to take on specific shapes. Lettuce was grown in a potsherd to encourage it to expand outward instead of upward and take the shape of a saucer, with many white leaves (XII.13). Lettuce seeds could also be blended with celery, rocket, and basil by putting the seeds of all these plants in a linen cloth and making them grow together (XII.13). This technique, which resulted in a salad of mixed greens, could also be applied to other species (XII.13).

The range of vegetables discussed was broad, but not excessively so, with the following kinds listed here in the order in the twelfth book, followed by the reference to the paragraph in which they were studied: mallow (12), lettuce (13–14), beet (15), cabbage (17), asparagus (18), gourd and melon (19), musk-melon (20), turnip (21), radish (22), celery (23), mint (24), rue (25), rocket (26), cress (27), chicory and endive (28), leeks (29), garlic (30), onion (31), hartwort (*Tordylium apulum*) (32), pennyroyal (*Mentha pulegium*) (33), dill (34), spearmint (35), grape hyacinth (*Muscari comosum*) bulb (36), squill (37), sorrel (38), artichoke (39), purslane (40), and mushroom (41). None of these vegetables was special in and of themselves, but their quality was improved, resulting in produce that might have been rare and, consequently, sought after. Beet, for example, which was by no means a delicacy, was grown in such way that it would be larger and whiter (XII.15), and celery, leeks, and grape hyacinth were increased in size (XII.23, 29, and 36, respectively). Garlic was made finer when grown in white earth (XII.30), and onions in red earth (XII.31). Asparagus was grown all year round (XII.18). In some cases, the taste of vegetables was modified through specific growing techniques. Cabbage was made salty (XII.17), musk-melon was flavored with roses (XII.20), and radishes, by contrast, were made sweet (XII.22).

Plants were also used to produce fine, flavored wines, which occupy most of Book VIII. Such wines were not a novelty of the Byzantine period, as several had already been described, with the recipes to make them, in *De materia medica* by Dioscorides (first century CE) and also in *De agricultura* by Palladius. However, some of those listed and described in the *Geoponica* did not appear in these earlier works and so might have been advances on them. Such wines in the *Geoponica* were many, ranging from those made with vegetables like parsley (VIII.12), celery (VIII.16), and dill (VIII.3), for example, to sweeter ones flavored with anise (VIII.4), fennel (VIII.9), pear (VIII.5), or quince (VIII.17), or bitter ones with pennyroyal (VIII.7), bay (VIII.8), or laurel (VIII.8). Hazelnut wine (VIII.6) might have been rare, and rose wine (VIII.2), colored with saffron and sweetened with honey, was probably an exquisite delicacy with a taste similar to that of the fine *rhodomel* made of juice of rose petals and honey, as described in the *Geoponica* (VIII.29).

Vegetables, too, were flavored, being sprinkled with vinegar infused with pepper (VIII.39) and even sweetened with must (VIII.36). A seasoning that was both sour and sweet was prepared with olives and grape lees (IX.31). Indeed, the whole garden was carefully tended so as to generate a complete sensory experience, which was certainly not common. It was fertilized to have lush vegetation (XII.6). Roses were bred to yield cultivars with a stronger fragrance (XI.18), and perfume was introduced into cuttings of grape with the (fanciful) idea that the vine exhaled its perfume across the land where it was growing (IV.9). Walking through such a garden before meals or dining in it by the light of candles on a fine summer night certainly would have added to the pleasure of the food and transformed the meal into a multisensory experience accessible only to a few (see Figure 2.1).

FIGURE 2.1 Grape vine in an Arabic manuscript (Istanbul, Topkapi Sarayi, MS Ahmet III, 2127, f. 252v, northern Iraq or Syria, 1229). Photo by SSPL. Courtesy of Getty Images.

Strangely enough, none of the spices that contributed to enhancing the taste of meals and adding to their exquisiteness in antiquity appears in the *Geoponica*. These spices were not native to the Mediterranean region and were imported from the Far East, sometimes at a very high cost, which doubtless contributed to their prestige and their status as luxury foods (see Chapter 2 of Volume 1 in this series). If their presence or absence in the available written documentation can be used as an indicator of their actual presence or absence in the local markets through what can be aptly called the international trade, their availability fluctuated through time. The many wars that Byzantium waged disrupted international trade and, in consequence, affected the availability of spices from afar. The conflict between the Byzantine and Sassanian Empires until the early seventh century was followed shortly thereafter by the rapid conquest of most of the Near East by the Arabs and the establishment of the Abbasid Empire. Only in the ninth century did the prevailing geo-political forces establish a new international equilibrium, with the consequence that the previously used Oriental spices reappeared in Byzantine records.

UNCERTAINTIES AND FLUCTUATIONS

In the meantime, the Arab world had developed a new agriculture that has variously been identified as the "Arab Agricultural Revolution" and the "Medieval Green Revolution" by historian Andrew Watson, in seminal studies published in 1974 and 1981, and completed and expanded in 1983 and 2008. Watson's thesis was that the expansion of the Arab world, which extended from India to Spain, acquainted those inhabiting diverse parts of that world with plants that grew abundantly in the whole Indian sub-continent. The Arabs then took these plants across their dominions. In just a few centuries, this diffusion, coupled with efficient water-engineering and improved land-use resulting in an early form of intensive agriculture, transformed agricultural practices, and increased crop-yields in an unprecedented way. Better harvests contributed to economic prosperity that, before long, was followed by demographic growth. This dynamic, which began shortly after the seventh-century military conquest, declined in the eleventh century because of major international instability caused by both the Crusades, which impacted the western part of the Arab Empire, and the Seljuk and Mongol invasions in the east.

After an initial period of wide acceptance, Watson's thesis has been revised and corrected. The resulting adjustment was usefully synthesized by historians Michael Decker (2009) and Paolo Squatriti (2014). The main problem with Watson's thesis was that the major crops used as indicators of this "new" agriculture had already been known around the Mediterranean in the pre-Arab period. This observation was true also of other plants supposedly introduced by the Arabs into the Mediterranean region, including those that constituted luxury foodstuffs. One of them is citron (*Citrus medica*). It was mentioned in Theophrastus, *Historia Plantarum* (4.4.3), and it was present in the Roman world. It appears in the *Georgics* of Virgil (2.126–7 and 131–5), in the *Naturalis Historia* of Pliny (12.15–16; 15.47; 16.107; 17.64; 23.105), and in Dioscorides, *De materia medica* (1.166). It seems to have been grown in places such as Pompeii, where it was represented on frescoes and, arguably, mosaics. It also appeared in a mosaic from the Baths of Diocletian in Rome (built late third to early fourth centuries).

Contrary to Watson's thesis, plants from India and farther afield were not all introduced into the Mediterranean world for the first time by the Arabs, but rather reintroduced by them after their production in the classical period had been abandoned, as the case of balsam (*Commiphora gileadensis* [*C. opobalsamum*]), for example, clearly indicates. The tree, which is native to the Arabian Peninsula and requires very specific climatic conditions and special arboricultural care, was cultivated in the eastern Mediterranean until the sixth century (see Figure 2.2). At that time, however, its plantations were abandoned. What the Arab Agricultural Revolution/Medieval Green Revolution really achieved was successfully re-introducing and re-acclimatizing species like balsam in a deliberate way, including bringing their use and cultivation to local populations under their control. Also, it resulted in generating intensive production of a certain range of plants, whatever the precise details of techniques employed and its economic and demographic impact might have been: Asiatic rice, sugar-cane, banana and plantain, lemon, lime, hard wheat, sorghum, watermelon, eggplant, spinach, artichoke, taro, sour orange, pomelo, mango, and coconut palm.

Returning to Byzantium, the empire benefited from both the cessation of the hostilities with the Arab world, and the ensuing reopening of trade routes, and the new agricultural dynamic generated by the Arab world. The first signs of the effects of these new developments can be perceived in the early tenth century in the so-called *Book of*

DEL DVRANTE. 65

BALSAMO.

Elidit menſes, partes ſimul atque ſecundas
BALSAMVM, *& exiccat; horrores febribus aufert;*
Calfacit ex oculis nubes & diſcutit; anguis
Aduerſatur item, cunctis ſimul atque venenis;
Vrinamq; ciet, tum concoquit, vlcera purgat
Sordida, tum laterum mulcet, capitiſq; dolores;
Iſchiadi, tuſſi, conuulſis, atque caducis
Proficit hoc morbis; ſimul & pulmone grauatis,
Anguſtæ vrinæ, crudis ſtomachiſq; & anhelis;
Vulneribus capitis confert, vertigini, & ipſi,
Torminibuſq; trahit ſquamma ex oſſibus inde,
Atque cicatricem vel citra vulnera iungit;
Corpora & halenti tandem putredine ſeruat.

NOMI. Gr. *Βάλσαμον ὀποβάλσαμον, ξυλοβάλσαμον; καρποβάλσαμον*, Lat. *Balſamū, oleū balſami, lignum balſami; Fructus balſami*. Ital. *Balſamo, olio di balſamo, legno di balſamo, frutto, ò ſeme del balſamo*. Arab. *Ba'eſem*. Franz. *Baume*. Ted. *Baſam*.

FORMA. La grandezza di queſt'arbore è come d'vn Melagnano, fatto di molti rami, le cui frondi ſi raſſembrano à quelle della ruta: ma più bianche, & ſempre verdeggiano, il ſuo frutto nella grandezza, & nel colore è veramente ſimile à quello del Terebinto, il quale ſpira di maggiore odore, che non fa il liquore, del quale non ſe ne porta à noi del ſincero; D però in luogo dell'opobalſamo ſi mette l'olio delle noci moſcate, ò quello della ſtorace, ò lo ſtate della mirra, ò delle bacche di ginepro, & in luogo del xilobalſamo, il legno aloe, & per il carpobalſamo, le cubebe vſuali.

LOCO. Naſce in Soria.

QVALITA'. Il vero balſamo è calido, & ſecco nel ſecondo grado, portaſi nuouamente dall'Indie occidentali vn liquore odoriſſimo molto ſimile alla ſtirace liquida, il quale coloro, che lo portano lo chiamano parimente balſamo per hauer egli alcune

A qualità ſimili al balſamo: ma queſta ſarà più toſto la vera ſtatte della mirra, ò il ſucco della ſtirace. Et cō buona ragione ſi può vſare in luogo del vero, ſe bene non è bianco: & ſe queſto non fuſſe alle mani; ne quell'altro ſi potrà vſar qualche balſamo artificiale, ſi come s'inſegnano eccellenti nel noſtro oracolo de gli ſpetiali.

VIRTV' *Di dentro*. Beuuto il liquor del Balſamo prouoca l'vrina, rompe le pietre, gioua alli ſtretti del petto, daſſi con latte à coloro c'haueſſero beuuto l'aconito: metteſi nelle medicine delle laſſitudini, & ne gli antidoti, & vale à i morſi de gli animali velenoſi. Il ſeme non hà tanta virtù, & manco d'ambedue il legno, daſſi commodamente à bere il ſeme ne i dolori laterali, ne i difetti del polmone, alla toſ-
B ſe, alle ſciatiche, al mal caduco, alle vertigini, & all'aſma, à difficultà d'vrina, dolor di corpo, à i morſi de i ſerpenti, beuuta la decottion del legno fatta con acqua vale alle crudità, à i dolori del corpo, allo ſpaſimo, & al morſo de i velenoſi animali: prouoca l'vrina.

VIRTV' *Di fuori*, Efficaciſſima virtù hà queſto liquore: leua via tutte quelle coſe, ch'offuſcan la viſta, & la pupilla de gli occhi, applicato con cerotto roſato gioua alla frigidità della madrice, prouoca i menſtrui, le ſecondine, & il parto, caccia vngendoſene il freddo, che procede dalle febri, & il tremore, purga le ſordine vlcere, matura & digeriſce le crudità: applicato il ſeme in profume è molto vtile alle Donne, & ſedendoſi nella decottion del ſeme apre
C l'oppilation della madrice, tirandone fuora l'humore. Il liquore oltre à ciò è efficaciſſimo rimedio per le catarratte applicato è meſſo ne gli occhi in poca quantità con acque di finocchio, & di eufragia: ma biſogna lauarlo prima molto bene con acqua roſa, & poi vltimamente con acqua di piantagine.

BAMBAGIA.

E vai-

FIGURE 2.2 Balsam in the *Herbario Nuovo* by Castore Durante (1529–90), 1636 edition, 65. Photo by DeAgostini. Courtesy of Getty Images.

the Eparch. At that time, trade activity in Constantinople was in the hands of guilds. It was highly specialized, with each guild handling either one type of goods or the goods related to a specific activity. All the trade was overseen by the Eparch, who was a high-ranking official, second only to the emperor. The *Book of the Eparch*, which had been promulgated (put into law and made publicly known) by the Emperor Leo VI the Wise (886–912) possibly in 911–12, contained all the articles of law that regulated trade. For each guild, it stipulated the products they were authorized to trade, the conditions of the trading, the places from where products could be acquired, and the intermediaries and buyers, in addition to the penalties for infractions and any other relevant matters. Spices were traded by the guild identified as the *myrepsoi*, who were not perfumers, as they are thought to have been according to the usual translation, but providers of the imported plants and products that were used for culinary, medical, and cosmetic purposes. The items that the *myrepsoi* were authorized to trade were not only spices such as pepper, spikenard, and cinnamon (with different provenances), but also several other products like aloe-wood (*Aquilaria malaccensis*), amber, musk, incense, balsam, indigo, lacquer, lapis lazuli, and sumac (*Book of the Eparch* 10).

The mention of Oriental spices in the *Book* should not necessarily be taken as a sign of their availability on the market and their presence on gourmets' tables in Constantinople. It might, instead, result from tradition, since these spices were already listed in the trade codes of the previous periods. A sign of this is provided by the textual tradition of *De materia medica*, where only two of the spices listed in the *Book of the Eparch* appear: cinnamon and spikenard. In effect, they were reintroduced into the text after they had been eliminated from some of its editions in previous centuries. These two plants do not reappear in the body of the text, but in an illustration on a flyleaf at the beginning of a mid-eleventh-century manuscript (Mount Athos, Lavra Monastery, MS Ω 75). This leaf seems to have been added after the manuscript was completed. It contains two representations of cinnamon (one as a leaf floating on a river and the other as a tree) and one of spikenard (as a low-growing bush, in the form of a small conifer). Interestingly enough, one of the representations of cinnamon (the leaf floating on water) closely corresponds to that of the plant in an Arabic manuscript, and the other two representations (cinnamon as a tree and spikenard as a bushy conifer) do not resemble in any way the plants they are supposed to represent. They manifestly were creations by an artist who had only seen the drug obtained from the plants and created these images on this basis: cinnamon sticks were drawn as the trunks of trees at the upper extremity of which some foliage was added, and a bunch of spikenard rootlets was represented upside down as a small conifer.

INTERNATIONALIZATION

Two treatises, almost contemporary and dating to the same period as the *Book of the Eparch,* allow one to complete the picture. They were by individuals coming from different places and stations in life: one was by the Arabic-speaking physician Ibn Butlan (d. 1066), and the other by the Greek scholar and writer Symeon Seth (eleventh/twelfth centuries). Ibn Butlan was born in Baghdad, where he practiced medicine at the hospital founded by the emir ʿAdud al-Dawla (936–83). In 1049 he moved to Egypt and, from there, he went to Constantinople. A Christian, he ended his life in a monastery in Antioch (now Antakya, Turkey). He is most famous for both a ferocious polemic with his contemporary ʿAli bin Ridwan (998–1061 or 1069), whom he met in Egypt, and his *Table of Health*. This work interests us here. According to its Arabic title, Ibn Butlan's treatise is the *Taqwim al-sihha* (that is, Tables of Health in an exact translation). It is a systematic presentation of 280 items that contribute to good health. Most of them are foodstuffs (nos 1–210). But several were activities such as singing, playing music, and dancing (211–13); traits of personality (214–16); consumption of alcohol, with drunkenness and vomiting (218–19); sleeping (221–4) and sexual activity (227–8); toothpaste (229); movement and sport (232–8); houses and their orientation (240, 265–6); water and bathing (239, 241–4); clothes and their fabrics (250–1); winds (267–8), seasons and places (271–4 and 275–8) with their climate, in addition to oils and their ingredients (247, 252, 257, and 253–6, respectively) along with syrups and other sugary preparations (258–64). The most original contribution of Ibn Butlan was his way of presenting the data he had collected: he renounced the discursive form of exposition that was traditional, and instead chose to tabulate data. The work's data took the form of 280 rows (each one devoted to an item) and 14 columns (each containing the data related to the *materia medica*: name; nature; degree of activity; usefulness; counter-effect and ways to counter-balance it; effect on health; usefulness according to the constitution of the patients, their age, the season, and the region; and comments by earlier physicians). The nature of the therapeutic action of each *materia medica* and its degree were expressed according to the system developed in antiquity by the Greek physician Galen (129–after ?216), which had been adopted by Arab scientists.

Symeon Seth is of importance comparable to that of Ibn Butlan, even though he is less known. Possibly born in Antioch, he traveled to Cairo, where he observed a solar eclipse in 1086. From *c.* 1071 on he was in Constantinople until he was exiled to Raidestos in eastern Thrace, possibly some time before 1112. He probably died there but when is unknown. He is best known among scholars for his translation of Arabic tales into Greek, and for a short treatise on the properties of 159 products (mostly plants). Although the treatise was written in Greek by Seth, it is usually known by the Latin translation of its title: *De alimentorum facultatibus* (On the Properties of Foodstuffs). The work was presented to the Byzantine Emperor Michael VII Doukas (r. 1071–8) and contains data coming from the Arab world and beyond, which might result from the translation of Arabic into Greek. Unlike Ibn Butlan, Seth presented data in a traditional, discursive way that followed the alphabetical order of the common name of the substances.

Returning to Ibn Butlan's *Taqwim* and its plants, the book opens with fruits: fig, grape, peach, plum, pear, pomegranate, quince, apple, apricot, and melon; it then continues with nuts, including jujube, which are followed by cereals including rice, and legumes: fava beans, lentils, and lupins; after this, it moves on to vegetables, including onions, garlic, asparagus, spinach, beet, and carrot. After a long section on meats, it returns to some plants, followed by perfumed waters and flavored wines. The treatise includes the plants from India that had been introduced into the Arab world and the Mediterranean, along with some of their derivatives: banana (20), coconut (21), rice (36) and preparations made of it (152), eggplant (67), citrus (202), pistachio (207), Indian melon (17), a preparation made of watermelon (166), sugar cane (204), sugar (169) and its derivatives (syrups [260–2]), distilled waters and other distilled products (camphor [255] and water with camphor [178]). Two substances grouped together in the *Taqwim* came from farther afield (the Himalayas): a plant (spikenard [180]), and parts of an animal (the glands of the musk deer [181]). All such substances had their nutritional and health properties defined according to the "mainstream" medical theory of that time, that is, Galenism, a fact showing that all of them had been fully assimilated into the contemporary medical theory and thinking in Ibn Butlan's time. They probably had lost their status as novelty and luxury foods, and were instead familiar in both cultivation and consumption.

Symeon Seth's *De alimentorum facultatibus* listed 159 substances, of which 100 were plants. Of the vegetables, almost all were common ones: cucumber (5), mushrooms (6), asparagus (11), basil (19), radish (25), parsnips (30), endive (45), cabbage (52), onion (58), lettuce (79), fennel (84), leek (113), horseradish (117), white beet (124), beetroot (125), and mustard (130). Some legumes were included: chickpeas (33), fava beans (145, 147), and lentils (146). There were several herbs: dill (9), basil (19), savory (41), caraway (61), coriander (64), nigella (black cumin, *Nigella sativa*) (81), rue (105), parsley (123), marjoram (132), and tarragon (138). Some species of nuts were included, none of which was rare: acorns (20), almonds (3), walnuts (49), chestnuts (51), pine nuts (53), and hazelnuts (74). There was a great variety of fruits: pear (4), apricot (17), quince (47), cherry (48), apple (78), medlar (90), pomegranate (115), peach (116), fig (119), grape (120) and raisins (121), and date (144), along with myrtle berries (88). This enumeration should not give the impression that Symeon Seth's treatise did not contain new data. It did list plants and other substances that were not those common in the Byzantine world. However, it is difficult to interpret *De alimentorum facultatibus* with regard to these plants and other products, and to determine its role as a possible indicator of the presence, and status as luxury food, of these plants and products in the Byzantine Empire. The problem is that many of the plants not included in the list above had already been known in antiquity. Citron is a good case. Seth's chapter reads as follows (Seth 1868: 52–3):

> Citron peel is intensely hot and dry to the first degree; its pulp is cold to the second degree. The inside is sour, cold, and dry to the second degree. The seed is warm and moist in the second degree.
>
> Its pulp is difficult to digest and produces a thickened phlegm, which is why it is usually eaten with honey.
>
> The peel aids in the digestion of food and is useful against black bile and the lack of tonus that comes from it. Taken in moderation it is good for the stomach. Taken in excess it becomes difficult to digest.
>
> Its preparation with honey and spices, which is called *diakitrion* [that is, made of citrus], warms the stomach, cleanses it of moisture, and digests food. The preparation without spices is difficult to digest.
>
> The seed is an antidote of poisons and bites by venomous animals, and helps particularly those stung by scorpions. Taken as a drink or in application, it provokes menstruation, and it harms the fetuses in the womb. Its acidity drastically stops yellow bile.
>
> It is surprising that hot bread put next to citrus makes it putrefy sooner.
>
> It is said that, when it is used also at night, it provokes a squint.

Since the citron had been introduced into the Mediterranean world in antiquity, Seth's chapter must be compared to ancient texts to assess its originality. Galen devoted a paragraph to citrus in his major treatise *De simplicium medicamentorum temperamentis et facultatibus* (On the Mixtures and Properties of Simple Medicines). Seth was familiar with Galen's immense oeuvre, so familiar, in fact, that he wrote a short refutation of some of Galen's theories. A comparison of the two is thus logical. The data in Galen's entry on citron can be summarized as follows: among substances that are both drying and cooling, its seed is sour and drying to the third degree. Its peel is drying without being sour, thus being of the second degree. It is not exceedingly cooling, but only moderately. The pulp makes bodily humors denser, and generates phlegm and cold. The internal part of the seed is bitter and therefore diaphoretic and drying to the second degree. Its leaves are drying and diaphoretic (8.12.19 [Galen, Gottlob ed. 1821–1833, 12.77]).

Seth and Galen's entries do not correspond, and Galen's cannot be considered to be the source of Seth's, not only because it does not provide much of the information found in Seth, but also, and particularly, because it identified the therapeutic properties of citron seed in an opposite way: for Galen, indeed, it was sour and drying to the third degree, whereas, for Seth, it was warm and moist to the second degree.

Searching for possible Arabic sources, Ibn Sina's *al-Qanun fi al-Tibb* (The Canon of Medicine) immediately comes to mind, as that was the major reference for medicine and *materia medica* in the Arab world in the period preceding that of Symeon Seth. Its entry on citron reads as follows (*Qanun* II, *līmūn*) (extracts):

> **Nature:** Citron is a well-known drug ... Yeminite lemon has a rind with fragrance similar to that of cinnamon.
>
> **Temperament:** Citron is hot in the first degree and dry in the last of the second. Its pulp is hot in the first degree with some moisture. There are some physicians who say it is cold and moist in the first degree tending to coldness. Its sourness is cold and dry in the third degree. The seeds are hot in the first degree and dry in the third ...

Visual Organs: A collyrium is made from the juice of the sour portion of citron. It helps in removing the yellow tinge from the eyes from jaundice. When taken orally, Yemenite citron rind is useful in treating conjunctivitis ...

Food and Alimentary Organs: Citron pulp is harmful to the stomach. It causes excessive gas in the digestive tract. Its digestion is slow. Therefore, it is important to use it in the form of a preserve, in particular one made in honey and used over a long period of time. The leaves strengthen the stomach and organs in the cavities of the body. When the buds and rind are used in foods in the way that spices are used, it helps in digestion. The rind itself, however, is not digestible due to its hardness ...

Excretory Organs: ... The pulp causes acute abdominal pain. The juice of the sour part produces stability in the stomach. Its sour part is useful in treating cholera. The seeds are laxative and also useful in treating haemorrhoids ...

Poisons: 2 dirham of its seeds, when converted into soothing or cleansing liquid rubbed on the skin or a drink with wine and hot water, can counteract all poisons. It is to be used orally or as a soothing or cleansing liquid rubbed on the skin. This is particularly the case with scorpion sting. The rind is similar. The extract of the rind is useful in treating snake bite when used as a poultice.

Ibn Sina's entry is much more complete than that of Seth, something that would not contradict its possible use by Seth, however, as Seth could have excerpted the *Qanun*. There are similarities that might favor this interpretation, specifically about the pulp, deemed difficult to digest, and the recommendation to use it in combination with honey. Both texts expanded on this, though in different ways. Finally, both recommended the use of citron in the treatment of poisoning and envenomation (injection of venom), especially by scorpions. But, aside from these common points, both substantially differed, especially in the manner of identifying the general therapeutic properties of citron and its parts. These differences can be summarized as follows:

Ibn Sina	**Symeon Seth**
pulp: hot, 1st degree, some moisture	**pulp:** cold, 2nd degree
sourness: cold and dry, 3rd degree	**sour part:** cold, dry, 2nd degree
seeds: hot, 1st degree, dry 3rd degree	**seeds:** warm, moist 2nd degree
whole fruit: hot, 1st degree, dry, 2nd degree	**peel:** hot, dry, 1st degree

As this comparison makes it clear, it is difficult to believe that Ibn Sina's entry might have been used by Symeon Seth as a source that Seth would have translated from Arabic into Greek. The differences in the definition of the properties of citron and its parts are too great for a possible relation between the two texts. As Ibn Sina stated, physicians identified the properties of citron in diverging ways, including with some of them considering that citron is cold and moist in the first degree.

A closer similarity is discovered when Symeon Seth is compared with his almost contemporary Ibn Butlan. Given the tabular presentation of the *Taqwim*, data from the latter need to be collected and organized as follows to allow for a comparison with those of Seth.

Ibn Butlan	**Symeon Seth**
general property: medium hot	–
peel: hot and dry, 2nd degree	**peel:** hot and dry, 1st degree, intense
seed: hot and dry, 2nd degree	**seed:** hot and moist, 2nd degree
pulp: cold and moist	**pulp:** cold, 2nd degree
juice: cold, dry, 3rd degree	–
sourness: dry	**sourness:** inside, cold, dry, 2nd degree
uses: facilitates digestion	**uses:** peel aids digestion
strengthens the stomach	good for the stomach
is difficult to digest	in excess difficult to digest
against poisons	against poisons and venoms
comments: Bakhtayshu' claims that eating it at night provokes strabismus	**comments:** "It is said that, when it is used at night, it provokes squint"

Symeon Seth's entry is close to that of Ibn Butlan, even though it does not completely correspond to it. Nevertheless, it might be considered to be a translation of a text very similar to that of the *Taqwim*, be it one of its sources or a derivative of it. One could imagine that Symeon Seth and Ibn Butlan crossed paths, either in Egypt or in Constantinople where both lived at about the same time. And, if they did not meet, they might have had access to the same literature.

If we thus posit that Seth's *De alimentorum facultatibus* is a translation and adaptation of contemporary Arabic textual material, instead of being a rearrangement of earlier Greek data, it can be taken as an indicator of the introduction of new plants and products from the Arab world into the Byzantine Empire. The items that Seth's treatise might have contributed to introduce to Byzantium could be the following by categories (items names are followed below by the reference to Seth's text [the page numbers in Langkavel's edition]): vegetables such as artichoke (52) and eggplant (70); fruits such as damson plums (34), jujube (40), citrus (52–3), and melon (84–5); nuts like pistachio (87); spices including ginger (40–1), cardamon (53), cinnamon (55), cloves (56), nutmeg (56), pepper (81), sesame (99–100), and spikenard (103); scented substances like balsam (28–9) and musk (66–7); medicines, among them those that contained camphor (58–9) and rosewater (111); recreational products made with hemp (60–1); ritual (church) and household products containing frankincense (62); and condiments containing sugar (96–7).

It is highly probable that the substances imported from the Arab world were luxury goods at Seth's time because of their novelty, possibly their high price, and certainly the originality of their initial uses. It is significant that some of the species that contributed to the Arab Agricultural Revolution/Medieval Green Revolution did not appear in *De alimentorum facultatibus*: banana, coconut, orange, and spinach. This could be the sign that the process of introduction of these species (or reintroduction for spinach) was still happening in Byzantium in the eleventh century, as the Athos manuscript of *De materia medica* already suggested.

Limitations in the range of products grown in, or traded by, the Arab world (including luxury items) that were present in Byzantium could be attributed to the obstacles to trade: not only the distance, but also regulations and taxation, a lack of response from the market after centuries of economic restrictions, and also a slow return to earlier consumption habits. However possible these explanations might be, they were probably not exclusive and do not take into consideration another factor possibly reflected by the documentation: a transformation of tastes and luxury.

A DIFFERENT *LUXE*?

Some time around the mid-twelfth century in Byzantium a vindictive individual who remains anonymous "usurped" the name of the well-known author Theodore Prodrome (*c*. 1100–*c*. 1170) while writing four poems in which he provided a detailed picture of luxury food in Constantinople. One of these poems, the third, was presented as if it had been written by a monk who identified himself as Hilarion Ptochoprodomos at the Philotheou monastery in Constantinople in the title and verse 387. The poem was presented as a complaint addressed to the emperor about the luxurious lifestyle of the abbot (*higoumenos*) of the monastery and his son. The author of the poem, whoever he was, compared their *grande vie* to his own humble one and the bad treatment he constantly received. Food made available was among the many differences between the superiors of the monastery and the author of the poem. With an extremely ironical tone, Hilarion repeatedly described the luxurious meals of his superiors. In a passage, one such extravagant meal consisted of four courses:

> First comes the boiled dish, a flounder in its broth. Second, a sauce made of fresh hake purée. And, third, a sweet-and-sour, of golden fish, with spikenard, valerian, cloves, cinnamon, mushrooms, and vinegar as well as unsmoked honey. In the midst of the dish lies a big, golden gurnard, and a grey mullet three palms-long, caught from Rygin waters, and a bream *à point* (perfectly cooked), a choice one ... Fourth, a good, well-roasted dish, and fifth the fried dish: the best parts (from the middle), moustached red mullets, a full double-pan of big "small-atherines," a flounder well cooked on its own, with garum and sprinkled from top to tail with caraway; and the fry from a big sea bass.
>
> (*Prodromic Poems* 3.147–54 and 158–63)

These several dishes were followed by a casserole, which arrived on the tables preceded by its aroma:

> After all the dishes that I have mentioned had been served came also—what a marvel!—the *monokythron* [a casserole], with a light steam diffusing its aroma. If you wish, know also the ingredients of this *monokythron*: four hearts of cabbage, dense and snow-white; the neck of a salted swordfish; the best cut [the middle] of a carp; twenty splendid *glaukoi* [a fish]; smoked flesh of sturgeon; fourteen eggs; Cretan cheese; four fresh cheeses; a bit of Valach cheese; a liter of holy olive oil; a handful of pepper; twelve little heads of garlic; fifteen dry mackerels, and, on top, a tumbler as large as a tray of sweet wine.
>
> (*Prodromic Poems* 3.174–85)

The luxurious food served to the superiors during "ordinary" days, was ironically contrasted to that of the fasting days:

> On Wednesdays and Fridays they stick with a strict regime: indeed, they do not eat any fish during these days, my lord, but only delicious small breads, lobsters and real crabs, and boiled crayfish, pan-fried shrimps, and vegetables and lentils with a medley of oysters and mussels, and [...], Lord, comb-shells, knife-shells, a purée of beans, rice with honey, unshelled beans, olives and caviar, and autumn botargo against lack of appetite, small apples and also dates, dried figs, and walnuts, and raisins from Chios, and the sweet made of lemon [...] To digest this strict regime, sweet wine, Gamitic, Cretan, and Samian ...
>
> (*Prodromic Poems* 3.273–86)

The contradistinction between the lifestyle of the superior and his son, on the one hand, and Hilarion, on the other, is made explicit in a passage where the components of the respective meals are compared one by one, with "they" referring to the superior of the monastery and his son, and "we" to Hilarion and his more ordinary companions:

They	We
Frogs	Holy broth
quantities of Chian wine to satiety	Varna wine cut with water
quantities of sweet wine in large cups	water in tumblers
refined bread	bran bread
yeasted bread with sesames	strained barley water the wheat filtered out
multiple beignets with honey	their pseudo-medicine (Holy broth?)
full plates of sweets	bitter castor oil seeds
...	...
several pieces of bass mullets, and squids	this smelly, smoky Holy broth
quantities of bluefish, catfish, and turbot	that thing again (What do they call it?)

(*Prodromic Poems* 3.311–22, 325a–d)

What Hilarion calls the *Holy Broth* is boiled water with onions:

> ... As for us, thus, we eat only the *Holy Broth* ... In a two-handled caldron of about four units, the cooks pour water up to the edge, and light an abundant fire underneath the caldron; they toss in about twenty onions ... The cook drops in three drops of oil, and he adds woody twigs of thyme as flavor, and he then pours the broth on pieces of bread. This is what they do give us, and we do eat it, and they do call it *Holy Broth*.
>
> (*Prodromic Poems* 3.287–95 and 298–301)

The luxuriousness of the meals described by Hilarion was evident to those who knew the dietary rules of Byzantine monasteries. These were stipulated in the foundation charters of monasteries (the so-called *typika*), which prohibited consumption of meat and authorized, instead, eating of vegetables and some species of fish. In the eleventh–twelfth centuries, *typika* generally prescribed two or even three courses per meal, authorized wine, and also allowed for more abundant meals of four courses during some periods (after Christmas) (Thomas and Constantinides-Hero 2000: 2.480–1, 533–4, 691–3, 746–7, 826). In all cases, however, food consisted mainly of fish and vegetables, and consumption should have been moderate.

It is relevant for the veracity of all the descriptions of luxury food that what the *Poem* calls the *Holy Broth* was not an invention of the author of the *Poem*; it was prescribed in the *typikon* of the twelfth-century monastery Kosmoteira located near Bera, Greece. It was indeed made of onions boiled in water and served with oil over a few pieces of bread (Thomas and Constantinides-Hero 2000: 826). Knowing this, the luxurious lifestyle of Hilarion's superiors can be better understood: it was quantitatively and qualitatively transgressive, characterized by an extraordinary abundance of food, and a fine selection of fish and shellfish, delicate sweets, and rare wines, probably expensive. Vegetables were almost absent, and spices did not contribute in any significant way to enhancing the taste of the ingredients. Luxury came firstly from the ingredients and secondarily from their preparation, but certainly not from vegetables that the Arab Agricultural Revolution reintroduced or introduced, or from spices traded by the Arab world that created an artistically orchestrated palette of tastes and fragrances harmoniously combined, possibly also with some touch of color. *Luxe* here resided in the pure taste of fine ingredients carefully selected and served in a well-crafted menu leading to a crescendo, in addition to variety and abundance. This would suggest that alimentary *luxe* in the late Byzantine world did not reside in previously unknown vegetables valued on the basis of their novelty and rarity. It did not reside, either, in expensive spices used alone or in combination to sprinkle on the food and create exquisite aromatic experiences arousing all the senses, in a synesthetic experience of smell, taste, and vision.

BACK TO THE WEST

In the West, by contrast, there seems to have been interest in exotic and rare vegetables and spices. A text attested in different forms from the end of the thirteenth century listed multiple plants used for medicinal and alimentary purposes. The text is traditionally identified with the generic title *Tractatus de herbis* (Treatise on [Medicinal] Plants). Its oldest surviving form is Egerton MS 747 held in the British Library, London, and possibly dates from the decades 1280–1310. Another significant member of this set of manuscripts is the codex Sloane 4016, also in the British Library and probably dating from the 1440s. These manuscripts and other derivatives of the *Tractatus* are lavishly illustrated with color representations of the plants and other *materia medica* studied in the treatise.

The most striking characteristic of these manuscripts and the variant forms of the *Tractatus* is that they include many, if not all, of the Far Eastern plants traded by the Arab world, in addition to the species cultivated by the Arabs and introduced/reintroduced by them to the West. There are more than three dozen of them. Traded species included spikenard, amomum, costus (*Saussurea costus*), cassia, aloe-wood, frankincense, mace, myrrh, pepper, and ginger. Cultivated plants were citron, banana, eggplant, oranges, pistachio, melon, and jujube. To these plants was added an animal product (musk from the musk deer), a distillate (camphor), and a mineral (Yemen alum). The illustrations of all the plants in these manuscripts are telling: traded commodities were represented in a way that made it clear that they were not known as living plants, but only as dried drugs. Cultivated plants, however, were figured as living ones, even though the representation of some of them was not faithful (see Figure 2.3).

The large number of these eastern plants and substances reveals a taste for exotic substances and new vegetables, all of which contrasted with both the earlier medieval alimentary customs and Byzantine culinary, as discussed above. The supposed place of origin of the textual corpus in which these substances appear is revealing: it has traditionally been identified as Salerno, south of Naples, on the shore of the Tyrrhenian Sea. At that time, Salerno was experiencing a major flowering of science and culture, characterized by exchanges between the different peoples inhabiting or frequenting the region: Latins, Arabs, and Jews, in addition to Byzantines. The interaction between these communities is thought to have been the origin of the local medical school known as the Salerno Medical School. This multiculturality is evident in the manuscript Sloane 4016 mentioned above, in which the representations of plants are not accompanied by any text, but only by captions providing the names of the plants in the different languages of the area. All such names are written in the Latin alphabet according to current conventions of transliteration. These names were mostly Greek, Latin, and Arabic, in addition to some dialectal and regional ones. The use of plant names in various languages facilitated

FIGURE 2.3 Nutmeg (on the left) in a copy of the *Tractatus de herbis* (Modena, Biblioteca Estense, Est 28 e M 59, f. 98v). Photo by DeAgostini. Courtesy of Getty Images.

a connection between the representations of plants and the texts in different languages owned and read by the people of diverse cultural origins active in the region. The multilingual nature of the manuscript is an indicator of a multiethnic and multicultural society in which plants and their uses circulated and possibly modified earlier alimentary and culinary uses. The resulting "new cuisine" generated a "new luxury," which might have evolved over time (see Figure 2.4). Whereas newly introduced vegetables probably were luxurious when they first appeared, these became part of the range of everyday produce once their cultivation intensified and made them more available. They then left the realm of novelty, which was reduced to imported and expensive spices sought after for the eastern taste they added to local or naturalized products.

Once this new luxury was established in the south of the Italian Peninsula, it moved northward as the Latin translation of the *Taqwim al-sihha* by Ibn Butlan showed. The work was translated into Latin in the second half of the thirteenth century. Its title, *Tacuinum sanitatis* (Tables of Health) corresponds to the Arabic title (*Taqwim al-sihha*), with a transliteration of *Taqwim* (meaning table) transcribed as *Tacuinum*, and a proper translation of *sihha* (meaning "health") rendered as *sanitas* in Latin (that is, "health"). The exact place, time, and name of the translator are uncertain, as various surviving manuscripts offer divergent information. According to one codex, the translation was made at the court of the king of Sicily Manfred (*c.* 1232–66), possibly in Palermo. But, in another manuscript, the translation is attributed to Farag ben Salem, known for his

FIGURE 2.4 A medieval kitchen from *Tacuinum sanitatis* (Latin manuscript, fourteenth century, Vienna, Österreichische Nationalbibliothek, MS Series Nova, lat. 2644). Photo by DeAgostini. Courtesy of Getty Images.

translation into Latin of Arabic medical literature. Of Jewish origin, he was active at the court of Charles I of Anjou (1226 or 1227–85), who opposed and killed Manfred in the battle of Benevento in 1266, and became king of Sicily in a change of dynasty from the Hohenstaufen to the Anjou. Whatever the place and time, and whatever the identity of the translator, the translation was made in a multicultural milieu not different from that of Salerno. Interestingly for the purposes of this chapter, it attests to the same taste for Oriental spices and vegetables introduced by the Arabs, even though the number of such products was slightly lower in the *Tacuinum* than in the *Tractatus*: two dozen in the *Tacuinum* instead of three in the *Tractatus*, with fluctuations in these numbers according to the manuscripts, since their writers took some liberties in transmitting the text, adding or deleting items. The nature of the substances of interest here was slightly different from the *Tractatus* to the *Tacuinum*, possibly as a result of the transfer of the text from southern to northern Italy, which also witnessed to a different kind of luxury.

The *Tacuinum sanitatis* is mostly known through a group of lavishly illustrated manuscripts produced in northern Italy (possibly Lombardy) between the very late fourteenth through the fifteenth centuries. The most salient trait of the illustrations in these manuscripts is that they "staged" each and every *materia medica* in an image that expresses magnificence, abundance, and luxury. Interestingly, the number of traded substances is drastically reduced from the *Tractatus* to the *Tacuinum*; it is limited to camphor, camphor and rose water, along with galanga, sandalwood, and musk. As for produce introduced/reintroduced by the Arabs, this consisted of eggplant and spinach for the vegetables, and banana, citron, dates, jujube, melon, coconut, and oranges for the fruits, in addition to sugar. This sharp limitation might have resulted from differences in the environment, since the northern plain of Lombardy did not offer the same climatic conditions as the southern part of the Italian Peninsula, coupled with a less central position with regard to trade routes. The selection of produce reveals a taste for warm-climate fruits that might not have been common in Lombardy and were most probably luxurious products accessible only to the wealthy minority who commissioned the large-size, lavishly illustrated manuscripts of the *Tacuinum*. Their luxury status did not reside only in their consumption—for example, of sweet, red- and yellow-colored, sunny, juicy fruits—but also in the enjoyment of the books that depicted these foods and extolled their benefits for health. The sensory experience triggered by the colorful, sunny, and warm foods and spices that probably enhanced tables loaded with venison, poultry, and common vegetables, was translated in the images of the *Tacuinum* manuscripts with a subtle, yet perceptible play on colors. Whereas the illustrations of meat, fish, and ordinary plant species are dominated by grey, light brown, matte, and cold colors, those featuring spices and Mediterranean vegetables are rendered with a light, bright, and warm palette with a wide range of vivid colors and tonalities.

As these illustrated manuscripts of the *Tacuinum sanitatis* indicate, consumption of luxury plants and spices and sensory delight in Lombardy in the late Middle Ages accompanied visual and intellectual enjoyment, and a feeling of belonging to an exclusive, and economically and politically powerful milieu. It was a synesthetic experience that in a certain sense bridged the gap created by history during the centuries from 476 to 1453–92, by bringing together the tasteful experience of the Mediterranean world and the rising power of the northern European nations in the making.

CONCLUDING REMARKS: LUXURY AND ITS FATE

Between 500 and 1400, luxury foods took on very different forms in the Mediterranean world, ranging from costly Oriental spices in the late Roman Empire to plants widely cultivated in the Arab world and transmitted to Byzantium in a renewal of lost traditions, to tables loaded with local fish, in Byzantium, and Mediterranean fruit species, along with some Oriental spices, in fifteenth-century northern Italy. Plants were not always the luxury food of choice, as the Byzantine case suggests, and distant origin was not necessarily what constituted luxury. This hints at a meaning of *luxe*—whatever the goods that contributed to it—during the period under consideration here, suggested by the *histoire vécue* (real-life story) of the mid-twelfth-century Byzantine author of the so-called *Prodromic Poems*: *luxe* did not reside in the inherent nature of the goods, but in the distance that these goods generated between an established group and any other group in the same context that did not have access to these goods for whatever reason. The multisensorial, synesthetic, and taste- and enjoyment-enhancing experience was secondary, coming after a feeling of uniqueness, possibly in combination with a sense of evanescence or, on the contrary, of permanence. It might be that these contradictory feelings best represent the essence of luxury plants.

CHAPTER THREE

Trade and Exploration

FEDERICA ROTELLI

Between the fourth and the sixth centuries CE, the cooling of the climate, the onset of epidemics, and the southward movements of the nomadic populations of the plains of Central Asia contributed to bringing about a deep crisis, particularly in the Mediterranean and China, and less so in India, where, however, the invasions of the Huns during the fourth century wrought havoc, as they also did in China. In northern India, the Guptas established a highly feudalized society between 319 and 543, and, in China, the Han Dynasty (206 BCE–220 CE) was followed by four centuries of great political and social unrest, concluded by the establishment of the Tang Dynasty (seventh to tenth centuries) and the reunification of the Empire. In Europe, the Germanic and Slavic populations pushed by Asian groups moved southwards, to France, Spain, and Italy. Between the fifth and the ninth centuries, China went first through a phase of demographic reduction and withdrawal from the countryside; then, it entered a phase of demographic and economic growth that lasted for five centuries, from the ninth to the fourteenth century with only a few interruptions. Europe followed. The new great crisis of the fourteenth and fifteenth centuries was the prelude to an unprecedented expansion of the Europeans, who occupied and transformed the Americas, Australia, and important parts of Asia and Africa in the space of four centuries (the sixteenth to nineteenth centuries).

Major movements of populations punctuated the period under consideration here, lasting until late in the thirteenth century with the Mongol invasion in Asia. By putting into contact peoples who had not been so before and might not even have known of each other, such movements deeply transformed the Old World, which became a highly multiethnic and multicultural region. Moving populations brought with them their habits and culture, including plants of their region of origin which modified not only the environment of their new location, but also the agricultural practices, the food, and the medicine of the populations with which they integrated. Taking advantage of the routes opened by these migrations and also going further, exploring the spaces newly opened to them, populations traded these commodities and created an international space in which peoples and plants circulated extensively. This chapter reconstructs the dynamics and itineraries of these exchanges and the modifications they introduced into the life of populations, the details of which are analyzed in more depth in the chapters on luxury foods and other plant products (see Chapters 2 and 4). It follows the chronology of these exchanges as far as the intertwined histories of these transfers allow.

CRISIS AND EMERGENCE OF THE EUROPEAN AND CHINESE ECONOMIC POLES

Before their expansion eastwards in the Mediterranean, the Romans knew and used some spices from India: cardamom, ginger, putchuk (costus; *Saussurea costus*), spikenard (*Nardostachys jatamansi*), sesame seed, and taro (*Colocasia esculenta*). Pepper, too, was imported from India as from two species: black pepper (*Piper nigrum*, from which white pepper was obtained) and long pepper (*P. longum*) (Young 2001). Cinnamon and cloves, which came from the Indian food tradition, were present in the Roman world where they were originally used for medicinal purposes.

During their conquest of the eastern Mediterranean from the second century BCE to the early second CE the Romans came into direct contact with several plant species previously unknown to them, such as spices like coriander and oregano in Egypt; saffron in Cilicia; caraway in Caria; cumin in Ethiopia, Syria, and Libya; and sumac (*Rhus coriaria*) in Syria (see Figure 3.1). They also discovered several kinds of foodstuffs which they cultivated themselves: legumes, among them broad beans, chickpeas, peas, and lentils, the latter being imported from Egypt and Syria, subsequently to be cultivated above all in Sicily; vegetables, like cucumbers, pumpkins, cabbages, asparagus, and broccoli; fruits, like plums growing in Syria, apricots originally from China via Media (Upper Mesopotamia), melons and watermelons from Egypt, in addition to peach and jujube (*Ziziphus jujuba*) trees, azaroles (*Crataegus azarolus*), medlars, figs, quinces, and pomegranates; and also nuts, among which were chestnuts and pistachios (André 1961; Lanfranchi 2006).

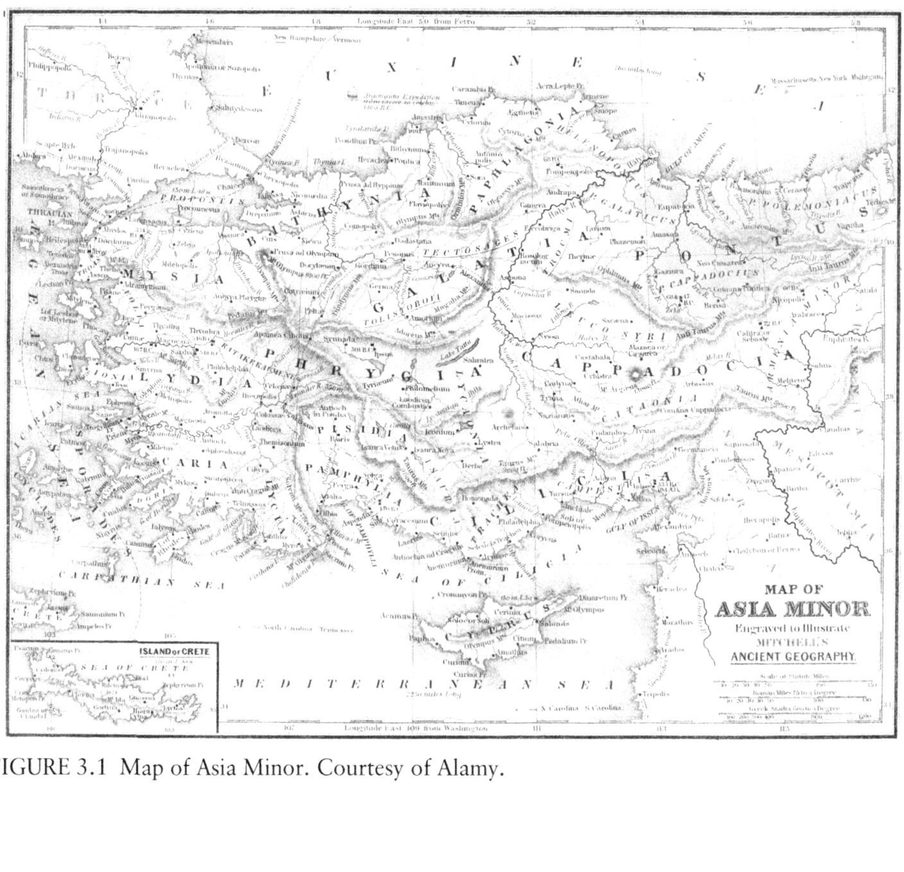

FIGURE 3.1 Map of Asia Minor. Courtesy of Alamy.

In the early second century CE, the Roman conquests were drawing to a close, and the contribution of the East to the Empire's diet was diminishing. The range of cereals, fruit, and legumes enjoyed by the populations of the Empire for two centuries greatly reduced in the early fifth century. Cereals were the first to lose the predominant position they had in the first-century CE production of Rome. The scanty yield of cereal harvest resulted not only from the unfavorable climate, but also (if not above all) from an agricultural decline. Cereal species that were less productive than wheat were cultivated because they could better withstand bad weather and could be consumed throughout the year (Fumagalli 1978). Rye and oats, which were known as wild grasses and had been selected and cultivated in the first centuries of the Roman Empire, became quite widespread in the Middle Ages. Millets and sorghum, too, which arrived from Africa between the second century BCE and the second CE, began to be cultivated at the beginning of the Middle Ages (Montanari 1984: 41).

The Germanic populations who settled in the western regions of the Roman Empire in the fourth and fifth centuries did not find a thriving economy (Fumagalli 1993). The wars that the Roman Empire waged against the peoples on its borders at that time contributed to bringing about a demographic crisis, further worsened by famines and epidemics that depopulated the countryside. Bubonic plague, which appeared repeatedly in the Eurasian continent from 180 onward, was most disastrous. The climate change, particularly toward cold, damp weather, reduced the harvests and promoted the already-advanced process of erosion of the earth, sparking off a slow, inexorable deterioration of the countryside, a crisis in farming production, and a general decline in the economy (Slicher Van Bath and Hendrik 1966). In the plains, the farmed land was taken over by woods and marshes (Montanari 1988: 13). Even oil production, which peaked in Italy during the first century, underwent a gradual reduction between the late Roman Empire and the Lombard period (fourth to eighth centuries) (Pini 1990). Many urban settlements of the Roman Empire were in a state of dereliction (Duby 1973; Lopez 1976: 10–13).

The food-production model of the Germanic world modified European agriculture. In continental Europe, the Germanic populations had developed a production based on cereals, chiefly barley, wheat, and oats in western Germany, and rye and millets in eastern Germany. This production was combined with hunting, sheep-breeding, and foraging for wild fruit in the woods and uncultivated areas. Alongside a rudimentary knowledge of the soil, these practices resulted in a sylvan-pastoral economy (Gasparri 1996). The transfer of this economy to the regions at the borders of the Roman Empire modified the status and perception of these regions in contemporary culture. In the Roman Empire the areas at the border were covered by woods and pastures. The uncultivated wild areas, which were marginal previously, became not only essential to the production system, but also culturally appreciated. Wild fruits were not only consumed as foodstuffs and flavoring agents, but also used as ingredients for the preparation of fermented beverages. The initial, radical difference between the agricultural tradition of the Mediterranean, based on wheat, wine, and oil, and the forest culture of the Germanic tradition resulted in a new type of foodstuff production and to an original consumption system, opening the era of the new Roman-Barbarian and Christian Europe (Duby 1973; Montanari 1988, 1990).

The reduction of land trade increased by the abandonment of many Roman roads at the height of the sixth century was further aggravated by the crisis of the artisanal and industrial production. In the sixth century, Rome lost its importance as a trade center, whereas Constantinople, the capital of Byzantium, emerged in the East as the main commercial center of the "Great Sea" thanks to its position between continents, the Mediterranean, and the Black Sea (Gasparri 1996: 330). The economic and social transformations of the Roman Empire at that time surfaced, only partly in different times, in most of the regions of the Eurasian continent. Unfavorable demographic and economic conditions affected Persia and Byzantium in the seventh century. Further East, China also went through a similar process after the fall of the Han Dynasty (206 BCE–220 CE). After its reunification under the Jin Dynasty (280–317), China was divided again between north and south. From the fourth century onwards, nomadic-pastoralist and hunter Huns and Turks coming from the north invaded its northern part (Von Glahn 2010: 140–5). There, the Turkish administration and constant civil wars transformed the countryside once thickly planted with wheat, millets, and sorghum, into uncultivated lands (Gourou 1972), whereas the south, under Chinese rule, experienced a surge in agriculture (Twitchett 1979). With the Sui Dynasty (589–618) shortly followed by the Tang Dynasty (618–907), China was reunified and started recovering from its demographic and economic crisis: in the eighth century, society picked up rapidly, more quickly than in Europe, soon attaining its full splendor, with a new economic expansion under the Song Dynasty (918–1279) (Corradini 1996). A series of farming and tax reforms boosted the agricultural, economic, and commercial system. Progress in farming techniques and hydraulic systems promoted a vast diffusion of wheat farming (*Triticum aestivum*), both in the north, particularly in the Shensi region where the Tang Dynasty had established its capital, and in the south, where rice continued to be the main crop for climatic reasons, even though its cultivation moved to northern China at that time (Sabban 2006a: 212–13).

Trade, which played only a secondary role in ancient China (Lopez 1976: 57), was taking place south of China, at Funan in the Delta of the Mekong (modern southern Cambodia and Vietnam), where pepper arrived from India, spices and medicinal herbs from Malaysia, and frankincense from Arabia (Von Glahn 2010: 151). From the end of the late seventh century through the eighth, however, the Chinese traded with both the Arab merchants, who had established commercial colonies in the ports on the southern coast of China, in Guangzhou and Quanzhou (Fujian), and the Uighur Turks, who created a military protectorate in northern China after 750. Trade with the West was carried out chiefly through caravans that crossed Central Asia via the Silk Roads, which constituted a two-way channel of exchanges, from China to the Mediterranean and also from Asia Minor and the regions crossed by the Roads to China. In Turpan, in northwestern China, the merchants coming from Sogdiana dealt in saffron from Asia Minor and northern India, sugarcane from India, and medicinal plants. Black pepper, which became a much-used ingredient of Chinese cooking in the late eighth century, came from southern India.

Plants from the Middle East and Central Asia that were consumed for dietary and medicinal purposes in these regions, were introduced into Chinese life between the seventh and the ninth centuries: yellow peaches arrived from Samarkand; grapes, imported from Qočo (Turpan) and acclimatized under the Emperor Tai Zong (626–49), were used for wine-making; aromatic jujubes and almonds came from Turkistan, with the latter cultivated in China during the last period of the Tang Dynasty; pomelos (*Citrus maxima*) and citrons (*C. medica*), myrobalans (*Terminalia chebula*), and the betel palm (*Areca catechu*) were introduced from India and the Malay Archipelago; sugarcane, which was already known in China, was more extensively cultivated thanks to the Emperor Tai Zong, who sent in 647 a legation to Magadha, in northern India, to study its production; the best pine nuts (*Pinus koraiensis*) were imported from Korea; sesame, native to India, was introduced perhaps before the seventh century; date palm, imported from the Middle East to Guangzhou, was cultivated there from the ninth century onward; pistachio trees from Persia grew in the area of Lingnan in the ninth century; and fig trees (*Ficus carica*), which also came via Persia, were cultivated in China as early as the middle of the ninth century.

Thanks to the intensification of trade in China and Southeast Asia, several spices gained in importance for both dietary and medicinal purposes: black, long, and cubeb (*Piper cubeba*) pepper; cardamom, nutmeg, cloves, and mustard (*Brassica nigra*), in addition to ginseng (*Panax ginseng*), imported chiefly from Korea, and saffron, from Kashmir and other Indian regions (Freeman 1977; Schafer 1977). Under the Emperor Chen-Tsung (968–1022) of the Song Dynasty, food production greatly increased thanks to the cultivation of a new variety of rice imported from Champa (Vietnam) in the Yangtze Valley. Its yield was much higher than that of the previously cultivated cereal crops in dry lands, something that stimulated new developments in the production of such other commercial crops as tea and sugar cane (Sabban 2006a).

BYZANTIUM BETWEEN THE WEST AND THE EAST AND THE TRANSFORMATION OF THE SPICE MARKET

During the reign of the Byzantine Emperor Justinian (527–65) and up to the first decades of the seventh century and the advance of the Arabs, the territory of Byzantium extended along the coasts of the Mediterranean and the Black Sea. The main ports and markets were Alexandria, Tyre, Antioch (near today's Antakya), Beirut, and Constantinople. On land, traders in the sixth and seventh centuries used the routes that had linked the main cities of the Empire to Asia since ancient times, even though the wars between the neighboring empires of Byzantium and Persia until the early seventh century, and Byzantium and the Arabic world later, interrupted these routes or required modification of their centuries-old itineraries in favor of two maritime routes: one to Egypt via the Red Sea, and the other to Syria via the Persian Gulf. The products were the same as in the Roman Empire of the previous centuries, with one major exception, however: in the fourth century if not earlier, papyrus was abandoned for parchment, which was not only more resistant, but also less expensive. With this change in book production, the Egyptian production of papyrus and its industry of book-making came to a close (Abulafia 2012: 247).

Together with Egyptians, Syrians, and Jewish merchants, the Byzantine traders played a decisive role in the westward exchanges between the eastern Mediterranean and southern Italy, southern France, Spain, and northern Africa. Established in Carthage, Cordova (where there was a Greek colony), Merida, Marseilles, and Arles, all of them imported spices from the Far East, wine from Gaza, olive oil and wheat from Africa, and also rice, dates, and figs from their native regions (Riera Melis and Soler Sala 2016). This is how Anthimus, a Greek physician who had taken refuge at the court of the Frankish king Theodoric I (*r.* 511–533/4), could include pepper, putchuk, spikenard, ginger, and cloves, the latter of which were an absolute novelty in cookery at that time, in the manual of dietetics that he completed around 515: *De observatione ciborum* (On the Observance of Foods) (Anthimus 1927: 5–13).

The Byzantines also traded with the East. In his *Topographia Christiana* (Christian Topography), the Alexandrian merchant Cosmas Indicopleustes reported his travels to East Africa, the Arabian Gulf, and the western coast of India between 535 and 547. He devoted the eleventh book of his work to the island of Sri Lanka (Ceylon, formerly Taprobane), which was then an important exchange point in the Indian Ocean, in addition to being a point of convergence between China and Southeast Asia and the West. Cosmas described the activity in Sri Lanka, where several products from China and Southeast Asia (including silk, aloe-wood [*Aquilaria malaccensis*], cloves and clove-wood, and sandalwood [*Santalum album*]) were loaded onto Indian and Persian ships that sailed along the Persian Gulf (Cosmas Indicopleustes, ed. McCrindle 1897/2010: 447). From the ports of Apologos or Charax Spasini (al-Muhammara) in the Persian Empire, which both functioned as almost tax-free centers of commerce, they proceeded up the Euphrates and reached Callinicum (al-Raqqa) or Nisībīs (Nusaybin), in Upper Mesopotamia. From there the products were transported by land to the commercial cities of Palmyra, Damascus, and Antioch in the Byzantine Empire (Heyd 1967: 5).

Trade between Byzantium and India was intense at that time. The ports of Barygaza (Bharuch) and Barbaricon (Bhambore) in northwest India were the starting point for Indian ships sailing up the Persian Gulf on the east coast of Arabia and to Adulis in Erythrea, on the west coast of Arabia. Seas were also sailed by Arabian ships heading for Cana, south of Arabia, and Muza (Al Mokha), and by Egyptian vessels that reached Alexandria via the Byzantine ports of Clysma (now the Egyptian city of Suez), Aila, and Berenice on the Red Sea coast. Byzantium traded not only spices from eastern India (bdellium [*Commiphora* resin], putchuk, spikenard, long pepper, and sandalwood), but also cotton fabrics from Ozene (Ujjain) and Tagira (Ter), from the western Indian hinterland (Miller 1969: 150–1).

Some goods were transported by land with several routes, particularly the different branches of the Silk Roads. One of the northern branches went through Central Asia to bypass Persia and its Empire, with which the Byzantines were at war; it brought Oriental spices (particularly turmeric), silk from China, and textiles to Petra, Antioch, and Constantinople (Boulnois 1963; Pelliot 1921). The other northern branch of the Silk Roads, called the Scythian Route, connected northwestern China and the Black Sea via the Altai Mountains, the Southern Urals and the Don; along it were traded ginger and rhubarb (*Rheum* spp.), but mostly gold (Miller 1969: 139–40). As for the southern branch of the Silk Roads, this conveyed amomum (identity disputed), bdellium, putchuk, long pepper, and spikenard from the center of Bactra (see Figure 3.2). Other land routes crossed southwestern Arabia up to the Gulf of Aqaba, and Palestine up to the Mediterranean Sea, or they proceeded to Transjordan, Damascus, and the Orontes Valley up to Antioch.

The tax codes regulating commercial activity in Byzantium listed the products imported into the Empire from the East. The so-called Alexandrian Tariff in Justinian's *Digest* (533) enumerated, under the title *Species pertinentes ad vectigal* (Types [of products] Subject to Taxation), goods (chiefly spices) that were subjected to a 25 percent tax in Alexandria. Although the Tariff does not cite all the spices traded from the East and Arabia to Byzantium, it is a good indicator of the substances in demand for medical and culinary uses in the eastern Roman Empire. Among the plants of Asian origin and their by-products, were cinnamon bark, long and white pepper, *folium pentasphaerum* (perhaps nard-leaf), *folium barbaricum* (an unidentified Barbary leaf), putchuk, *costamomum* (an ointment prepared with *costum* and *amomum*), spikenard, Turian cassia and cassia bark (*Cinnamomum aromaticum*), myrrh, amomum, ginger, *malabathrum* (possibly cinnamon leaves), *aroma indicum* (an unspecified Indian product), galbanum and asafoetida (*Ferula* spp.), aloe-wood, *sarcocolla* (*Astragalus gummifer*), cardamom, and *xylocinnamomum*, that is, cinnamon-bark. Besides these spices, Byzantine trade with the East also included linen, raw cotton fabrics, and *opia indica vel serica* (unspecified Indian or Chinese drugs) (Justinian, ed. Scott 1932: 651).

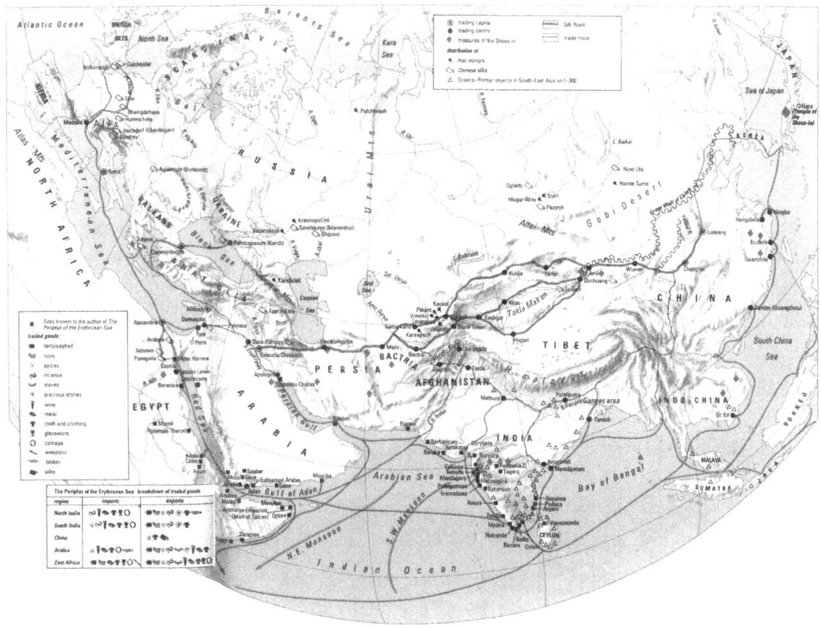

FIGURE 3.2 The Silk Road and the Trans-Asian trade, from *Past Worlds. The Times Atlas of Archaeology* (London: Times Books Limited, 1988), 190–1. © Times Books, HarperCollins Publishers (1988), used with permission.

Among the products traded with Byzantium that were not included in the Tariff were benzoin (*Styrax* spp.), camphor, cloves, galangal (*Alpinia* spp.), nutmeg and mace (both *Myristica fragrans*), and turmeric imported from southern Asia; aloe-wood, bdellium, sandalwood, sesame, sweet flag or calamus (*Acorus calamus*), and sweet rush or ginger grass (*Cymbopogon martini*) from India; sagapenum (*Ferula persica*) from Persia; balsam (both *Commiphora kataf* and *C. opobalsamum*), and frankincense (*Boswellia carteri*) from Arabia; and ammi (royal cumin; [?] *Cuminum cyminum*), other kinds of balsam (*Commiphora abyssinica*) and frankincense (*Boswellia sacra* and *B. frereana*), and myrrh (*Commiphora* spp.) from northeastern Africa.

South of the Red Sea, the Christian kingdom of Ethiopia dealt with India in products from China, Southeast Asia, and Sri Lanka, such as cassia (Chinese cinnamon) and cinnamon, aloe-wood, benzoin, camphor, cloves, ginger, lakawood (*Dalbergia parviflora*), mace, nutmeg, and sandalwood. Ships from Sri Lanka arrived directly at the port of Adulis (now Zula, Eritrea) in the kingdom. The cinnamon from eastern Indonesia reached Rhapta on the African coast via Madagascar along the route to East Africa (Miller 1969: 145–7). After the expansion of the Arabic world, trade shifted from Ethiopia to Muza in the south of the Arabian Peninsula on the Red Sea, and goods also went by land and down the Nile to Alexandria.

Throughout the seventh century, Byzantium was under constant pressure on its eastern borders: first, the Persians occupied Syria and Egypt, then, after their defeat by the Byzantines, the Arabs conquered Syria, Egypt, and northern Africa in a very short period of time. The main Byzantine cities fell into their hands: Antioch in 636, Alexandria in 641, and Carthage in 698 (Kazhdan 1996). By 702, the Arab cavalry had subjugated the entire Maghreb and its coasts, and, in 711, Cordova, in the Iberian Peninsula, also fell (Chaunu 1969).

The Arab presence throughout the Mediterranean world deeply modified trade activity. Initially, trading relationships between the Byzantine and the Arab empires were few, although Egyptian cotton still reached Constantinople. The Oriental spices traveled via the northern land route of the Silk Roads, through Afghanistan and Persia, thanks to Armenian merchants, who carried the products to Trebizond (Trabzon), where the Byzantine ships loaded them to sail to Constantinople. In the western Mediterranean, the few cities on the French coast that had managed to preserve a feeble, yet uninterrupted, trading activity during the seventh and eighth centuries gradually disappeared. Marseille was cut off from Africa after the Arab conquest of Carthage, and its importation of African oil and wheat stopped, whereas it was still an active business center during the sixth century and the beginning of the seventh.

Decrease in activity around the Mediterranean had deep repercussions on the Continent. North of the Alps, Oriental spices became dramatically rare, without disappearing, however, in the ecclesiastic milieu. At the beginning of the eighth century, the monks of the Corbie Abbey in northern France, taking advantage of the privileges granted by the Frankish king Chilperic II (715–21), purchased a considerable quantity of such Oriental spices, mainly pepper and cumin, at Fos-sur-Mer, in the Rhone Delta (Abulafia 2012: 247). About two centuries later, in the tenth century, the same abbey imported from Cambrai, some 70 km (approximately fifty miles) away, another substantial quantity of spices, including some that had not been present in the eighth-century order: galangal, ginger, rhubarb, turmeric, and lentisk (*Pistacia lentiscus*), besides a smaller quantity of pepper and cumin (Laurioux 2005: 176–7).

After the conversion of the Franks to Christianity under King Pepin II the Short (714–68) in 751, and the long reign of his son Charlemagne (*r.* 768–814), founder of the Carolingian Dynasty, the liturgical use of frankincense, myrrh, and balsam coming from the Holy Land or Arabia was boosted. Later, during the eighth and ninth centuries, Oriental spices were in demand at the court of the Carolingian kings in Aachen, where they were appreciated for their culinary and therapeutic value. The Roman Catholic Church consumed spices coming from southeast Asia and China, such as cinnamon, Indian pepper, costus from Kashmir, and aromatic resins and frankincense from Arabia (McCormick 2001).

The almost simultaneous rise of the Abbasids (750) and the Carolingians (751) completed the transformation of the trade geography. The ancient sea routes that linked Rome, Constantinople, and Syria, and the centers of the international trade between the Byzantine, Arab, and Carolingian worlds shifted toward peripheral areas. In *c.* 762 the Arab Caliphate moved from Damascus to Baghdad, and the trade activity of the Syrians in the West gradually declined. Between the middle of the eighth century and the end of the ninth, some Italo-Byzantine ports (Venice, Naples, Amalfi, and Gaeta) started replacing the Byzantine cities of Greece in the trade between the East, Egypt, Syria, and the West, while Constantinople increasingly concentrated its foreign trade within its own boundaries (Duby 1973).

THE DIFFUSION OF ARAB CULTURE AND ORIENTAL PLANTS

After their conquest of the Persian Empire (636–42) and their expansion into Central Asia (up to Kabul, Bukhara, and Samarkand), the Sindh region in India (712), the

Maghreb (702), and the Iberian Peninsula (711–18), the Arabs dominated a vast territory with a great variety of flora and specialized local cultivation as, for example, apples in Upper Mesopotamia, Syria, Lebanon, and Palestine; figs, grapes, and plums in Cis- and Trans-Jordan; and olive in Palestine and Iraq. Beyond their empire, the Arabs were also in contact with India, which cultivated citrons, bitter oranges (*Citrus* x *aurantium*), and lemons (*C.* x *limon*); with China, which produced sweet oranges, trifoliate oranges (*C. trifoliata*), and some types of mandarins (*C. reticulata*); and with Indonesia, which cultivated pomelos and limes (*C.* x *aurantiifolia*), the latter of which grew also in Sindh (Riera Melis 2002; Watson 1998: 99–115).

In terms of staple foods in the Arab Empire, wheat was the main cultivated cereal under the Abbasids (750–1258), together with barley, which, however, was chiefly fed to animals. Upper Mesopotamia (the northern part of Iraq) was the breadbasket of Persia and supplied southern Iraq with wheat via the Tigris and Euphrates rivers. Syria too, from the Negev to Eilat, and Egypt, particularly its southern provinces, grew and exported considerable quantities of wheat. Rice and sugarcane reached Persia from India between the fifth and the seventh centuries (Rosenberger 1997: 271). The cultivation of rice increased in Lower Mesopotamia (southern Iraq) and Egypt after the Arab conquest in the 630s (Ashtor 1976). It intensified along the Euphrates Valley and the tributaries of the Tigris River, on the southern shores of the Caspian Sea, along the Volga Valley, on the banks of the Jordan River, in Yemen, in southern Morocco, and, from the tenth century onward, also in Ethiopia and around Mogadishu (Somalia). Sugarcane expanded into Mesopotamia between 633 and 642, and was cultivated in Syria and Palestine around 700, and in Egypt a few years later (Lagardère 1994: 337). Sorghum, in the form of *Sorghum bicolor* Durra Group from India, was cultivated for a long time in the Near East, northern Africa, and Europe. In the tenth century it was grown in the coastal area of the Persian Gulf, Persia, the southern part of the Arabian Peninsula, Egypt, and in the coastal areas of Northern Africa (Watson 1998: 38–42).

The vegetables and fruits known and consumed in the Arab Empire came from India, China, Tibet, Central Asia, Turkistan, and Ethiopia. Most, however, came from India, where they had been cultivated since ancient times. Some were native to the East (Southeast Asia, the Malay Archipelago, and Myanmar), and included bitter oranges, lemons, limes, coconuts, bananas, and mangoes (Watson 1998: 165). Others like tamarind, instead, had come to India from Africa through the ancient Saba Route, which linked the two continents via the eastern coast of Africa and the southern part of the Arabian Peninsula (Watson 1998: 166).

Traders came from all across the Arab world and neighboring regions. The Indian merchants of Dabul and Persia established settlements on the islands of Zanzibar, Pemba, and Kilwa between the eighth and ninth centuries. Those from Yemen, Oman, and Siaf were active in the islands of the eastern coast of Africa (Zanzibar, Pemba, Madagascar), and those of Oman and Siraf, on the Persian Gulf, which was one of the main ports of call in the journeys from India and China, traded mostly between India and the Persian Gulf. All of them actively contributed not only to the introduction into the Arab world of the plants they were trading, but also to their naturalization and their further distribution to the West. Some species of citrus, such as bitter oranges, lemons, and citrons, were introduced into Oman in the ninth century. Bitter oranges were further diffused to Iraq, Syria, Palestine, and Egypt. Through northern Africa, they reached Spain and Sicily (Watson 1998: 168–71). Citrons were intensively cultivated in Palestine and Egypt from the tenth century onward. Mango, native to India, appeared at the end of the ninth century in Oman, and from there expanded into southern Iraq.

Interestingly enough, Arab trade also expanded towards Africa. Mango, for example, was seen in the fourteenth century in Mogadishu by the Arab geographer Ibn Battūta (b. 1304). Taro, native to tropical Asia, reached eastern Africa south to Madagascar, and was spread by the Arabs into the tropical and subtropical regions of the African continent (Watson 1998: 159–61; 147–52). Limes were grown near Basra, in Egypt, as well as in Yemen, and they expanded in Saharan and sub-Saharan Africa.

Some species were introduced into the Islamic world by the Arabs themselves during their conquests. This is the case of coconut, for example, which is probably native to the western Pacific. It was "discovered" by the Arabs themselves in the eighth century, when they conquered the provinces of Sindh, in northeastern India, and it was imported into Iraq in the tenth century (Watson 1998: 126–7).

Trade was not limited to edible species and their products. Linen, for example, was produced in the Egyptian countryside, while the so-called "cotton of the Old World," perhaps native to Asia or Africa, was probably cultivated at first in the northwestern part of the Indian Continent. During the eighth century, it was cultivated in Egypt where it became a major crop in the thirteenth century. From the tenth century onward, it spread to the entire Arab world, where it gave rise to a prosperous manufacturing activity which had its most important trade in Baghdad. Bahrein, Yemen, and Socotra too, became famous for their textile production (Ashtor 1976).

The products imported by the Arab world from northern Africa, China, and India to Mesopotamia are documented by Amr ibn Bahr (d. 869), better known as Al-Jāhiz, in his *al-Tabassar bi'l-Tijara* (The Investigation of Commerce). Al-Jāhiz showed that the Arabs imported more products from the Far East, India, and China into Mesopotamia than from northern Africa and Egypt. He mentioned sandalwood and coconuts from India; cinnamon and rhubarb from China; frankincense, *Khitr* leaves (likely a form of indigo, in which case *Indigofera tinctoria* or *I. argentea*), and turmeric from Yemen; musk and sugarcane from Zhwarizm; grapes from Balkh; pomegranates from Gurgan; plums, quinces, and pears from China; apples and saffron from Isfahān; indigo and cumin from Kirman; rose-water and ointments made out of waterlilies and jasmine from Fars; pistachio from Fasa; sugar and dates from Ahvāz; citrons from Susiana (formerly Elam); figs and vinegar sauces from Hulwan; and only papyrus and balsam from Egypt (Lopez and Raymond 1955: 27–9).

The plants and other natural species found by the Arabs during their conquest of large regions of the world of that time and further traded by merchants and introduced into the Arab world generated a true scientific interest among Arab people. Sources dating from the eighth to the ninth century mentioned that foreign plants were sent to the caliphs or governors of the eastern Arab Empire. This is how the governor of Egypt, 'Abd Allāh Ibn-Thāir (798–844), for example, received in 825 some melon plants from Jurásan. Local species could also be used in the conquered provinces to pay tribute to the governors who appreciated their dietary, ornamental, and medicinal value, or to enrich the gardens of their palaces.

Interest among the Arabs was sustained by the translations into Arabic of the Greek botanical, botanico-medical, and agronomic literature that were made from the late eighth century onward, particularly during the ninth century (see Figure 3.3). The most significant translated treatises were *De materia medica* by Dioscorides (first century CE); *De simplicium medicamentorum temperamentis et facultatibus* (On the Mixtures

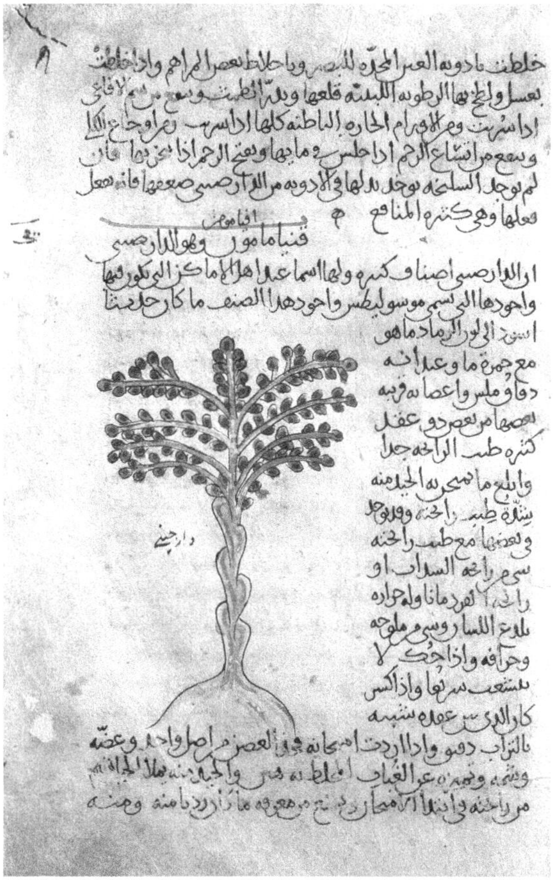

FIGURE 3.3 Cinnamon. Illustration in the Arabic version of Dioscorides *De materia medica*, (Leiden, Universiteitsbibliotheek, MS Leiden Or. 289, f. 9r, dated 1083 CE). Photo by Werner Forman/Universal Images Group. Courtesy of Getty Images.

and Properties of Simple Medicines) by Galen (129–after ?216); the *Geoponica* (Farm Work), which was translated as early as 795 and was known in the Arab world under the title *Kitāb filhāt al-ard* (Book of Agriculture), to which should be added the *Nabatean Agriculture* translated from Syriac into Arabic in 902 and further known as *Kitāb al filāha al-nabitiyya*. These translations provided a basis for the scientists of the Arab world to study botany and *materia medica*. For example, the Arabic version of Dioscorides' *De materia medica* reached Persia, where al-Bīrūnī (973–1048) studied and used it to describe and analyze the local flora, together with its therapeutic properties. In the West, a manuscript of the Greek text of *De materia medica* was donated by the Byzantine Emperor to the Caliph Abd-al-Rahmān III (929–61), and a Byzantine monk named Nicolas arrived in the city from Constantinople in 951 to help local scientists understand the Arabic version made in the eastern part of the Arab Empire (Touwaide 2013a; Zangheri et al. 2006: 60–1).

Interest in botany was encouraged, if not promoted, by the civic authorities. During the caliphate of Al-Mahdī (775–85), emissaries were sent to India to study medicinal plants, and some of their species were sent to Baghdad. The Caliph Al-Ma'mūm (813–33) transferred some bitter orange trees from northern Persia to Rayy, while the Caliph Al-Qāhir (932–4) in Baghdad built a patio with orange trees that had come from India and had crossed Oman, down to Basra and the capital (Watson 1998: 186). The species from abroad introduced into the Arab world by trade were the object of intense horticultural activity. Farmers of Indian, Persian, and Syrian origin contributed to their introduction, developed their propagation, and transmitted their production, and related new farming techniques from the eastern Arab world to its western regions.

Gardens, both royal and private, played an important role in the diffusion of exotic plants throughout the Arab world. Under Persian influence, the Arabs, who wished to learn and appropriate the cultures of the populations with which they came in contact, showed great interest in garden art (Gothein 1979: 195). Exotic plants, which were cultivated in gardens or kitchen gardens at first for ornamental purposes, not only helped enhance the royal palaces with their beauty and rarity, but also encouraged the study of their acclimatization in environments that were different from their native ones. The medieval geographers are clear witnesses to this true interest. In his *Mu 'jam al- Buldān* (Dictionary of Countries), the Syrian Yāqūt al-Hamawī (1179–1229) described Basra, Nisībīn, in Mesopotamia, and Damascus as cities rich in gardens and kitchen gardens where plants coming from the Far East grew. In the second half of the ninth century, dwarf palms, date palms, roses of various colors, almond trees, saffron, and other plants imported by the Sultan Jumārawayh (884–95) from the East were cultivated at the castle of Tulunids in Cairo (Gothein 1979: 201–3).

Many sovereigns were directly interested in exotic plants from a scientific point of view. In Cordova, plants from Syria and farther east, such as date palms, arrived at the garden of 'Abd al-Rahmān I (756–88), who founded the Umayyad Dynasty of Spain, and in the tenth century this garden became the first Andalusian botanical garden to receive seeds and roots of plants from all the Arab world (Gothein 1979: 204). Indeed, royal gardens became real scientific laboratories: this was the case of that of the Taifa King al-Mu'tasim (1051–91) in Almeria, which was planted with bananas and sugarcane. They were often run by botanists, who supervised their maintenance, studied and described the characteristics of the plants newly introduced, wrote agronomy treatises, and oversaw the arrival of the exotic plants to be acclimatized in the gardens. In Toledo, Al-Munya al-Nā' ūra (The Water Wheel Orchard), better known as Huerta del Rey (Garden of the King), was run by two famous Andalusian agronomists: Ibn Wāfid (1008–74) and Ibn Bassāl (*c.* 1050–*c.* 1100). After the capture of the city by the Christians in 1085, Ibn Bassāl moved to Seville, in southern Spain, where he designed a botanical garden, Hā'īt al-Sultān (Garden of the Sultan), for the Taifa sovereign Al-Mu'tamid (1069–91) (Watson 1998: 251).

The Arab presence in the Mediterranean world contributed to the diffusion and consumption of the spices used in cooking in the medieval West, mostly cinnamon, ginger, cloves, and nutmeg, which were not much used in Roman times and came to the fore during the Middle Ages, over other already-widespread ones: pepper, Chinese cinnamon, camphor, and musk (Rosenberger 1997).

By contrast, the plants newly introduced, or re-introduced, into the Mediterranean world by the Arabs seem to have entered European agriculture rather slowly. Their presence was not registered until several centuries after the Arab conquest of the Iberian Peninsula. Around the middle of the ninth century, sugarcane was cultivated in the western Maghreb and the Iberian Peninsula, and somewhat later, around 878, in Sicily.

However, it is not mentioned in literature until such eleventh-century authors as Al Bakrī (1028–94) and Al-Idrīsī (1100–66), who attested to their production and processing in certain parts of the Maghreb and Andalusian Spain (Riera-Melis 2002: 7). Spinach was grown in al-Andalus from the thirteenth century onwards; lemons and bitter oranges began to be grown in Spain and Italy as late as the thirteenth and fourteenth centuries. The cultivation of rice, which was widespread in Sicily and al-Andalus from the tenth century onwards, did not reach the plain of Pisa before 1468 and the Po Valley before 1475 (Watson 1998: 175).

Among the citrus fruits, citrons were the only ones that were known to, and cultivated by, the Romans, particularly in the countryside around Naples and at Pompeii. They were not reintroduced on the coasts of the Mediterranean Sea by the Arabs before the ninth century onward, citrons arriving in al-Andalus in 961 (Zangheri et al. 2006: 54–5). From the tenth century onward, bitter oranges gradually reached the Maghreb, Al-Andalus, and Sicily. Lemons were widespread in northern Africa, Sicily, and al-Andalus more or less in the same period, before the Crusades. Among the other citrus fruits introduced by the Arabs during the Middle Ages there were bergamot oranges, mandarins, trifoliate oranges, and pomelos. Sweet oranges were probably introduced in the West by the Portuguese, that is, not before the fifteenth century (Watson 1998: 110). The cultivation of aubergine was spread by the Arabs all across the Mediterranean region, where previously it was unknown (Riera-Melis 2002: 14–16).

Bananas and plantains (*Musa* spp.), native to South and Southeastern Asia to the Pacific, were introduced in the Middle East and Egypt from the ninth century onward, and probably reached al-Andalus from there. Watermelon, the original form of which spread from the equatorial regions of Africa to northern and western Africa, expanded slowly in Egyptian agriculture, where its presence was reported only between the twelfth and the thirteenth centuries, whereas it had certainly been introduced into al-Andalus in the tenth century. Artichokes (*Cynara carduncellus*), probably native to the Maghreb area, were introduced to Europe as late as the end of the Middle Ages (Bolens 1997; Watson 1998: 144–6). All these plants newly introduced into the western Arab world (above all citrus, sugarcane, and rice) converged with those of the earlier Roman-barbaric dietary tradition that resulted in a new gastronomic identity.

As for plants utilized for "economic" purposes, the cotton plant was probably transferred from Asia Minor to Greece first and then, from the twelfth century onwards, to Cyprus and Italy. In regions further west, travelers such as Al-Idrīsī, Al-Bakrī, and Al-'Umarī (1301–49) witnessed its cultivation in Tunisia. From there it was exported to al-Andalus and Italy, and also to Sicily, where it probably arrived from Tunis. The Andalusian world began to domesticate and export it into the coastal cities of northern Africa from the ninth century onward (Watson 1998: 77–97).

Whereas the Arabs' use of safflower (*Carthamus tinctorius*) and lentisk resin was inherited from the Mediterranean area, they should get credit for transmitting new knowledge to the world of Western medicine by introducing new medicinal plants. Cubeb (*Cubeba officinalis*), for instance, appeared for the first time in Western medicine thanks to Constantine "the African" (d. before 1098/9). A native to Kairouan, he moved to Salerno in the second half of the eleventh century and translated several medical treatises from Arabic into Latin. Through his translations and many others subsequently made, the West reconnected with the medical traditions going back to ancient Greece and, at the same time, discovered the new therapeutic uses of plants developed by the Arab physicians in the previous centuries (Laurioux 2005: 180).

THE RISE OF ITALIAN MERCHANTS

At that time, the trade of Oriental products in Europe was controlled by the Arabs. The most important market places were Zaragoza in Aragon and Cordova in Andalusia, and Tunis, Kairouan, Alexandria, and Old Cairo in North Africa. During the period spanning the ninth to twelfth centuries Arab trading activity in the Mediterranean was at its peak. Nevertheless, however flourishing it might have been, it never reached a level comparable to that of the centuries before the Arab conquest of the Mediterranean. Whatever the volume of traded goods, the Arab zenith was soon followed by a decline. With the First Crusade (1096–9) and the affirmation of such Italian merchant cities as Venice on the Adriatic side of Italy, and Genoa, Pisa, Gaeta, and Amalfi on the Tyrrhenian side, the Arab trade domination in the Mediterranean started to decline (Tangheroni 1996).

While the Byzantine and Arab Empires maintained their supremacy on the East–West trade during the tenth century, Europe was establishing the foundations of a commercial revolution that changed the economic geography two centuries later and resulted in putting Italy in a position of commercial supremacy in western Europe (Bautier 1970: 263–5). Life conditions started to change in Europe at that time. The great epidemics disappeared, the climate became milder, and populations grew. In the countryside that had been deserted during the Early Middle Ages, swampy areas began to be reclaimed, drained, and brought under the plow (Chaunu 1969), as were wooded lands, leading to an expansion of the cultivated area and a growth in the cereal production between the tenth and the twelfth centuries (Duby 1962; Montanari 1996: 409). Improvement in farming techniques increased productivity, which, in turn, led to population growth. As a consequence, populations also increased in the urban centers, which promoted the development of construction and such manufacturing activities as textile weaving.

By the tenth century, Amalfi, Gaeta, Salerno, and Venice traded with Constantinople, which they supplied with several goods from within the Byzantine Empire: wheat from Asia Minor, mastic from the Island of Chios in the Aegean Sea, saffron from Cilicia, linen fiber from Asia Minor, Thrace, and Bulgaria, and cotton from the western Peloponnese and the Aegean coasts of Asia Minor (Laiou 2002: 725). Various texts attest to the flow of Oriental products arriving in the Byzantine capital. The treatise *De alimentorum facultatibus* (On the Properties of Foodstuffs) by Symeon Seth, physician to the Emperor Michael VII Doukas (1071–8), mentions several spices to be used as remedies: ginger, balsam, cinnamon, camphor, cumin, aloe-wood, pepper, and sugar. *The Book of the Eparch* (or *Prefect*) (*To Eparchikòn Biblìon*), a prescriptive text possibly promulgated by the Byzantine Emperor Leo VI (886–912) between 911 and 912 and addressed to the city guilds, mentions pepper, frankincense, and cinnamon among the products sold by the apothecaries of Constantinople. Cotton appears among the textile plants that were imported (Patlagean 1993).

Besides markets and bazaars within the city walls of Constantinople, fairs that attracted merchants from many countries were held throughout the year in all the Byzantine Empire. The largest one in the eastern Mediterranean was that of St. Demetrios in Thessaloniki, in northern Greece. It was frequented by Italian and Muslim merchants who imported, among other products, cotton (Laiou 2002: 725–8). At the margins of the Byzantine Empire, Trebizond, on the Black Sea, was the crossroads of the trade routes from the Byzantine Empire, the Muslim world, and the Caucasus, and Attalia (Antalya), on the southwest coast of Asia Minor, was the entry point of the merchants coming from Syria.

At the end of the eleventh century, however, the Byzantine trade began to decline. After the First Crusade (1096–9), some small Christian states created on the Syrian and Palestinian coasts enabled the Western merchants from Venice, Genoa, Barcelona, and Marseille to transport Oriental spices from Antioch, Tripoli, and Acre to the ports of the western Mediterranean Sea. But, although these merchants managed to monopolize the sea trade from the eastern Mediterranean toward Europe and northern Africa, they were never able to gain access to the sea routes that went from the Red Sea and the Persian Gulf to the Far East. The trade with Asia through these channels remained an exclusive privilege of the Arab merchants through almost the whole Middle Ages (Bautier 1970).

Around the middle of the twelfth century, Constantinople was frequented by Persian, Egyptian, Palestinian, Arab, Russian, Hungarian, Bulgarian, and Spanish merchants, and above all Italian ones who permanently settled in the city. At the end of the century, the latter controlled all the ports of the Empire and the trade of the Oriental spices loaded in the ports on the coasts of the eastern Mediterranean Sea (Lopez 1966: 321). In 1204, however, this prosperous, multinational, economic activity was brought to an end as a result of the conquest of Constantinople by the troops of the Fourth Crusade, which founded the Latin Kingdom of Constantinople. Even though the Byzantine Empire was restored in 1261, it never recovered its past grandeur and started a slow, yet unavoidable decline. By the end of the twelfth century the Italo-Byzantine merchants, assisted by the favorable position of their ports, had replaced the Oriental, Greek, and Syrian merchants in the trade between the eastern Mediterranean world and the West, besides extending their supremacy to the Arab markets of Asia and Africa. Thanks to the trade relationships their city then had with the Arabs in Sicily, the merchants of Amalfi had access to the coastal markets of Syria and Palestine, where they gained some commercial predominance (Abulafia 2012: 267–70). In the first half of the thirteenth century, Genoa established settlements in the part of Spain reconquered by the Castilians, and in 1277 the galleys of the Genoese merchant Benedetto Zaccaria (d. 1307) crossed the Strait of Gibraltar in order to bring Oriental products to England and Flanders (Chaunu 1969).

The spice trade was one of the main driving forces of the Western economy and one of the main economic goals pursued by the growing naval powers (Freedman 2009). The demographic increase and the growth of farming production stimulated the demand for products from the Middle and Far East in Europe, especially within the aristocracy and the new wealthy classes there. Pepper, saffron, cinnamon, and ginger were most in demand on the international market, while cloves and nutmeg, cardamom, cubeb, coriander, turmeric, galangal, and malagueta pepper (*Aframomum melegueta*) occupied an important place among luxury goods (Laurioux 2005).

FROM THE APOGEE OF THE MONGOL EMPIRE TO THE CRISIS OF THE EURASIAN TRADE

The Mongol attacks in Asia, which culminated with the capture of Baghdad in 1258, interrupted the Asian land routes through which spices traveled from the Far East to the Black Sea and the eastern Mediterranean. As a result, the trade in aromatic plants shifted to the ports of Alexandria, Alep, and Beirut (Dyer 1996: 496). Between the middle and late thirteenth century, however, the Mongol Empire unified a vast territory from China to the Russian shores of the Black Sea and the Mediterranean coasts of Asia Minor under its rule. Thanks to the so-called *Pax mongolica* (Mongol Peace), the commercial routes were reopened. The Western merchants (Genoese and Venetians) reached the production centers of Oriental spices, thus avoiding the intermediation of the Arabs, who still controlled them, however (Lopez 1976: 108–9).

During the reign of Kubilay Khan (1260–94), the grandson of Genghis Khan (r. 1206–27) and the unifier of China, the Mongol Empire was at its zenith. The Mongols, who had to administer a huge territory, opened the kingdom to foreign travelers, merchants, officials, and religious dignitaries, entrusting them with administrative and diplomatic posts (Dainelli 1960: 104–5). With the Yuan Dynasty (1279–1368), the Mongol domination connected the Far East and Europe in an unprecedented way, through a dense network of trade exchanges, which also allowed for cultural contacts (see Figure 3.4).

Khanbaliq (Peking, modern Beijing), which became the capital of the Chinese Empire, was a cosmopolitan city then: the diet of its population acquired an international quality, and the introduction of spices supplied by foreign trade was one of its main features. The treatise *Yinshan Zhengyao* (The True Principles of Eating and Drinking), written by the dietician Hu Sihui (d. 1330) for the Emperor Tuq Temür (1328–32), to whom it was presented in 1330, illustrates the role of spices in Mongol medicine and diet. Besides the spices that already belonged to the Chinese tradition (pepper, ginger, and coriander), this treatise mentioned foreign ones that were previously unknown to the cooks then, such as saffron, turmeric, long pepper, asafoetida, lentisk resin, rose-water, fenugreek (*Trigonella foenum-graecum*), and spikenard. The presence of spices from the Middle East, Arabian legumes such as chickpeas, and also syrups and preserves made with sugar, attested to the influence exerted by the Arab tradition on China at that time. Venetian merchants traded in the city of Urgench (western Uzbekistan) and, in Peking, Franciscan

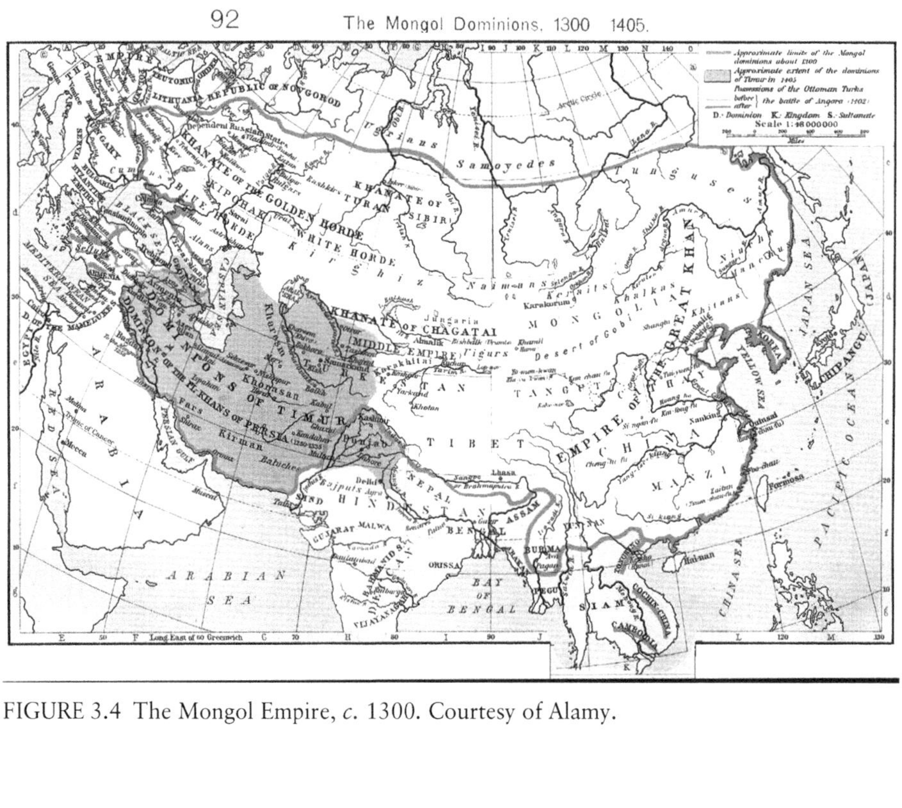

FIGURE 3.4 The Mongol Empire, *c.* 1300. Courtesy of Alamy.

missionaries built warehouses to accommodate the merchants, while Chinese travelers, soldiers, and missionaries moved to various parts of the Empire, from Moscow to Tabriz (Sabban 2002).

The first real exploration of the Asian continent by Europeans dates from this time. It is attested by several historical documents written by authors with very different backgrounds. The best known of these documents is the work that became known as the *Milione* (The Million), by the Venetian traveler and merchant, Marco Polo (1254–1324). Its reports and descriptions of the customs and products of the Far East contributed enormously to the knowledge of China in Europe (Verdon 2004). Marco Polo set out in 1271 with his father Nicolò and his uncle Maffeo, reaching the court of Kubilay Khan, in Peking, via the Silk Road. During the seventeen years he spent in Asia, Marco Polo was appointed by Kubilay to government posts and official missions that allowed him to become acquainted with the local dietary traditions. Up to the end of the thirteenth century, Europeans had known Oriental spices only as dried products. The possibility to observe their cultivation directly in their natural habitat was stimulating for Polo. In Xandu (Shang-du), he noticed that aloe-wood was used as a perfume and a medicine; in Cianfu (Ph-Zhou-fu), he found many native plants, such as ginger, galangal, and *spigo* (*Nardostachys jatamansi*) used for perfume; and, in Java, he saw pepper, nutmeg, galangal, cubeb, and cloves (Polo 1559).

Another explorer was the Franciscan Friar Odoric of Pordenone (*c.* 1265–1331). Unlike most Christian missionaries who ventured into Persia—which was a popular destination because it was close to the Mediterranean, Odoric went as far as China. He set off from Venice in 1317–18, passing through Constantinople, Trebizond, and Tabriz down to Baghdad and Basra, and then from Hormuz reaching Tana and the Malabar Coast. In his *Descriptio orientalis partis* (The Eastern Parts of the World Described), which was already in wide circulation at the time of its writing, Odoric included several references to medicinal and food plants (Melis 1984). He described the Malabar Coast as the main source of pepper, while in Colombo (Quilon), he saw ginger, which was used as both a medicine and a spice. In Java, he observed many spices, such as nutmeg, cloves, and aloe-wood. He proceeded to the important Chinese ports of Canton (Guangzhou) and Zailon, and, in 1325, he reached Khanbaliq, the capital of the Great Khan, where he stayed for three years. Then he returned to Italy by a different route, through Tibet, India, Afghanistan, and Persia, up to the Mediterranean Sea and Venice (Odoric of Pordenone 1513).

Explorers of a different kind were the many Western merchants who traveled to the East for personal interests, or as agents of trade companies. One of them was Francesco Balducci Pegolotti (1310–47) from Florence, an agent of the Bardi trade company. He compiled the most complete manual of commercial practice and history of the East–West trade of his time. He first stayed several years (1324–9) in Cyprus and in the commercial colonies along the Asian coast where he collected the data that he later incorporated in his *Pratica della mercatura* (Practice of Commerce), compiled around 1340 (Balducci Pegolotti, ed. Evans [1936] 1970). The fact that he included information about the trade routes to the Mongol capital via Asia Minor attests to the interest of the Italian traders in the Asian continent. Balducci described the northern route from Crimea to Peking, via southern Russia and Turkistan, as a safe itinerary that could be completed in nine months and through which it was convenient to export French and German linen to China and to import silk. For other goods, he advised taking shorter routes, such as that from Asia Minor (Trebizond) to India and China via Persia and Afghanistan, or the sea route, which was less expensive but took two years, from the Persian Gulf, along the coasts of the Indian Ocean, to Peking. Under the heading "spices," he gave an alphabetical list of 288 products coming not only from the Arab countries, the Byzantine world, Asia, and Africa, but also from Italy and more generally from Europe. Not all are strictly spices; as a rule, they are the non-perishable products of the great trade with the East. Among the spices some were culinary; others, such as cotton (including *bambagio d'oltre mare*, from Syria, *bambagio di Romania*, from the Eastern Roman Empire, i.e. Byzantine cotton, *bambagio di Puglia*, from Apulia), were used in textile crafts; others in medicine, such as mastic (*mastico*); and still others, such as balsam (*balsimo*), were used for cosmetic purposes. The numerous other "spices" included oranges (*aranci freschi*), cubeb (*cubebe dimestiche* and *cubebe salvatiche*), cloves (*gherofani)*, malagueta pepper (*meleghette*), coconuts (*noce d'India*), and rice (*riso d'oltre mare* and *riso di Spagna*) (Balducci Pegolotti, ed. Evans [1936] 1970: 293–8).

The internationalization of activity in Peking did not compensate for the damage to farming production caused by the Mongol invasion, which generated hunger and epidemics, further aggravated from 1330 onwards by the repeated floods of the Huang He River. From the 1340s onwards, unrest among the Chinese and revolts led to the fall of the Yuan Dynasty (1368). The Ming Dynasty that then took control was economically unstable, with a decreasing population (Chaunu 1969). The founder and first emperor of the Ming Dynasty, Zhu Yuanzhang (Emperor 1368–98), promoted the repopulation and reforestation of the ravaged areas and the restoration of farming as the main activity of Chinese society (Sabban 2006b).

In western Asia, Osman I (1299–1326), the founder of the Ottoman Dynasty, extended the Turkish conquests into Anatolia. In 1363, the Ottomans established their capital in Hadrianopolis (Edirne), in Thrace, and they began their penetration into Europe. Shortly afterward, in 1380, Tīmūr (Tamerlane [1336–1405]), from the Chagatai Khanate, conquered India, Mesopotamia, and Persia. In the second half of the fourteenth century, the Asian Continent was almost impenetrable to Westerners, and the trade routes that linked the East to the West through the Russian steppes of the Volga, the lands of Xinjiang, in northwestern China, and Mongolia, were cut off (Dyer 1996: 501). Trebizond, which was still one of the main ports of the trade between the West and Persia, lost its importance when it was conquered by the Ottomans. The only routes accessible to the Western merchants were those branching out from the Syrian coast toward Mesopotamia and the Persian Gulf, and the one that reached the Red Sea and the Indian Ocean via the Lower Nile. Trade of goods from the Far East required the intermediation of the Arab world.

A similar crisis was hitting Europe. After a maximum at the end of the thirteenth century, its population decreased by more than 30 percent. A major cause was the so-called Black Death, which ravaged Europe, Anatolia, and Constantinople up to Crimea from 1348 onwards. It was not the only cause of the European decline of that time, however. During the first half of the fourteenth century, the European climate was characterized by cold and rainy spells that affected farming production, which was already barely sufficient to feed the population when it still was in a growing phase. The untilled lands and the sylvan-pastoral agro-economy typical of the Early Middle Ages reappeared in Europe (Montanari 1996: 411–12). Epidemics hit Germany during the last decades of the fourteenth century, and France, which was fighting the Hundred Years War, went through a demographic decline. The eleventh-century expansion begun in Europe was slowing down, and manufacturing activities underwent a decrease in production and sales. The European economy had entered a recession (Chaunu 1969).

In the second half of the fourteenth century, the volume of Genoa's trade with the Arab East, which peaked between the second half of the thirteenth century and the beginning of the fourteenth, began gradually to diminish. After the Ottoman presence in the Middle East had cut off the routes of the East–West trade, Syria and Egypt became the main places for the exchange of products from Arabia and the Far East between the Italian coastal towns and the Arab countries. The register of the Genoese customs of 1376 and 1377 contains information about the spices and other products imported and exported by Italian and European cities. The direction of trade was inverted again, and Genoa played a major role in the export of Spanish and Italian oil to Egypt, and also, though in smaller proportion, of hazelnut (*Corylus* spp.). Lorenzo Raynaldi's cog (vessel fitted with a single mast and a square-rigged single sail), for example, which set off from Apulia heading for Alexandria, carried oil, as did also Borellus de Nigro's cog from Gaeta to Alexandria, which had also hazelnuts. Antonio Grasso and Basilio Lomellino's cogs from Spain carried Spanish oil to Egypt (Ashtor 1986: 391–432). Whatever the products and the direction of the trade, the data on the ship loads arriving at Genoa from Egypt and Syria between 1367 and 1377 indicated a decline of the Levantine Genoese trade.

FROM OLD WORLDS TO THE NEW: CONCLUDING REMARKS

The fourteenth century ended in a climate of pervasive crisis which lasted at least a century. But, by that time, the wealth promised by the trade with the East made it more desirable than ever to find ways to increase trade by bypassing the intermediation of the Arabs and the Turks. After the brief interlude of the Portuguese exploits, this prospect motivated Christopher Columbus (1451–1506) to set off in search of the land of spices. The global market of agricultural and medicinal products that resulted from this expedition moved the center of gravity of the spices and plants trade from the Mediterranean to the New World. For the first time in its long human history, the Mediterranean was to lose its supremacy.

CHAPTER FOUR

Plant Technology and Science

Medieval China

HUAIYU CHEN

INTRODUCTION

Plant science and technology in medieval China cannot be separated from the developments in agriculture, economics, and medicine, as well as cultural practice. The Chinese empire ruled most of East Asia in the medieval period. Numerous species of plants were observed, cultivated, harvested, and used in the vast land of China that spanned a wide range of biomes from boreal through to temperate and tropical, with most regions classed as subtropical. Besides indigenous plants, many plants from West, Central, South, and Southeast Asia were introduced into China and East Asia in general. This chapter focuses on particular food plants, fabric plants, flowers, trees, and wood which played vital roles in the socio-economic life of medieval Chinese people. It highlights various scientific and technological aspects of these plants, such as how medieval people knew about them, cultivated them, grew them, harvested them, and used them for socio-economic purposes. These plants were also important in the political, cultural, and religious life of medieval Chinese people, and some of these plants were introduced to Korea and Japan to become East Asian plants in general.

From the perspective of plant science and technology there were several important changes and breakthroughs in medieval China, particularly between the sixth and fourteenth centuries CE. First, there was a spatial shift in terms of the main agricultural production areas from the north region centered on the Central Plain to the lower Yangtze Valley and Sichuan. In south China, rice-farming spread and multiple-cropping systems which yielded high rice production were used. In north China, by contrast, the more prolific and superior winter wheat which could be harvested three times in two years replaced millet to become the major food crop. These new developments made it possible to provide sufficient food supplies to the fast-growing population of the Chinese Empire; in the twelfth century China surpassed a landmark when its population grew beyond a hundred million. The development and expansion of agricultural production in south China might have been due to several factors. Politically, there were three big waves of southward immigration that occurred: in the fifth century when the Western Jin court lost control of the Central Plain and moved to Jiankang (modern Nanjing, Jiangsu Province);

in the ninth century when the Tang Empire collapsed; and in the thirteenth century when the Mongols defeated the Southern Song Dynasty. In terms of technical innovation, as early as the sixth century, the Sui regime started building a canal system across the empire which became a major transportation route facilitating economic connection between the Central Plain and the lower Yangtze Valley. In the eighth century, after the An Lu-shan Rebellion, the Tang Empire was forced to shift its economic productivity to the southern regions, especially the lower Yangtze Valley and Sichuan areas. Agriculturally, in the tenth and eleventh centuries, more rice fields were developed in the Jiangnan area, especially the lower Yangtze Valley (Golas 1980: 291–325; Hartwell 1982: 365–442; Liu 2015: 141–96; von Glahn 2016: 208–54; Zeng 2015: 351–429).

Second, there was a long-term change in Chinese sericulture. Traditionally, the upper class valued silk textiles more than hemp and ramie fabric textiles. Hemp and silk were two of the most common fabric source materials. Mulberry (*Morus alba*), as the main source for providing leaves to feed silkworms, was a practical and symbolic plant for Chinese sericulture. Hemp (*Cannabis sativa*) was widely used for making oil, fabric cloth, and paper. Although silk production reached maturity in the Tang Dynasty and flourished, cotton was introduced to China in the eleventh century and its cultivation grew gradually. By the fourteenth century, cotton eventually became a significant fabric plant for the Chinese textile industry.

Third, with the development of agricultural production in south China, the cultivation and processing of tea (*Camellia sinensis*) began to increase. Tea culture appeared to become a significant part of daily life across the boundaries of various social classes. The established religious traditions such as Buddhism and Daoism played vital roles in this social and cultural change. Both literati and members of the clergy contributed to the rise of tea culture in the Tang and Song Dynasties. They discussed philosophy, religion, and literature while drinking tea. Numerous well-known poems on tea were composed. South China became the most critical region for cultivating tea trees. Various tea fields covered hills and mountains across the southern regions.

Fourth, in the Tang Dynasty, the expansion of the Chinese Empire and the development of its tribute system motivated gift exchange activities between the Tang court and other regimes as well as religious exchanges. These activities brought many exotic materials to Central China from abroad, including jewels and medicines, animals and plants, and science and technology. Fruit such as litchi (*Litchi chinensis*) and grapes (*Vitis vinifera*) and wood such as sandalwood (*Santalum* spp.) and rosewood (*Dalbergia* spp.) were introduced to China and became favorite luxuries of the upper class. And, lastly, expansion of religious monasticism and private gardens, religious buildings, temples, offices, houses, pavilions, bridges, boats, and other buildings occurred to a significant extent in the Tang and Song Dynasties not least because of the demands of the growing population. For instance, many medieval Buddhist monastic establishments received gardens and fields as well as animals and plants as donations; therefore, monks had to handle the classification and acceptance of these donations (Chen 2009: 31–51). Many gardens and buildings increasingly used wood rather than stone for construction in the Tang and Song Dynasties. Wooden architectural technology became sophisticated in the Song Dynasty, which was mainly manifested by the appearance of a technical manual entitled *State Building Standards* (*Yingzao fashi*) compiled by Li Jie (1065–1110) in the twelfth century.

It should be noted that the expansion of agricultural cultivation toward the lower Yangtze River region and the west resulted in large-scale deforestation. Further, as already noted, wood was also extensively used as architectural building materials. Some of the earliest extant Buddhist temples and pagodas in north China are made of wood, such as the East Hall of the Buddha's Light Temple in Mt. Wutai, Shanxi Province (857), the Southern Chan Temple (782) and the Wooden Pagoda in Ying County, also in Shanxi, and the front façade of Mogao Cave 444 in Dunhuang, Gansu Province (926 CE). Parts of these structures have been renovated and amended extensively over the years, but their main wooden structures and frameworks have endured for more than a millennium (Steinhardt 2004: 228–54). The maturity of wooden architectural technology resulted in demands from both government and private projects for large amounts of timber, spurring the rapid deforestation in the Song and later periods. Mark Elvin (2004: 84–5) lays out three phases of deforestation in Chinese history. He notes that deforestation was caused by various factors: clearcutting to create fields in which to cultivate crops and vegetables; harvesting timber for constructing houses, boats, carts, coffins, and religious artifacts as well as for use as fuel in cooking, heating, and industrial processing, and so forth. The first phase took place in the north during the second half of the first millennium BCE when some measure of public exploitation could be detected. The second phase happened in the medieval period when the economic revolution began in the Yangtze River region where the cultivation of paddy rice was tremendously increased. It was during this time that severe wood shortages were first reported in the lower Yangtze River Valley and some parts of the western Chinese Empire. This crisis was characterized by bitter quarrels between local communities over the use of forested areas. The third phase started in the seventeenth century and continued to the modern period. Overall, natural disasters, military campaigns and wars, social and economic developments, the explosion of population, and the expansion of agricultural activities, as well as the extension of religious monastic establishments led to extensive deforestation.

Finally, in terms of knowledge of plant science, it was remarkable that during the medieval period, Chinese writers and scholars produced a large number of agricultural and biological works that recorded the cultivation of various domestic and imported plants, the socio-economic functions of different parts of plants, and the technology for processing plants to produce food, oil, and fabric materials.

KNOWLEDGE OF PLANTS

In medieval China, knowledge of diverse plant species is attested not only in numerous works on particular flowers, fruits, and trees, but also in agricultural works. One example is the *Compendium of Important Rules for Four Seasons* (*Sishi zuanyao*) by Han E (*c.* late ninth and early tenth centuries), which contains details about numerous agricultural technologies as well as a large amount of biological knowledge. Some ethnobiological works focused on geographical regions with exotic animals and plants based on the authors' own traveling experiences or local plants that they observed and encountered. In the medieval period, many learned individuals traveled afar for the purposes of governing, exploring, or leisure. They left numerous records on local animals and plants. They expanded a genre of writing about exotic animals and plants which was called the record of exotic things (*yiwu zhi*). For example, the *Record of the Land of Northward Facing Doors* (*Beihu lu*) by Duan Gonglu (*c.* late ninth century) and the *Record of Exotics in Lingbiao* (*Lingbiao luyi*) by Liu Xun (*c.* late ninth century) both concern animals and plants in the Guangdong area. The former contains details about more than twenty plants, and the latter discusses more than ten plants and many types of fruit native to the Lingnan region. In the Song Dynasty, Fan Chengda (1126–93) wrote *Treatises of the Supervisor and Guardian of the Cinnamon Sea* (*Guihai yuheng zhi*, 1175) which describes plants in Guangdong and Guangxi (Hargett 2010). Further, Song Qi (998–1061) wrote *Short Notes on Local Products from Chengdu* (*Yidu fangwu lueji*) which contains details about plants in Sichuan. Li Gefei (?–1106) prepared a book entitled *A Record of Celebrated Gardens in Luoyang* (*Luoyang mingyuan ji*) which discusses nineteen gardens in Luoyang and the numerous species of flowers they contained (de Pee 2009: 85–116; Yang 2004a: 221–55).

More works on fruit and flowers appeared in the Song Dynasty. Some of the most prominent works are as follows: *A Record of Peonies in Luoyang* (*Luoyang mudan ji*) by Ouyang Xiu (1007–72) which focuses on *Paeonia suffruticosa* and was written in 1031 when the author traveled to Luoyang; *A Register of Peonies of Tianpeng* (*Tianpeng mudan pu*) by Lu You (1125–1210), composed in 1178 and containing details about thirty-four different peonies in Tianpeng, Sichuan; *A Register of Chinese Herbaceous Peonies* (*Shaoyao pu*) by Liu Ban (1023–89), who wrote about Chinese peonies (*Paeonia lactiflora*); *A Register of Plum* (*Meipu*) by Fan Chengda (1126–93) who discussed five categories of flowering plum (including *Prunus mume* 'Alboplena,' *P. mume* 'Viridicalyx,' and *P. mume* 'Bungo'); *A Register of Chrysanthemum* (*Jupu*) by Liu Meng (*c.* twelfth century) which was penned in 1104 and focused on 160 kinds of locally cultivated chrysanthemum flowers (*Chrysanthemum morifolium*); *A Register of Orchids from Jinzhang* (*Jinzhang lanpu*) by Zhao Shigen (*c.* thirteenth century), written in 1233 about twenty kinds of orchid; *A Register of Litchi* (*Lizhipu*) by Cai Xiang (1012–67) in 1059 on thirty-two types of litchi from the Fujian area; *A Register of Citrus* (*Julu*) by Han Yanzhi (*c.* twelfth century), penned in 1178 on twenty-seven kinds of citrus (such as *Citrus x aurantium* Sour Orange Group, *C. reticulata* (mandarin), and *C. x aurantium* Sour Orange Group); and *A Register of Tong* (*Tonglu*) by Chen Zhu (1009–61) which was released in 1051 and was concerned with the cultivation and use of *Paulownia* trees. These works detailed traditional Chinese botanical classifications of flowers, fruit, and trees, rather than embodying any modern knowledge of plants. However, their meticulous descriptions of each kind or category (rather than the modern sense of species) of these plants can still help us understand and identify them by modern standards. More importantly, these books preserve extensive information about the cultivation and distribution of these plants across many regions of the Chinese Empire in the medieval period.

The knowledge of herbal plants has a long history in China and flourished in the medieval period as the government started issuing the official versions of *materia medica* texts. The Tang court actively encouraged learning about traditional herbs. For example, in 659, the imperial court ordered Su Jing (*c.* seventh century) to organize a team to compile the *Newly Revised Materia Medica* (*Xinxiu bencao*). Literati who benefited from both traditional herbal learning and religious practice also extended their plant knowledge. Some of them devoted themselves to learning from the *materia medica* and left substantive written accounts. In the eighth century, Zhang Ding compiled one of the most interesting *materia medica* in the Tang Dynasty entitled *Materia Medica for Successful Dietary Therapy* (*Shiliao bencao*), which appears to inherit a lot of information from the medical herbal learning of Sun Simiao (581–682), a noted Daoist hermit and medical expert, and the prescriptions of Meng Shen (621–713), a medical expert and nutritionist who once served as a drafter for the emperor briefly. A fragment of this work was discovered in 1907 by Aurel Stein (1862–1943) from the cave library of Dunhuang, an oasis city in northwest China. It describes the medicinal and nutritional value of more than 160 food crops, oil, vegetables, and fruit as well as other plants (Engelhardt 2001: 184–7).

In the Song Dynasty, new pharmacopoeia with illustrations were compiled. Both medicinal and biological knowledge benefited from such endeavors. Further, traditional herbal texts continued to be written, annotated, compiled, and edited. Some writings on specific vegetables, fruit, and flowers appeared in the Tang and Song Dynasties, such as records of citrus, peony, bamboo, litchi, mushrooms, paulownias, and tea. Biological writings flourished in the Song Dynasty since many new works that included herbs and trees or flowers in their titles appeared, recording numerous plants in particular villas and private gardens across the country (Liu 1994: 35–59; Luo 2015: 446–53). Since other chapters will cover medicinal plants, the main focus herein is on food plants, fabric plants, flowers, trees, and wood in the medieval period, particularly those that were more visible and remarkable than others.

SCIENCE AND TECHNOLOGY OF FOOD PLANTS

In the medieval period, Chinese agriculture witnessed the dominant rise of winter wheat (*Triticum aestivum*) and paddy rice (*Oryza sativa*) in the cultivation and production of grain crops. Before the Tang Dynasty, millet was the dominant traditional grain crop, which included two species: foxtail millet (*Setaria italica*) and broomcorn millet (*Panicum miliaceum*). Wheat gradually replaced millet and became the most important grain crop in north China after the mid-Tang period. Wheat was also cultivated in the southern regions. In the Yunnan area, there were even two harvesting seasons for wheat since the climate was warmer and more humid than in north China. In the moist regions of the Yangtze Valley, wheat was planted in the eighth month and harvested in the fourth or fifth month of the following year, same as in the north and the Central Plain. However, they could harvest three times in two years via a rotation system, which dramatically increased production. Paddy rice was also cultivated in north China. Still, the semi-arid climate placed high demands on the artificial irrigation system, which impeded its popularity and sustainability, unlike its cultivation in south China (Zhang 2016: 233–9). In the Song Dynasty, even in the Yangtze Valley, technological innovation meant that wheat could be harvested twice per annum. In addition, both wheat and rice were cultivated respectively in different regions and seasons in the south, which enhanced the ability to support the rapidly growing population of the Chinese Empire.

In the Tang Dynasty, two issues seem to have contributed to the development of wheat and rice cultivation in the south. First, large-scale immigration from north China brought abundant agricultural labor to the south. Numerous fields were developed and expanded. Second, in the Sui and Tang Dynasties, the cultivation of wheat and rice first developed along with the canal system that linked the Central Plain and the lower Yangtze Valley. The skilled immigrants settled in the lower Yangtze Valley propelled the cultivation of these grain crops. According to the *Old History of the Tang Dynasty* (*Jiu Tang shu*), in the mid-Tang period, ten times more rice was transported from the south via the grand canal than in the early Tang era. Farmers in the south also developed more diverse rice fields to ease demands on irrigation. For example, the technology of the ladder fields (*titian*) probably first appeared in Yunnan in the Tang period and then flourished across all of south China in the Song Dynasty (Bray 1984: 126) (see Figure 4.1). Technically, the expansion and improvement of rice cultivation in the Song Dynasty was due to many factors such as water control, farm implements, draft animals, planting methods, weed and pest control, manuring, seed selection, and human innovation.

FIGURE 4.1 Ladder fields in Yunnan Province, China. Courtesy of JialiangGao/Wikimedia commons.

Paddy rice became the principal grain crop in the Song Dynasty. It should be noted that the introduction of Champa rice from Southeast Asia into the Fujian area in the eleventh century became a game-changer. This rice can be double-cropped in the southern regions of China. Because of its early ripening, it also matured much faster than traditional paddy rice. Champa rice offered better drought resistance qualities, enabling it to grow in deficient soil. Its flexibility to cope with flooding and salinity also helped it tolerate more abiotic stresses for surviving, maturing, and harvesting. Therefore, its cultivation was swiftly extended to the lower Yangtze Valley (Barker 2011: 184–6). In the twelfth century, paddy rice eventually took the leading role across the country. The increasing production of both winter wheat and paddy rice laid the foundation for supporting a large population in the twelfth century when the people of Song China probably just passed the landmark of a hundred million.

In the medieval period, soybean (*Glycine max*) was another important food plant, which often served as a supplement to winter wheat and paddy rice. However, it was also important for oil production. The technology for extracting oil from soybeans was already established in medieval China and there were two major plants for extracting oil for domestic uses which were widely cultivated across the country: sesame (*Sesamum indicum*) and choy sum (*Brassica rapa* Chinensis Group, Chinese name: *youcai*). Sesame was the principal oil plant until the thirteenth century when choy sum became the chief choice for farmers to plant as the winter vegetable. Sesame oil was widely used for daily cooking and medical healing in the medieval period. It seems that, although choy sum was first cultivated in the northwestern region for its resistance to cold winters (Chinese farmers also called it "*hancai*," cold vegetable), it was introduced into the Yangtze Valley and was expanded in the southern regions in the Song Dynasty. Its leaves were used as food, but its seeds were pressed for making oil. Farmers used the oil for lighting lamps and also washing hair, and in this respect it seems to have been a better option than sesame oil. From the thirteenth century, it became the dominant oil plant.

The rise of tea culture in the Tang and Song Dynasties marked one of the most important developments in the history of Chinese plants, economically, socially, culturally, and diplomatically. The cultivation and production of tea trees originated in southwest China, especially Yunnan and Sichuan. In early medieval China, Daoist clergy started consuming tea as a drink made from young leaves of tea trees. Aristocratic literati accepted this beverage and introduced it to the Chinese court. Even aristocratic women enjoyed it. Tea then became a typical drink for the upper class in the early medieval period. In most cases, tea was made from the leaves of wild tea trees that grew naturally in the mountains. Later in the Tang Dynasty, most tea producers grew domesticated tea trees. Therefore, the cultivation and production of tea flourished across the country, especially in the Yangtze region.

In the Tang Dynasty, the technology of cultivating tea trees was also improved. Around 760–2, Lu Yu (733–804) wrote a book entitled *The Classic of Tea* (*Chajing*) offering a detailed account of the cultivation of tea trees and the production of tea beverages as well as the fashion of tea drinking as a culture. His horticultural description touched on the climatic, soil, and water conditions for cultivating tea trees. His book also discussed the technology of transplanting tea trees and harvesting tea leaves. Lu Yu listed fifteen tools for picking, steaming, pressing, drying, and storing tea leaves and cakes and the methods of making tea cakes. This allows us to learn many specific details about how tea beverages were made in the eighth century (Benn 2015: 21–41; Mair and Hoh 2009).

In the ninth and tenth centuries, the technology of cultivating tea trees and water irrigation became more mature and sophisticated, and tea cultivators even started harvesting tea leaves in the tenth month of the lunar calendar year, besides the second, third, and fourth months. Tea production increased to meet the demand of a large population of consumers. Eight percent of the local prefectures in the Tang Dynasty cultivated tea trees and produced tea in the eighth century. In the Song Dynasty, more farmers were involved in growing and producing tea and the government instituted tea households as a new category in its household registration system. In the Tang and Song Dynasties, the Yangtze and Southeast regions such as Zhejiang, Fujian, Jiangxi, Hunan, Hubei, Anhui, and Sichuan were the main places for cultivating and producing tea. In the ninth century, there were probably four kinds of gardens for growing and producing tea: government tea gardens, private tea gardens of landlords, tea gardens of Buddhist monasteries, and self-supported tea gardens of farmers. The quality of these tea products varied depending on the region, technology, and the skill of the cultivator. The popularity and success of tea resulted in the government imposing a tax on tea production and trade (Wang 1987: 14–21). The rise of tea culture certainly laid the foundation for the status of tea as an iconic product of the Chinese Empire in later periods.

PLANTS FOR SEASONING AND AROMATICS

As Edward H. Schafer noted (1963: 155), there was little clear-cut distinction among drugs, spices, perfumes, and incenses, because all of them were used to nourish both body and spirit. Many plants were used as sources for making seasoning and aromatics. Medieval Chinese people enjoyed cooking their food using various types of spice and seasoning, which came from a broad array of aromatic plants such as ginger (*Zingiber officinale*), green onion (*Allium chinense*), Sichuan pepper (*Zanthoxylum simulans*), garlic (*Allium *sativum*), Chinese parsley (*Coriandrum sativum*), garlic chives (*Allium tuberosum*), and many others (Li 1984: 130–40). *The Monthly Ordinances for Four Classes of People* offers detailed descriptions of the stages and skills in growing these vegetables, including the selection and planting of seeds, transplanting, irrigation, and fertilizer use. In the thirteenth century, the technology of cultivating chives in greenhouses was improved by introducing horse manure to fertilize soil.

In the medieval period, aromatics were widely used among royal families, bureaucrats, religious adherents, businesspeople, and the military. Different types of incense commonly appeared in political, military, and religious rituals, especially sacrificial and funeral ceremonies. According to Schafer (1963: 158), among numerous native plants and animals that the Chinese used for producing perfume and incense, cassia (*Cinnamomum cassia*, also called Chinese cinnamon), camphor (*Cinnamomum camphora*), and liquidambar (*Liquidambar acalycina*) were commonly used. However, other aromatic plants were introduced to medieval China from abroad as exotic indigenous tribute. One of the most popular incenses (Schafer 1963: 163–5) was imported agarwood (*Aquilaria* spp.) from Southeast Asia. Both Buddhist and lay communities highly valued this type of fragrant wood in their daily use of incense. Aristocratic people even developed a fashion for wearing a bag containing this incense. Agarwood was used for making statues of Buddha saints for its fragrance, as well as highchairs for eminent monks to sit on whilst they preached Buddhist teachings. Clove (*Syzygium aromaticum*) from Indonesia was used as both a culinary spice and incense in the Tang Dynasty. In the underground chamber of Dharma-Gate Temple near Xi'an, Shaanxi Province, archaeologists discovered that clove and incenses were offered. Other exotic aromatic plants included lakawood (*Dalbergia parviflora*) from Indonesia, elemis (*Canarium* spp.) from Southeast Asia, storax (*Storax balsam*) from Asia Minor, benzoin (*Styrax*) from Indochina and Indonesia, frankincense (*Boswellia* spp.) from Arabia, and putchuk (*Saussurea costus*) from Sri Lanka. Nevertheless, these perfumes and incenses might not have been known to all social classes, but they were favored by the upper classes.

TEXTILE PLANTS

In traditional China, many plants, including hemp, ramie (*Boehmeria nivea*), chingma (*Abutilon theophrasti*), jute (*Corchorus capsularis*), and flax (*Linum usitatissimum*) were used as fabric materials. It is beyond doubt that hemp was the most commonly used plant for cloth-making and paper-making in medieval China. As Joseph Needham noted, the "hemp" or the Chinese "*ma*" nexus might include more than twenty plants (Needham 1986: 170–4). Although the identification of hemp in ancient Chinese sources was often a puzzle due to the ambiguous meaning of the Chinese character "*ma*," scholars have combined textual sources and extant species from manuscripts found in Dunhuang for tracking it down. In medieval China, the flax was often called "barbarian flax (*huma*)" in Chinese sources. There was both cultivated hemp and wild hemp in China.

The cultivation of hemp has a very long history as it was a popular plant in the Shang and Zhou dynasties (eleventh–third centuries BCE). At that time, it was commonly cultivated in the middle and lower regions of the Yellow River. But around the common era, its cultivation was expanded to the southern regions of the Yangtze River. In the medieval period, hemp was cultivated throughout the entire Chinese Empire. It continued to be cultivated in other frontier regions until the thirteenth century when the central government began to encourage the cultivation of cotton. Some early agricultural sources including the *Book of Si Sheng* (*Sisheng zhishu*) and *Essential Techniques for the Common People* (*Qimin yaoshu*, *c.* 535) noted methods for cultivating hemp to ensure hard stalk, thick bast, and more internodes in order to harvest strong hemp raw materials for textile production; indeed, these sources offer many processual details covering seed selection, seed soaking, intercropping, sowing, and harvesting. The hemp seed and oil occasionally served as food and medicine for some families in north China (Li 1974a: 293–301; 1974b: 437–48; Liu et al. 2017: 235–42). According to some Dunhuang manuscripts dated to the ninth and tenth centuries, hempseed was used for oil pressing. Although hemp oil was used for both cooking and lighting, the quality was poor (Trombert 2004: 62).

Hemp was the main source for making cloth and paper in the medieval period. In early medieval China, the fiber made from hemp even functioned as money for paying taxes to the state government in northwestern and southern China (Wang and Wang 2013: 263–80). Chinese families, mostly women, converted the hemp fiber into a textile through spinning and weaving. Knitting and felting might also be used for making textiles. Traditional spinning and weaving technology first involved warp and weft threads bound together with cloth beams made on a loom. There were usually three basic structures to this cloth, namely tabby (*pingwen*), twill (*xiewen*), and stain (*duanwen*) (Chen 1984: 41–3). Even as early as the Warring States period (475–221 BCE), both silk and hemp cloth were accepted as tax payments. This tradition continued, with various changes experienced during the early medieval period. In 737, in compiling a new code, the Tang Government regulated hemp cloth as an alternative tax payment if silk cloth was not an option. An adult was required to pay 77.7 cm (31 in.) length of hemp cloth and three pounds of raw hemp fiber per annum (Sheng 2013: 182–4).

Hemp was the most important raw material for making cloth in traditional China. Many scholars have noted that hemp remained as the generic name for plant fibers before cotton was introduced from Central Asia into China around the tenth to eleventh centuries. In the Tang Dynasty (618–907), four main types of establishment worked on textile production: rural households, official workshops, private urban workshops, and weaving households. In the eighth century, they contributed about 7.4 million bolts of tabby and more than 1.85 million hanks of silk floss to the state. As a typical labor division in medieval China, male peasants focused on cultivating hemp, ramie, and mulberries in the field, and female household members played the most important role in yarning and weaving for producing textiles. According to Hong Mai (1123–1202), when a young lady died at twenty-one, she was buried with thirty-three bolts of open weave, seventy bolts of plain silk, and about 50 meters (about 164 ft) of coarse silk. The inventory demonstrated her production in her lifetime. In addition, it should be noted that the ancient Chinese cultural tradition required family members to dress in raw tabby clothing while mourning the death of a relative (Bray 1997: 189–200; Liu et al. 2017: 239–40; Sheng 2013: 182–4).

Hemp was also a common raw material used for paper-making and printing technology in medieval China. Traditional Chinese historiography regards paper-making in proud terms since its history can be traced back to the inventor Cai Lun (48–121). Although many textual sources have described the process of papermaking technology in ancient China, early paper products have not survived to the present day. However, various manuscripts written in multiple languages have been found by Aurel Stein, Paul Pelliot (1878–1945), and other Western explorers in the early twentieth century in the extraordinary cave library in Dunhuang and other sites of ruined cities in Chinese Central Asia. Many of these manuscripts are made from hemp, which was called small hemp paper, for it might have been produced locally and was different from the paper used by the central government of the Tang Empire. Even though the manuscripts were found in Dunhuang, numerous documents were originally issued by the central government, and some religious manuscripts were distributed from the capital city of the Tang Empire, Chang'an. The paper from the central part of the Tang Empire, especially Chang'an, was better made with higher quality than what might have been made locally in Dunhuang.

Official documents issued from the central court of the Tang Empire were prepared on paper that was called "*huangbo*" hemp paper; this was more durable than regular hemp paper. The *huangbo* paper was produced from hemp paper that was completely soaked in oil extracted from the Amur cork tree (*Phellodendron amurense*) (Drège 1986: 403–15; Rong 2013: 485). Along the Silk Roads, many manuscripts were made of paper that was locally produced. As Agnieszka Helman-Ważny's fiber analysis, based on samples from 130 manuscripts has shown (Helman-Ważny 2016: 127–40), there were four distinct groups of rag paper, which came from textile remnants. Among these 130 manuscripts, only 4 were made using pure hemp. The majority of these manuscripts, 84, were made from ramie, some with a hemp addition. In the third group there were 8 cases where manuscripts were made of ramie and hemp with the addition of paper mulberry (*Broussonetia papyrifera* or a *Morus* sp.). The addition of paper mulberry fibers might have been due to a shortage of ramie and hemp in the northwestern region. The fourth group includes fifteen samples made of ramie and hemp with the addition of other, singular, fibers such as flax, jute, silk, or cotton. Clearly, hemp was the most important raw material for producing paper across the Tang Empire, with the addition of other fibers varying region by region depending on the supplies of available plant materials.

Flax was another important fabric plant in medieval China. It was called "barbarian hemp" in traditional Chinese literature, which indicates that it was an exported plant from foreign lands. According to Chinese sources such as the *History of the Han Dynasty* (*Hanshu*), it was first brought to China by Zhang Qian from a mission to Central Asia. However, some modern scholars (Liu et al. 2011: 561–6) question the veracity of this claim. Instead, they suggest that flax was first cultivated as a fiber crop and then an oil crop in India and China. The third-century agricultural work entitled *The Monthly Ordinances for Four Classes of People* (*Simin yueling*) offers information about planting flax. In medieval China, flax became one of the most popular plants along the Hexi Corridor, the pathway region linking the central part of the Chinese Empire with its northwestern frontier regions. It was used by local people for making oil, paper, and ropes.

Two species of cotton were introduced into medieval China through different routes. One was the African-Asian species (*Gossypium herbaceum*) that came to China mainly through the Silk Roads from Central Asia. Stephen Dale even suggested that the Silk Roads should also be regarded as the Cotton Road (2009: 79–88). In early medieval Chinese sources such as the social and economic documents among the manuscripts uncovered from Chinese Central Asia, cotton was called "*baidie*." In 1959, Chinese archaeologists discovered two fragments of cotton cloth from a second-century tomb near Niya in Xinjiang. As early as the sixth century, a local regime centered on an oasis city, Gaochang Kingdom (modern Turfan), was known to produce cotton cloth. In the ninth and tenth centuries, this area often sent cotton cloth as a tribute to the central court of the Chinese Empire. However, cultivation of this species never became widespread due to limitations of both environmental conditions and political control. The second species was the South Asian species (*Gossypium arboreum*), which came to southern regions of the Chinese Empire from India (especially modern Assam and Eastern Bengal states) via Myanmar. After the fourteenth century, this species flourished across southern China. In the medieval period, mostly in the Tang and Song Dynasties, cotton was not as popular and essential as in modern times. On the one hand, it was not a plant indigenous to China. On the other hand, when some Chinese residents in the Tang Dynasty started to learn about cotton and its production, the use of cotton garments was regarded by the government as a threat to preserving social status markers, since luxury silk and common hemp already occupied dominant positions to serve the main sources for making different garments to distinguish between the upper and lower classes. Moreover, official financial administrators also viewed cotton as a potential complicating hindrance to the fiscal regime. In the Song Dynasty, the cultivation of South Asian species of cotton was growing fast, especially in the lower Yangtze River region. In 1273, the government even issued a significant agriculture handbook entitled *Fundamentals of Agriculture and Sericulture* (*Nongsang jiyao*) detailing the cultivation of cotton. In the thirteenth century, cotton cultivation and cotton textile production were encouraged by the government. Cotton cloth was adopted as an important material for making both military uniforms and civilian garments (Zurndorfer 2011: 705–6).

Arguably, silk was the most reputable fabric material in medieval China, especially before the fourteenth century when cotton took over the dominant position in the textile production of China. Although silk came from silkworms and was thus not produced directly from plants, the leaves of a plant were used as the indispensable food for feeding silkworms. This plant was the white mulberry tree (*Morus alba*), one of the most extensively cultivated plants across the Chinese Empire for four thousand years. The Chinese were devoted to developing their sericulture by cultivating mulberry trees, raising silkworms, weaving silk cloth, and producing silk textiles. Silk cloth was used not only for making textiles but also served as money (legal tender) and a means of paying taxes along the Silk Roads, thus ultimately becoming a significant financial mainstay of the Chinese government. According to the *Tang Code* that was made law in 737, in silk-producing areas, the per annum tax for a household was about 62.2 cm (about 24.5 in.) silk plus 85 g (about 3 oz) of silk floss. For those areas that did not produce silk, the tax was about 77.7 cm (about 30.6 in.) of hemp cloth plus 1.3 kg (about 2.86 lbs) of hemp fiber. It appears that in northern China, silk was more favored than coins and hemp cloth, yet in southern China, coins and hemp were more common as currency, but silk was more valuable (Sheng 2013: 175–95). Expensive silk was often dyed by using plants such as safflower (*Carthamus tinctorius*), indigo (*Indigofera tinctoria*), and purple gromwell (*Lithospermum erythrorhizon*).

Chinese sources often used the characters "*nong*" (agriculture) and "*sang*" (mulberry) in conjunction with each other, and used the phrase "*sangtian*" (mulberry fields) for referring to cultivated lands, which highlighted the preeminent role of mulberry cultivation in traditional Chinese agriculture. The central court required local officials to take action to encourage farmers to cultivate mulberry trees and household women to weave silk and produce silk textiles. For instance, in the thirteenth century, each male adult of a household was required to plant twenty mulberry (or) jujube (*Ziziphus* sp.) trees each year (Kuhn 1988: 286–8). However, occasionally, in the medieval period, due to a shortage of mulberry leaves, silkworms could be fed with the leaves of oak trees which resulted in the production of coarser silk.

Besides serving as feedstock for silkworms, it should be noted that mulberry leaves were also used to feed cattle as well as being used as a convenient herbal medicine for healing wounds in medieval China. Mulberry trees also served as raw material for producing paper along the Silk Road (Helman-Ważny 2016: 132).

CULTIVATION OF FLOWERS

Flowers were cultivated and used for many purposes in the medieval period. Traditional Chinese sources often mention ornamenting and viewing activities centered on the flowers. Both textual and visual sources manifest the vibrant life of flowers among all social classes. Besides botanical records on flowers, medieval literati also wrote prose, poems, and lyrics about flowers they cultivated, observed, and discussed in their daily life. Visual materials such as murals in Buddhist caves and wall paintings in tombs also depict flowers. Some flowers seem to be more visible than others. For example, peonies were famous for their popularity in both large gardens and private villas in the medieval period (de Pee 2009: 85–116; Yang 2004a: 221–55). Other flowers of fruit often appeared in poems too, such as those of the pomegranate (*Punica granatum*) (Harper 1986: 139–53).

Li Deyu (787–850), a high-ranking official in the late Tang Dynasty, developed his own private villa in Pingquan, about 16 km (10 miles) south of Luoyang, modern Henan Province. Originally, he acquired a garden estate for commemorating his father for his filial piety, but soon he became obsessed with enjoying his reclusive residence surrounded by flowers and mountains. During his long career in the civil service and moving from place to place across the Tang Empire, he collected numerous plants from locations where he was stationed and cultivated them in his Pingquan Villa (Yang 2004b: 45–88). Not all of these plants survived. However, textual sources indicate that at least sixty-nine plant species did survive there, among which modern scholarship has identified about fifty-nine, including both trees and flowers that vary from tropical (39 percent) to temperate plants (3 percent). Based on the descriptions from numerous poems by Li, some flowers can be identified as follows: dwarf sedge (*Acorus gramineus*), Tartarian aster (*Aster tataricus*), buddleja (*Buddleja officinalis*), camellia (*Camellia japonica*), Japanese glorybower (*Clerodendrum japonicum*), eupatorium flower (*Eupatorium fortunei*, Chinese name: *peilan*), Korean sweetheart tree (*Euscaphis japonica*), hibiscus (*Hibiscus mutabilis*), rose of Sharon (*H. syriacus*), coastal rose of Sharon (*H. hamabo*), big leaf hydrangea (*Hydrangea macrophylla*), Chinese green-flower lily (*Lilium fargesii*), the flower of Kaido crab apple (*Malus x micromalus*), the flower of Magnolia (*Manglietia insignis*, *Magnolia biondii*, and *M. pilocarpa*), white lotus (*Nelumbo nucifera*), oleander flower (*Nerium oleander* [*N. indicum*]), ormosia (*Ormosia semicastrata*), sweet olive (*Osmanthus fragrans*), photinia (*Photinia serratifolia*), plum (*Prunus mume*), the flower of David's peach or Chinese wild peach (*P. davidiana*), Nanking cherry (*P. tomentosa*), the flower of Chinese white pear (*Pyrus bretschneideri*), rhododendron (*Rhododendron mariesii*), cloud-pattern silk rhododendron (*R. fortunei*), multiflora rose (*Rosa multiflora*), broadleaf lilac (*Syringa oblata*), sapphire berry (*Symplocos paniculata*), coltsfoot (*Tussilago farfara*), sea bilberry (*Vaccinium bracteatum*), the Chinese wisteria (*Wisteria sinensis*) (Feng 2016: 60–72). This long list illustrates the wealth of flowers that Li cultivated in ninth-century China. Many of them served as ornamentals for displaying and viewing, but some could produce fruit or herbs for medical use.

Among numerous flowers that were cultivated in medieval China, safflower occupied a prominent position for its use of multiple purposes. The discovery of numerous manuscripts from Dunhuang helps us understand that safflower played a sophisticated role in medieval Chinese economic, social, and cultural life. In 1993, Chinese scholar Wang

Kexiao first identified it in a careful reading of some manuscripts. Then Eric Trombert (1997: 509–47; 2004: 59–72) offered a comprehensive study of this flower. Together they demonstrate that safflower was used to serve as a medical herb, to dye various textiles, to produce rouge (*yanzhi*), and to extract oil for cooking purposes. Egyptians first cultivated safflower, followed later by the Persians. Eventually, it reached China via the Silk Road and flourished in northwestern regions. As early as in the *Essential Techniques for the Common People* in the sixth century, it was stated that oil could be extracted from safflower seeds to make candles and a lubricant for cartwheels. It also suggested that once the safflower was harvested in the summer, it should be processed immediately. In the ninth and tenth centuries the use of safflower seeds and petals expanded. Besides serving medical and cooking purposes, safflower was also used for making dyestuff. According to Trombert's study of a letter in the early tenth century, rouge was listed as the most expensive tribute that was sent by local people in Dunhuang to the khan of the Uighur Kingdom, along with jewels, silk, cotton, and hemp cloth. His study of economic documents from the cave library in Dunhuang also revealed that safflower and purple gromwell were the two most popular materials for making dyestuffs since most silk, woolen, and hemp textiles were dyed bright red or purple; here, safflower was used to produce a bright red color. Purple was often used for dyeing monastic robes to offer to high-ranking monks in the Dunhuang area. Safflower was also used for making rouge, a common cosmetic to wear on the cheeks for making them look bright red. Both textual and visual sources from Dunhuang illustrate the fashion of local women for wearing rouge. Nevertheless, it seems that processing technology limited the production of rouge, and it remained an expensive product from safflower, compared to its wide use for making dyestuff.

TREES AND WOOD TECHNOLOGY

In the Tang and Song Dynasties, one of the most remarkable accomplishments was the appearance of works on specific trees that illustrate meticulous knowledge about these trees. In particular, in the Song Dynasty, Chinese literati developed a predilection for planting trees and flowers in their private gardens and became interested in natural history. They were the first generation of Confucian scholars to focus their interest on nature rather than political and social order. Although the Chinese had already acknowledged the aesthetic and economic values of paulownia more than three thousand years ago, a specific book on this plant did not appear until the eleventh century (see Figure 4.2). Chen Zhu (1009–61) was a native of Tongling County (modern Anhui Province). While being a hermit who did not serve in the government as most Confucian scholars did, he devoted himself to cultivating bamboo and paulownia. He practiced the technology of cultivating paulownia and observed the growth of this plant. Eventually, in 1049, he wrote a book entitled *Record of Paulownia* (*Tongpu*) which offers a very detailed account of the history of this plant in China. He first traced the dispute on various names of this plant and identified them (including *wutong* [*Firmiana simplex*], paulownia, and *tung* oil tree [*Vernicia fordii*]) by analyzing their tree forms, leaf forms, habitats, colors, flowers, and functions, then discussed its species. Some of these species can be identified more specifically according to modern plant science. For example, from his description, the "white-flowered tong" (*baihua tong*) is *Paulownia fortunei*, while the "purple-flowered tong" (*zihua tong*) is *P. tomentosa*. Chen Zhu focused on the extraction of oil from the seeds of *Vernicia fordii*. One of Chen's descriptions can also help identify *Erythrina variegata*, a tree that gave its name to Quanzhou as "Citong" in the medieval period. Another species might be *Clerodendrum japonicum* (Haw 2017: 224–5). Chen Zhu also described the methods of planting, cultivating, maintaining, and processing. He provided a detailed account of the environmental conditions necessary for cultivating paulownia such as the required geomorphology, soil condition, sunshine, drying and soaking, temperature, and humidity. He even elaborated on a method for controlling harmful insects. According to his description, the planting of paulownia was very popular in the lower Yangtze River region. However, at that time, the most well-known place for producing paulownia was the Sichuan area. He demonstrated his extensive knowledge of paulownia based on both his textual reading and personal experience (Luo 2002: 173). Paulownia wood was widely used for various purposes including making shoes and musical instruments. Chinese monks and nuns used wooden rice pots, bowls, and spoons (Hu 1961: 11–27). According to Stephen G. Haw's study, two Chinese species *Vernicia fordii*, and *V. montana*, were important trees for extracting oil from their seeds. This *tung* oil was extensively used as paint and varnish in the Song and Yuan Dynasties. This oil became the chief paint oil for all exterior woodwork throughout the Chinese Empire in the early twentieth century. The oil was also used for making Chinese ink in the Song Dynasty (Haw 2017: 215–36).

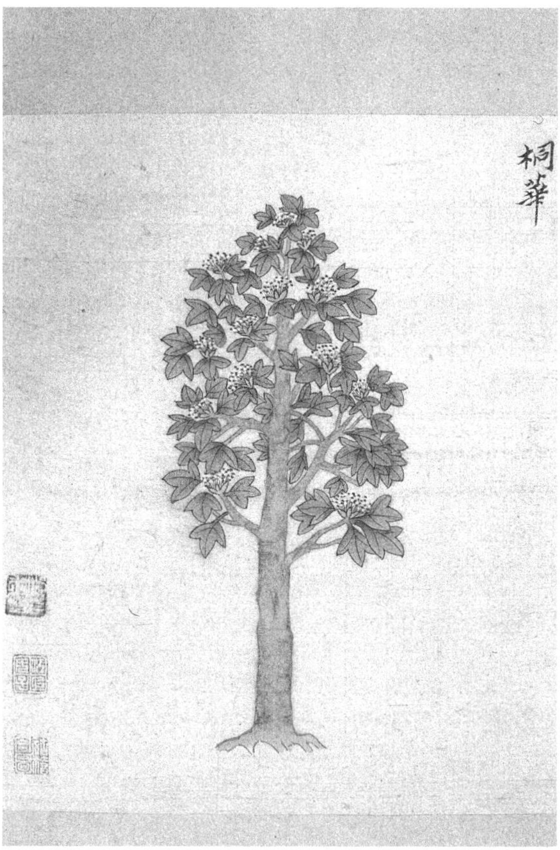

FIGURE 4.2 Painting of the paulownia from *Bencao tupu* (Illustrated Herbal) by Zhou Hu and Zhou Xi, 1644. Credit: Wellcome Collection. Attribution 4.0 International (CC BY 4.0).

In the medieval period, it was noteworthy that technological advances in the use of wood in architectural applications flourished. Building science and technology in China experienced a sort of revolution and became much more mature and sophisticated than in previous dynasties; this can be marked by a meticulous manual entitled *State Building Standards* (*Yingzao fashi*), a monumental work compiled in 1100 by Li Jie (1065–1110), a Superintendent of State Buildings, and released in 1103 by the Song Government when Li Jie was promoted to Director of Palace Buildings. The appearance of this work should be attributed to the Wang Anshi Reform (1069–74) launched by Wang Anshi (1021–86) who served as a chancellor for the socio-economic reform movement and was an advocate for specialized professions. Wang urged the compilation of a guide to building construction. After Wang's death, the emperor commissioned Li Jie to formulate a new manual. This manual consists of thirty-four chapters divided into five parts, offering detailed technical guidelines for building palaces, mansions, ordinary houses, and pavilions. The first part interprets forty-eight building technicalities, mathematical formulae, building proportions, and information on construction. The second part lists standards and regulations for design, construction methods, working procedures, and manufacture of bricks, tiles, and glazed tiles. The third part focuses on work norms and costs as well as personnel management (Guo 1998: 1–13). This manual provides highly reliable and specific information about the building technology in the Song Dynasty as well as previous dynasties. The major components required for constructing a building are blocks (*dou*) and brackets (*gong*), which means that timber was first processed for making these blocks and brackets. Flower-shaped bracketing seems to have been a typical technology from the ninth to the thirteenth centuries when wood-framed architecture extensively used brackets for structural and aesthetic purposes (Feng 2007: 369–432). This particular wooden architectural technology, which originated in medieval China, demonstrated its power and was also diffused abroad to medieval Korea and Japan, especially in those traditional palaces and religious temples which are still in existence. This work also briefly mentions the following trees and woods for making various parts of medieval buildings: sandalwood (*Santalum album*), Siberian elm (*Ulmus pumila*), pagoda tree (*Styphnolobium japonicum*), pine (*Pinus*), cypress (*Platycladus orientalis*), zhennan or black wood (*Phoebe zhennan*), and Chinese fir (*Cunninghamia lanceolata*).

Architecture and Building

The East Hall of Buddha's Light Temple and some nearby temple structures in Shanxi Province, which can be dated back to the ninth century, are generally cited as the earliest extant examples of wooden architecture in China (see Figure 4.3). However, the original wood source is not always identifiable. Most likely, they were constructed of Prince Rupprecht's larch (*Larix gemlinii* var. *principis-rupprechtii*), and, later, other materials from different trees such as Chinese red pine (*Pinus tabuliformis*) and Chinese willow (*Salix babylonica* [*S. matsudana*]) were used for repair and replacement (Tsinghua 2011: 49). Interestingly, these trees can be found among the plants cultivated in Pingyuan Villa, the ninth-century private garden of Li Deyu in the Sichuan Province. It seems that in the Tang Dynasty, Chinese red pine and willow were commonly used for architectural purposes. The wooden pagoda in Ying County, Shanxi Province, regarded as one of the oldest extant wooden buildings in China can be dated back to the eleventh century. The main parts of this building such as pillars, beams, arches, and blocks were made of Prince Rupprecht's larch, which is native to the Taihang Mountains in northern China, especially across the boundary between Shanxi and Hebei Provinces. Some other parts came from the Siberian elm and other pine trees (Shen 2006: 85–6).

FIGURE 4.3 Roof beams of the Grand East Hall of the Foguang Temple in Shanxi Province, China, built in 857 during the Tang Dynasty (618–917), one of the oldest wooden buildings in China. © Fan Minda/Xinhua/Alamy Live News/Alamy.

Shipbuilding

Shipbuilding technology experienced something of a renaissance in medieval China, becoming a more sophisticated and large-scale enterprise during this time. The construction of the great canal from Luoyang to Yangzhou and socio-economic development in south China during the Sui and Tang Dynasties created a high demand for shipbuilding in Sichuan and other regions. Since the Tangut Empire blocked the Silk Route linking Central China and the Western Regions in the eleventh century, trade between China and other parts of Asia often took place by sea, which provided the impetus for some major advances in shipbuilding technology in the Song Dynasty. After the Jurchen Dynasty occupied north China, the Southern Song Dynasty devoted itself to the building of a large navy to protect itself against enemies from the north along the Huai (across Anhui and Jiangsu Provinces) and Yangtze Rivers. Also, trade with Korea, Japan, and other regions needed to occur via sea; indeed, the Song Dynasty has been called the beginning of the maritime Silk Road era. In the Song Dynasty, technological innovations resulted in the construction of curved side boards, broad lateral beams, and lofty superstructures with mortise-and-tenon joints and iron nails being employed to make boats stronger, safer, faster, and more powerful.

Both textual sources and archaeological discoveries provide some explicit information about shipbuilding technology in the Tang and Song periods. In the Tang Dynasty, most river boats had flat bottoms and square prows. They were mostly made of zhennan or black wood (*Phoebe zhennan*), camphorwood, and Chinese fir (*Cunninghamia lanceolata*), as well as Chinese red pine (*Pinus tabuliformis*). Although these woods were the most commonly used in the shipbuilding industry, some small boats might have been constructed using other types of wood or other materials, depending on convenience and availability. Some fragments of canoes have been found in the Great Canal region near Liuzi (Huaibei, Anhui Province). Archaeologists identified one fragment, uncovered from Ningbo in Zhejiang Province in 1973, as a dragon boat from the Tang Dynasty. It was about 11 m (38 feet) long and 0.95 m (3 feet) wide, and cut from a single red pine timber. In the Song Dynasty, the most popular materials for building ships and boats were Chinese fir and Chinese red pine, which were widely available in south China. The Jiangxi region became the largest producer of river boats and warships. Zhejiang and Fujian served as the principal places for building sea boats (Huang 2017: 16–18), for there were abundant and good-quality wood materials available. For example, in 1974, the wreckage of a large sea boat from the Song Dynasty was uncovered in the Quanzhou Bay area, Fujian Province. The remaining structure of this boat was about 24 m (80 feet) long and 9m (30 feet) wide. There are thirteen chambers inside. The keel is made of two red pine timbers. The cant frames were constructed from solid and durable camphorwood while other parts were made using Chinese fir as major materials. It has been discovered that in the process of construction, this boat was thoroughly painted with paulownia oil, that had been obtained from the seeds, to solidify the structure (*Fujiansheng Quanzhou haiwai jiaotongshi bowuguan* 2017: 21–2).

In medieval China, the government established numerous large shipbuilding yards for constructing river boats, sea boats, and warships across south China, especially in the Jiangsu and Zhejiang areas. Some other places in Hubei, Jiangxi, Fujian, and Guangdong also had shipbuilding yards, producing small numbers of boats. As early as in the sixth century, the government built a large five-story warship which was nearly 55 m (180 feet) long. It can be estimated that this would have had the capacity for eight hundred military personnel to defend it against enemies in the Yangtze River. In the Song Dynasty, a type of warship called the cart ship was equipped with two giant turning wheels on each side to power the ship. Four soldiers were required to move each wheel-board mechanical system. This warship could move much faster than its contemporaries. Some large warships might have been more than 100 m (325 feet) long with 1,500-ton capacities. In the early eleventh century, the government shipbuilding yards in south China produced almost three thousand river boats and transport ships per annum. In 1257, in the Ningbo area alone, there were almost eight thousand boats. In one military campaign against the Jurchen invasion in the early thirteenth century, the Southern Song navy mobilized more than seven hundred warships. For trade purposes, boats made in China undertook voyages to Korea, Japan, India, Sri Lanka, and the Arabian Peninsula. Some sea boats were too large to enter the Persian Gulf so merchants had to switch to smaller boats (Feng 1957: 10–14).

Carts and Sedan Chairs

In terms of transportation in the medieval period, the Chinese tended to favor riding horses and using ox carts. A wooden model of an ox cart was discovered in an ancient tomb in Astana near Turfan, and was dated from the Tang Dynasty (618–907). It is only 53 cm (20.8 inches) high but clearly shows that an ox carried a black-painted cart for a single occupant (Liu 2002: 146–8). However, in the Song Dynasty, sedan chairs made of wood and bamboo gradually became more popular for both officials and civilians. Images often depict carts and sedan chairs in the Tang and Song Dynasties, yet archaeologists rarely discover remains of medieval carts and sedan chairs, which makes it difficult to determine the original materials that were used to make them (see Figure 4.4). Many sedan

FIGURE 4.4 Depiction of a wooden cart, from the *San li tu* (a design book of ritual objects) authored by Chongyi Nie (active tenth century) and preserved in an edition printed by Tong zhi tang, 1676, Beijing, China. Courtesy of The Metropolitan Museum of Art, New York. Purchased with income from the Jacob S. Rogers Fund, 1940, Accession Number: 230.51 N55.

chairs seem to have been shouldered by two people and to have carried one occupant, though luxury sedan chairs for the upper classes might have required more people to lift. Pine and pear (*Pyrus* spp.) wood might have been used since they were popular materials for making furniture in the Song Dynasty. The famous painting *Along the River during the Qingming Festival* (*Qingming shanghe tu*) by Zhang Zeduan (1085–1145) depicts numerous sedan chairs, and is commonly believed to reflect daily life in the capital city Kaifeng in the Song Dynasty.

Furniture in Medieval China

In the medieval period, grass, wood, and bamboo were often used to make furniture. Textual sources such as poems and prose from the Tang period mention a lot of furniture yet often do not specify the materials used for their construction. Medieval Chinese furniture can be categorized as follows: furniture for sleeping and crouching (beds and mats), furniture for sitting (chairs and stools), furniture for supporting objects and items (stands, tables, and desks), and furniture for storing objects and items (chests, boxes, cabinets, cupboards, and wardrobes). Archaeological findings of Tang furniture are very rare. However, some pieces preserved in Shōsōin, a treasure house in Tōdai-ji in Nara, Japan, are believed to stem from the Tang Dynasty. They are made of narra (*Pterocarpus indicus*). According to the *Comprehensive Statutes* (*Tongdian*) compiled in 766–801 by Du You (735–812 CE), many prefectures in North China (modern Shaanxi, Gansu) offered sabaigrass (*Eulaliopsis binata*) mats as local tributes to the Tang court, which seems to indicate its popularity in the northern regions. In other parts of China, sea water grass (*Cyperus malaccensis*) was used to make mats (Liu and Xiong 2017: 33–4).

In the Song Dynasty, various texts and archaeological discoveries provide information about furniture. Wood and bamboo continued to serve as two of the major materials for making furniture, though cane and grasses as well as other materials were also used. It appears that much furniture was made of wood from local trees (Shao 2010: 161), and Chinese users recognized three wood categories: softwood (*ruanmu*), wood of inferior quality (*chaimu*), and hardwood (*yingmu*). The softwood in this, rather than the modern sense, included Chinese aspen (*Populus adenopoda*), Chinese fir, and paulownia (*Paulownia tomentosa*), for example. In the Song Dynasty, Chinese aspen was commonly used for making screens, chests, boxes, and stands in North China. In the tomb of Yan Deyuan near Datong, Shanxi Province, illustrated screens and stands made of Chinese aspen were discovered. Chinese fir was a common tree in South China. A desk and a chair made of the wood of fir were found in the tomb of Sun Siniangzi in Jiangyin, Jiangsu Province, and dated from the Northern Song Dynasty. Paulownia was a popular choice for making musical instruments, such as Chinese zithers and drums (Shao 2010: 163).

In the medieval sense, wood of inferior quality included Siberian elm (*Ulmus pumila*), jujube (*Ziziphus mauritiana*), Manchurian catalpa (*Catalpa bungei*), Chinese catalpa (*C. ovata*), apricot tree (*Prunus armeniaca*), and zhennan or black wood. Siberian elm and Manchurian catalpa commonly appeared as major materials for making furniture in the Yellow River region. The wood from these trees is durable in wet conditions, so it was also used for making agricultural tools. Manchurian catalpa was also a good choice for making Chinese *go* (*weiqi*) game boards and scales. Chinese catalpa could be found in both the Yellow River and Yangtze River regions, and was used for making tables, chests, and storage items, as well as the outer layer of coffins. Apricot tree wood mostly appeared in the Yellow River region. The chair, bed, screen, towel stand, table, and zither stand found in Yan Deyuan's tomb were all made of apricot wood (Shao 2010: 163). Southwest China was the major place for producing black wood, which was often used to make chests and coffins.

The hardwoods included sandalwood (*Santalum* spp.) and amboyna wood (*Pterocarpus indicus*), which were imported from Southeast Asia and often used for making luxury furniture. One poem by the Northern Song Emperor Huizong seems to indicate that he possessed a sandalwood game board. A sandalwood container for storing Buddhist scriptures from the Northern Song was found in Rui'an, Zhejiang Province in the 1970s (Shao 2010: 165). Besides wood, bamboo was the second most popular material for making furniture. There are numerous different species of bamboo in south China. Due to its easy availability and the less sophisticated processing technology, it was widely used for making everyday furniture, especially ordinary chairs, armchairs, beds, tables, mats, and sedan chairs. Various species of canes and grasses across the country were also used to make mats and stools (Shao 2010: 161–8).

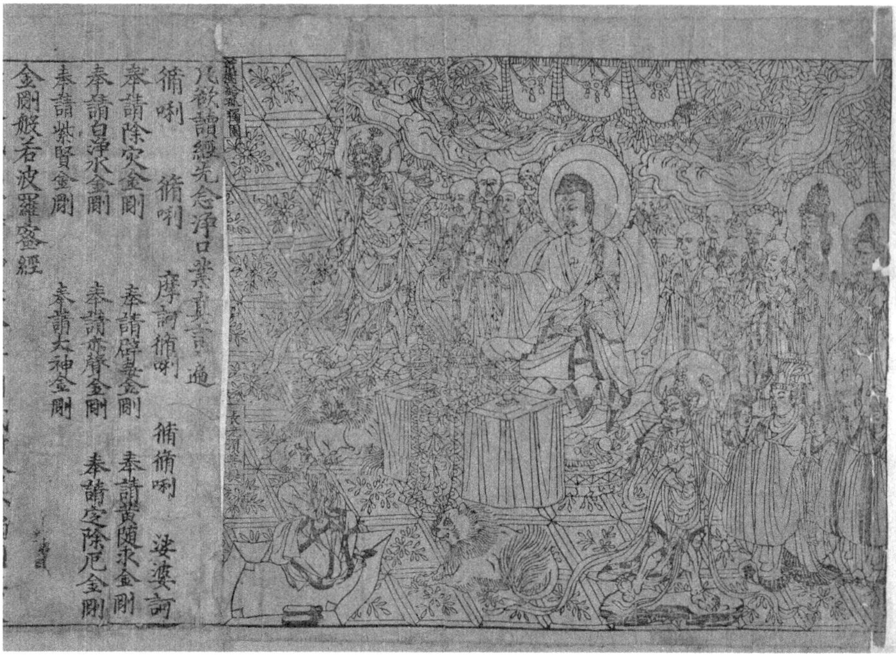

FIGURE 4.5 Frontispiece of the *Diamond Sutra* from Cave 17, Dunhuang, China (868 CE); the earliest known dated, printed book. Currently located in the British Library, London. Courtesy of Jingangjing/Wikimedia commons.

Woodblock Printing

In the Tang Dynasty, the Chinese invented the technology of woodblock printing. They used the hard, close-grained wood from pear (*Pyrus* spp.) and jujube (*Ziziphus* spp.) trees for making blocks (Barrett 2008; Twitchett 1983: 70). The carvers first engraved Chinese characters on the blocks and then printed text by pressing paper on the blocks. It is well known that the earliest dated printed book can be traced back to the *Diamond Sutra*, dated 868, and discovered in the Dunhuang cave library (see Figure 4.5). From evidence in historical sources, scholars have acknowledged that during the ninth century, some major urban centers of northern and central regions of the Tang Empire developed woodblock printed materials. For instance, in Sichuan, Chang'an, Luoyang, and Jiangxi areas, private printing activities might have produced medical books, poetry collections, reading primers, character lists, and divination books (McDermott 2006: 11–12). However, as early as the eighth century, some Buddhist seals, even if they were often not counted as books, might have been printed.

CONCLUDING REMARKS: EXOTIC PLANTS FROM ABROAD IN TANG CHINA

In the medieval period, the Tang Empire conquered many places in Central Asia and claimed numerous territories. As an open empire, it received various kinds of tribute from neighboring states and other regimes from afar. Persians, Indians, Sogdians, Turks, and Uighurs, as well as other peoples, often introduced exotic fruit to the Tang court.

The Silk Road linking East Asia with West, South, and Central Asia became a pathway for constructing a Eurasian network of exchanges of religions, medicine, science and technology, animals and plants, as well as other materials. Many foreign plants were introduced into the Chinese Empire. In his monumental work entitled *The Golden Peaches of Samarkand: A Study of T'ang Exotics 618–907*, Edward H. Schafer (1963: 117–32) discussed some of the most interesting exotic plants that were presented as royal gifts from abroad by drawing upon particular sources from Tang literature. He also noted that some of these exotic plants later became common domestic plants in China. He first described the administration and technology of preserving fresh fruit such as litchis from Lingnan (modern Guangdong Province) and mare-nipple grapes from Central Asia (modern Xinjiang Province). The Chinese Empire established an office called the "Office of His Highness' Forest" in charge of receiving, transporting, and preserving perishable fruit, flowers, and seedlings from afar. This office ordered local administrators to cut ice blocks from the mountains for storing these perishable tributes. The imperial court of Tang China also built Tabooed Park near the imperial palace in Chang'an to serve as a nursery for planting exotic fruit and vegetables. A medical branch under the central government was responsible for planting medicinal plants. Some exotic plants from abroad discussed in Schafer's work include the date palm (*Phoenix dactylifera*) from Persia, peepul tree (*Ficus religiosa*) and sal tree (*Shorea robusta*) from India, saffron crocus (*Crocus sativus*) from India and Bukhara, snake flower (*Nagapushpam* in Sanskrit) from India, Narcissus from Persia, many species of lotus flowers from India, and water lilies (*Nymphaea caerulea*) from India. Although not noted by Schafer, many of these plants may also have been introduced by religious adherents of Zoroastrianism, Buddhism, Manichaeism, and Nestorianism since some of these plants served in liturgical rituals and daily practices. In Schafer's account, some exotic wood, mostly from India and Southeast Asia, also came to the Tang Empire. For example, sandalwood (*Santalum* spp.) was imported from India and Southeast Asia. Rosewood (*Dalbergia hainanensis*) was from the island of Hainan, and probably other rosewood came from Southeast Asia. A kind of blackwood from India, identified as ebony (*Diospyros* sp.) by Schafer, was also valued in Tang literature (1963: 133–8).

CHAPTER FIVE

Plants and Medicine

IOLANDA VENTURA, TONY HUNT, AND JOHANNES GOTTFRIED MAYER†

During antiquity and the Middle Ages the approach to plants both in Europe and the Near East, the regions which are the focus of this chapter, varied not only according to periods and places, but also—if not primarily—according to the perspective of scholars and professionals who handled, studied, and interpreted plants both *realistically* as living organisms with a view to their nature and potential uses and qualities, and *metaphorically* or symbolically. Plants were dealt with by natural philosophers, physicians, apothecaries, agronomists, and architects as well as interpreters of the Bible, religious authors, and preachers, together with poets and writers in search of the right metaphor to communicate an idea. As in the case of other natural objects, discussion of plants raised important linguistic issues in the Middle Ages. In a time characterized by the interaction and the overlapping of several languages and various forms of exchanges of knowledge between linguistic areas, language played a crucial role, as it guaranteed—or should have done so if it was correctly used—the right transmission of plant knowledge across cultures, including accurate identification across time and space.

A CONFLATION OF SCIENCES, PERSPECTIVES, AND APPROACHES

The nature, names, and uses of plants formed a complex interplay of data, forms of knowledge, and approaches reflected by a number of authors who either dealt directly with the plant world, or extracted some of its concrete or symbolic features for use in their writings. In such a fluid epistemological and textual context, botany was a conflation of various sciences related to the world of plants rather than a single branch of knowledge. It was a complex scientific system that included natural philosophy (dealing with the general characteristics of the plant world), lexicography (concerned with plant nomenclature and the process of finding reliable and generally acknowledged synonyms in the main languages in which medical and pharmacotherapeutical works were written), and phytopharmacy (plant-based pharmaceutics). Horticulture, though less written about, also received significant attention.

This chapter deals with the scholarly approach to plants, specifically in medicine and phytopharmacy. Its perspective is chronological and typological. After offering a chronology of the main phases of the development of medicinal botany and introducing a presentation of the most representative authors and texts dealing with plants, it focuses on the role played by plants in medicine and therapeutics across the Middle Ages. Our survey of the history of medieval phytopharmacy distinguishes four major phases: the heritage from antiquity; the practice of medicine in monasteries from the eighth to the twelfth centuries; the "medical revolution" triggered by the reception of some major Arabic works in Western culture; and the development of universities, academic and scholastic pharmacotherapy and pharmacology (viz., the branch of knowledge dealing with theoretical notions of pharmacotherapy such as "therapeutical properties") during the last centuries of the Middle Ages, accompanied by the production of a wide body of vernacular translations that contributed to circulate data and information about plants originally accessible only to people versed in Latin among a larger public. This attempt to offer a pan-European overview includes authors and texts ranging from England to southern Italy, from France to Germany, without this geographical division corresponding to set borders, for medical and pharmacotherapeutical literature did not know any such geographical delimitations; it traveled across cultures and languages, and conveyed data and information to different geographical and cultural milieus, allowing for the exchange and development of knowledge.

THE BACKGROUND OF BOTANY

The renewed flowering of natural philosophy and botanical knowledge in the post-classical period was heralded and largely mediated by a modest and somewhat confused general account of the nature of plants: the treatise *De plantis* (On Plants) which, though attributed to Aristotle (384–322 BCE), was possibly written by Nicolaus of Damascus (*c.* 64 BCE–after 14 CE) and which was translated into Latin and commented on *c.* 1200 by Alfred of Sareshel (Shareshill, England; Alfredus Anglicus, twelfth–thirteenth century) in Spain. The main botanical treatise from Aristotle's school, later translated as *De plantis* (On Plants), by Theophrastus (*c.* 371–*c.* 287 BCE), remained untranslated and virtually inaccessible to the Western world until the Renaissance. The Pseudo-Aristotelian text, however, was very successful. It was commented on not only by Alfred, but also by an unknown English author who compiled the so-called *Glossa anglicana* (*c.* 1230), and by both Albert the Great (1200–80) and Peter of Alvernia (d. September 9, 1304). Among these commentaries, Albert the Great's *De vegetalibus et plantis* (On Vegetation and Plants; *c.* 1260) deserves special attention because of its particular structure: Books I–V contain the commentary on the Pseudo-Aristotelian texts, and Books VI–VII a description of plants (*c.* four hundred species including trees), that focused on the medicinal uses of the plants and the growing of trees. In this way Albert provided medieval readers with a wide, encyclopedic view of the plant world that drew on a large library of sources and paid particular attention to theoretical knowledge, practical experience, and uses of plants, resulting in connecting natural philosophy, science, and medicine.

As for plant nomenclature, many inquiries into the linguistics and lexicography of phytonyms developed, providing a rich source of material for study. Plant names, both Latin and vernacular, were frequently treated as interchangeable and were contextually determined. The same name might be used for a range of plants, not necessarily ones resembling one other, and, conversely, a single plant might have been known by several names. Some names were authentic indigenous, even regional, forms, while others were merely "literary" names resulting from literal translation. Writers of medico-botanical treatises often provided a series of names—*synonyma herbarum*—as an aid to clarifying plant nomenclature, as was the case with Dioscorides (first century CE) and many subsequent herbals, though clear distinctions were not always achieved.

There was a great variety of medico-botanical glossaries. In the eighth–ninth centuries, the most representative kind was the so-called *Hermeneumata medico-botanica* (Medico-botanical Explanations) aimed at providing a good Latin equivalent of Greek plant names. At the same time a large number of plant glossaries were being compiled in Anglo-Saxon England. It has been calculated that there are some six thousand Latin-Old English plant glosses, many practical and others more erudite. The *Laud Herbal Glossary*, for example, which is a twelfth-century manuscript (Oxford, Bodleian Library, MS Laud Misc. 587), contains over 1,500 entries based on a wide range of sources, and incorporates some two hundred Old English and six French words. These glossaries represent a valuable source for the study of medieval botany, because they transmit a rich nomenclature and, together with medical texts, help medical historians and ethnobotanists to identify the array of plants used in Anglo-Saxon medicine and, not infrequently, confirm the therapeutic properties attributed in the past to specific plants.

A second sort of glossary was produced in what may be called the golden age of medieval medicine, namely the time of the Greek/Arabic-Latin translations (eleventh–thirteenth centuries). During this time, the most advanced Arabic medical manuals reached Western medical culture and were translated into Latin. The *Kitab al-Malaki* by al-Majusi (d. *c.* 994) was translated twice: first by Constantine the African (d. 1098/9) as *Pantegni* (Medical Art), and then by Stephen of Antioch (*c.* 1127) as *Regalis dispositio* (Royal Regulation) in the first half of the twelfth century. The *Qanun fi al-Tibb* by ibn Sina (Avicenna; 900–1037) was translated by Gerard of Cremona (1114–87) as *Liber canonis* (Canon). Of Rhazes (Abu Bakr Muhammad ibn Zakariyya al-Razi, 854–925), his *Kitab al-Mansuri* was translated by Gerard of Cremona as *Liber ad Almansor* (Alamansor's Book of Medicine), and his *Kitab al-Hawi* as *Continens* (Collection of Writings) by Faraj ibn Salim in 1279. These and other similar works increased the need for reliable glossaries that could offer a good Latin equivalent for the names of medical remedies left without translation, or crudely transliterated, in the manuscripts. Different authorities compiled concordances of names and plants. Among such works can be mentioned the so-called *Synonyma Rhazes*, the *Synonyma Stephani* (that is, by Stephen of Antioch), as well as the *Salernitan* glossary *Alphita* (twelfth–thirteenth centuries), the *Clavis sanationis* (Key of Healing) by Simon of Genoa (twelfth/thirteenth centuries), and Matthaeus Silvaticus' (*c.* 1280–*c.* 1342) *Liber pandectarum medicinae* (Complete Book of Medicines) completed in 1332. Finally, a large number of *Synonyma apothecariorum* (Synonyms of Apothecaries) were produced during the late Middle Ages. They connected Latin names to both other Latin synonyms and vernacular (especially German) translations.

The practice of horticulture is less well documented, at least in written sources. In the absence of a comprehensive "flora" (treatise on plants), scholars largely depend on written documents like building accounts, charters, and surveys; archaeology; and painting. Horticulture in the Middle Ages was practiced on a small scale, often confined to the garth (garden) or yard. Special, more sophisticated, gardens were created for medical purposes in abbeys and monasteries, as shown by the first-hand account provided in the *Hortulus* (Little Garden) by Walahfrid Strabo (808/9–849; see below), whose book examines twenty-three plants in botanical and medicinal detail. Charlemagne (742–814) attempted to standardize the structure of private gardens through his *Capitulare de villis et curtis imperialibus* (Ordinance Regarding the Royal Villas and Farms), in which he prescribed some seventy-three herbs for such gardens. It is striking that the *Glossary* by

Aelfric (*c.* 995–*c.* 1025), written in the last decade of the tenth century, listed sixty-eight names of herbs. Combined with the evidence of his *Colloquia* (Discourses), it covered some 140 plants, employing about 170 Old English Names.

Further attempts to regulate agriculture were made throughout the Middle Ages, under the influence of *De agricultura* (On Agriculture) by Palladius (end of the fourth–first half of the fifth century), for example in the *Praecepta cistercensia ruralis oeconomiae* (Cistercian Instructions on Rural Management) (preserved in five manuscripts of Cistercian origin) by Alexander Neckam (1157–1217), the Abbot of Cirencester, England, who in *De naturis rerum* (On the Natures of Things) describes roughly 166 plants that he would like to see in the garden, and in *De laudibus divinae sapientiae* (On the Praises of Divine Wisdom), in which he devotes *distinctio* 7 (372 lines) to medicinal herbs and their uses, or in *Liber Floridus* (Floriferous Book) (*c.* 1120) of the French Benedictine, Lambert, canon of St Omer (*c.* 1061–1150).

Greater horticultural display was evident at larger abbeys and the palatial manor houses of their abbots, somewhat in the manner of the Roman villas (on whose ruins they were often constructed) such as the Fishbourne Roman Palace dated to 75 CE and situated in Chichester (West Sussex, England), which is often compared to the palace of the Roman Emperor Domitian (51–96 CE) on the Palatine Hill in Rome. Viticulture, too, was a regular feature of land-use, especially in royal gardens, but despite the fact that some English wine was rated to be as good as French, generally the vineyards produced verjuice (grape vinegar) for culinary purposes.

The monasteries made an important and long-lived contribution to techniques of land-management, crop-rotation, irrigation, drainage, production, and distribution of seeds, not to mention horticultural terminology. Monasteries typically had a *vinea* (a small growing area for vines), a *pomerium* (orchard), a *gardinum* (kitchen garden), an *herbarium* (for medicinal plants), and a *viridarium* (verger, often a modest pleasure-garden like the "paradises" found in monastic grounds). The Cellarer's (responsible for the provisioning of food and drink) garden needed to be extensive. For example, in his herbal *Aaron Danielis*, Henry Daniel (1315/20–after ?1385) listed 252 species that could be found in his garden in Stepney, London, in the mid-fourteenth century.

Grander gardens may have been inspired by the extremely popular treatise (over a hundred manuscripts identified so far) of *Ruralia commoda* (Book of Rural Benefits) (written by Piero de' Crescenzi (*c.* 1233–*c.* 1320), around 1305), Book 6, which deals with herbs and is indebted to Dioscorides and Pliny. There appears to have been an impressive physic garden in Venice in 1334 that followed a planting scheme proposed by a Dr. Gualterus, and one at Salerno, founded by Matthaeus Silvaticus (1280–*c.* 1342), who taught there and combined classical and Arabic materials in his celebrated *Liber pandectarum medicinae* (Complete Book of Medicines), which includes 487 plant names. In 1447 Pope Nicolas V (1397–1455; elected pope in 1447) set aside a plot in the grounds of the Vatican for the establishment of medicinal plants. By *c.* 1400 some physic gardens could boast of as many as 250 different kinds of plants.

The fourth, and by far the largest and most important, area of medieval botany can be identified in phytopharmacy—"green medicine." This branch of knowledge is usually defined and described according to, firstly, the main authoritative sources from antiquity and the Middle Ages, and secondly, the main steps of its development.

FIGURE 5.1 Ambrogio Lorenzetti, *The Effects of Good Government on the Country Side* (Siena: Palazzo Pubblico, 1338–40). Photo by Art Media/Print Collector. Courtesy of Getty Images.

PHYTOPHARMACY IN ANTIQUITY AND THE MIDDLE AGES

Phytopharmacy developed in the Middle Ages as a result of the reception of three major works dealing with *materia medica*—animal, vegetable, and mineral—which privileged the association of botany with medicine. The fountainhead of such activity was the work of Dioscorides, written in Greek, but commonly known by its Latin title *De materia medica* (On Medical Matters). In the Middle Ages, it was known in various translations (four only in Latin, produced between the sixth and the eleventh centuries) and forms (e.g. in the alphabetical version known as *Dioscorides alphabeticus*, possibly put together in southern Italy during the eleventh century). This remained the principal pharmacopoeia of western Europe until well beyond the Renaissance. A further reason for its fame is represented by the presence, in some manuscripts, of a rich corpus of illustrations which attests to the interest in, and study of, nature during the Middle Ages. In total, Dioscorides dealt with *c.* seven hundred plants, not all of which can be exactly identified.

A second great source of botanical knowledge, accompanied by a fair amount of superstition and anecdotes (by no means valueless from an ethnobotanical point of view) was the encyclopedic *Naturalis Historia* (Natural History) of Pliny (23/4–79 CE) in thirty-seven books, which treats trees, other plants, and medicines. As in the case of Dioscorides, Pliny's work retained an unbroken popularity with respect to botany and the study of nature generally until the sixteenth century, albeit rapidly losing authority in the field of phytopharmacy after the introduction of Arabo-Latin medicine in the West. In his coverage of some one thousand trees and other plants (Pliny had a personal interest in plants), he emphasizes his belief that their nature can only be properly understood by studying their medicinal effects (*Naturalis Historia*, XIX, 189). He illustrates this in a very large number of recipes, prescriptions, or remedies, which further demonstrate that he understands what the Swiss Paracelsus (Theophrastus von Hohenheim, 1493–1541) later developed as his "doctrine of signatures," an idea suggested by Dioscorides, a development of the principle which governs homoeopathy—"*similia similibus curantur*" (like is cured by like). An external feature of a plant, be it leaf, root, or flower, is perceived to resemble a part of the body or some symptom of human disorder and hence to be efficacious in the treatment of that disorder or body part. In consequence of that, some plants with yellow flowers, for example, are prescribed for jaundice and those with red flowers (or other parts) for staunching blood, the spotted leaves of lungwort (*Pulmonaria officinalis*, the Latin name reflecting the belief) suggesting its usefulness in treating diseased lungs and so on. Over two hundred manuscripts, counting excerpts and fragments, of Pliny's work survive.

Significantly, the works of Dioscorides and Pliny, together with treatises deriving from them, are well represented in monastic library catalogues in the period 600–1100, alongside a wide range of short anonymous works: recipe collections; botanical treatises; *dynamidia* (treatises, recalling Hippocrates' *Regimen*, dealing with the effects of a variety of substances, elements of diet, and weather among other things); treatises on weights and measures (often with a regional flavor); others dealing with the substitution and interchangeability of medieval substances (especially plants) commonly known as *Quid pro quo*; and notably, the *Medicina Plinii* and *Physica Plinii*, two collections of recipes mostly consisting of excerpts derived from the *Naturalis Historia* and integrated with other sources (viz. Dioscorides), written in the early fourth century.

Compared with these pillars of medieval Latin phytopharmacy, the theoretical background provided by the Greek, Galen (Aelius, or Claudius, Galenus [129-after [?] 216]), had less influence until the thirteenth century; it was mostly confined, in the absence of a Latin translation, to small compendia related to the *Corpus Oribasianum* (the corpus of texts attributed to the fourth-century CE Byzantine physician Oribasius). The influence of Dioscoridean and Galenic *materia medica* was, it should be stressed, confined to, and determined by, their precocious translation into Latin, possibly before the sixth century CE, that is, before the decline and the disappearance of the knowledge of Greek in western Europe after the fall of the Roman Empire.

Finally, very important are some Latin herbals among the most influential sources of medieval phytopharmacy. One of them is the *Herbarium* (Herbal) attributed to the Latin writer Apuleius of Madauros (*c.* 124–*c.* 170), certainly not written by him, but by an unknown fourth-century author traditionally identified as pseudo-Apuleius (Apuleius Platonicus). The *Herbarium* covers *c.* 130 plants and some 200 medical conditions, the entries of which are structured in the form of short recipes and therapeutical instructions, which further inspired the *St. Gall Botanicus* (see below) in the ninth century. Other influential collections were the *Compositiones*, a collection of medical excerpts, put together by Scribonius Largus (first century CE), and the anonymous treatise the *Ex herbis femininis* dating back to the end of the fifth century and based on Dioscorides, but integrated with new material, which deals with seventy-one plants. This last treatise was very popular; it is preserved in several manuscripts, many of which are illustrated. These sources (Dioscorides, Pliny, pseudo-Apuleius) constitute the basis of monastic medicine, which constituted medical science and praxis before the reception of Arab medicine in Western culture.

FIGURE 5.2 Portraits of Galen, Avicenna, and Hippocrates, the three leading authorities of ancient and medieval medicine (woodcut, 1511). Photo by Bettmann/Getty Images.

MONASTIC MEDICINE

"Monastic medicine" refers to medicine between the late eighth century and the birth of Arabic-oriented, scholastic and university medicine in the twelfth and thirteenth centuries. In this period monasteries were the only places where the medical literature of antiquity had survived the collapse of the Roman Empire and the Migration period (476–800) in western Europe. This is important because of a number of texts dealing with Monastic medicine (see below) and, from a strictly medical perspective, because of the connection evidenced between indigenous plants and Asian spices. The effectiveness of some of its remedies, established by recent tests, should also be emphasized (e.g. in the case of an eyesalve described in *Bald's leechbook* [Bald's Book on Medical Prescriptions], written in England in the tenth century, which has been shown to effectively kill a range of microbes, including *Staphylococcus aureus* [MRSA], *Klebsiella pneumoniae*, *Escherichia coli* [*E. coli*], and *Leishmania*), which shows the deep knowledge of therapeutic effects possessed by monastic physicians.

The preconditions of monastic medicine were established in the *Regula Benedicti* (Rule of Saint Benedict) by Benedict of Nursia (*c.* 480–550), who became the founder of the Order of Saint Benedict and of Western monasticism. The *Rule* was written as a guide for individual, autonomous communities. In chapter 36 it deals with the care of the sick and aged members of the community. "Before and above all things, care must be taken of the sick, that they be served in very truth as Christ is served," is how the first sentence of the chapter reads. Benedictine monasteries were engaged in medicine and played a decisive role in the implementation of the medical system stipulated by Charlemagne as part of the reform of the Frankish Empire known as the *Carolingian Renaissance*. But these monasteries were affected by the patchy transmission of Greek medicine after the decline of the knowledge of Greek. Available medical texts were reduced to few representatives which were the main elements of medieval medical monastic libraries and were complete works or small compendia that transmitted content originally derived from (or belonging to) Late Antique medicine. These texts were frequently used as handbooks in monastic infirmaries.

The most representative texts of monastic medicine are the *Lorscher Arzneibuch* (Lorsch Medical Handbook), a medical compendium formerly preserved in the Abbey of Lorsch (Germany) and now in the Staatsbibliothek, Bamberg; the *Plan of St. Gall*; Walahfrid Strabo's *Hortulus*; the text known as *Macer floridus* (viz. the poem *De viribus herbarum* [On the Properties of Plants] by the French Odo of Meung, probably written in the eleventh century); and the *Physica* by the German abbess and polymath, Hildegard of Bingen (1098–1179). The so-called *Lorscher Arzneibuch* was written between 790 and 795 in the monastery of Lorsch near the city of Worms (or, perhaps, at least in part in Italy). This collection of texts is well known because of the *Defensio artis medicinae* (The Defense of the Medical Art; ff. 1r–5r) it contains. The bulk comprises 482 recipes gathered in five books, among which should be pinpointed the second, derived from the *Physica Plinii*, and the fourth and fifth, transmitting complex recipes, such as the *theriaca* healing oils at the end of the latter work. The *Lorscher Arzneibuch* displays a rich array of plants (more than six hundred), the most frequently mentioned of which are fennel (*Foeniculum vulgare*), ginger (*Zingiber officinale*), and pepper (*Piper nigrum*), some of which were imported from the Orient. Indeed, in the eighth century medicinal plants were traded from India to Europe, as the chapter "Trade and Exploration" in this volume explains in great detail. Furthermore, the text is the first medical book to recommend St. John's wort (*Hypericum perforatum*) for *melancholia* (quite possibly depression), and valerian (*Valeriana officinalis*) in a remedy for better sleep and a better condition during the day.

The second important document related to monastic medicine is the famous *Plan of St. Gall*, a medieval drawing of a monastic compound known through a manuscript preserved in the Stiftsbibliothek Sankt-Gallen (Switzerland) with shelf-mark MS 1092. The plan was made in the monastery on the island of Reichenau, in Germany, between 818 and 835, that is, while the monastery of St. Gall was building a new church (830–5). It perhaps represents an attempt to create a monastery where all the prescriptions of the *Regula Benedicti* could be implemented. An infirmary, a house for the physician, and a house for bloodletting and bathing can be seen on the top left of the plan. In the corner on the top left, a small garden for medical herbs is mapped, with sixteen beds, each bed for one kind of plant.

The so-called *St. Galler Botanicus* (*Botanicus Sangallensis*), which is a compendium of medicine and phytopharmacy influenced by the *Herbarium* of the Ps.-Apuleius, was compiled in the same abbey of Sankt-Gallen, around 800. It is preserved in the Sankt-Gallen manuscripts bearing the shelf-mark MS 271. At the same time the Benedictine monk Walahfrid Strabo, the abbot of Reichenau and a typical writer of the Carolingian era, described the management of an herb garden containing twenty-four kinds of herbs in his poem *Liber de cultura hortorum* or *Hortulus* (Little Garden), which is largely influenced by Pliny (Stoffler 1996). More than half of these plants can be found also in the plan of Sankt-Gallen, and among the most important plants cultivated in monastic herb gardens are fennel (*Foeniculum vulgare*), lovage (*Levisticum officinale*), sage (*Salvia officinalis*), rue (*Ruta graveolens*), mint (*Mentha* spp.), and pennyroyal (*Mentha pulegium*). The *Plan of St. Gall* and Walahfrid's *Hortulus* show the execution of Benedict's idea of monastic medicine and Charlemagne's regulation of it.

The real "beststeller" of monastic medicine was Odo of Meung's poem *De viribus herbarum* (On the Properties of [Medicinal] Plants), written at the end of the eleventh century in France. The poem, often known as *Macer floridus,* was wrongly attributed to the Roman poet Aemilius Macer (first century BCE) and deals with more than seventy medicinal herbs with the addition of twelve spices from Asia like pepper, ginger, lesser galangal (*Alpinia officinarum*), zedoary (*Curcuma zedoaria*), clove (*Syzygium aromaticum*), and cinnamon. These entries, derived from Constantine the African's *Liber de gradibus* (On the Degrees of Medicines), reveal Odo's debt to the Arabic-Latin medical translations that paved the way for the development of Salernitan medicine, as well as the renewal of the use of Eastern spices in Western medicine. Odo's work became a definitive textbook in herbal medicine and was translated into several vernacular languages. Some one thousand manuscripts containing copies in both Latin and the vernacular (among others German, French, Italian, English, and Spanish) survive.

As early as 1118–35 the historian Henry of Huntingdon (*c.* 1080/90–*c.* 1157) had produced an impressively ambitious and formally original herbal in verse, the *Anglicanus ortus* (The English Garden), dealing with 162 herbs and spices with an obvious dependence on both *Macer floridus* and Walahfrid. It also incorporated materials from Constantine the African's *Liber de gradibus*, and practically merged the traditions of Monastic and Early Arabo-Latin phytopharmacy into one text. Henry's poem enjoyed only limited success (only three complete manuscripts are known so far), but its value as a link between medical traditions cannot be denied.

The last, and most famous, representative author of monastic medicine is Hildegard of Bingen (1089–1179). Hildegard founded her own monastery very close to the town of Bingen on the river Rhein *c.* 1147–50. Beside three books based on visions she had had since the age of four, she wrote two books on nature and medicine, the *Physica*, and the *Causae et curae* (Causes and Treatments), respectively. The former consists of nine books, the first of which describes *c.* 220 herbs and the third deals with trees. The latter is a scientific and medical text in six books. Hildegard gives no physical description of the plants, but focuses on their culinary and medicinal uses, and, in the case of trees, on their uses in building. The sources she used are still to be precisely identified. She seems to echo some of the definitions of the nature and the therapeutical effects of plants, as well as prescriptions, circulating during the twelfth century. Hildegard's medicine depends heavily on her original idea of nature and medicine. One of her main contributions lies in its description, in the *Physica*, of medical plants that had not been mentioned earlier in medical texts, such as marigold (*Calendula officinalis*), lily of the valley (*Convallaria majalis*), and garden angelica (*Angelica archangelica*), and in the use, in both works, of remedies featuring spices from Asia, such as ginger, gangal, zedoary, clove, nutmeg (*Myristica fragrans*), and cinnamon.

Monastic medicine became outdated, and partly faded into oblivion (with the exception of the *Macer floridus*) for two reasons: the Arabo-Latin translations of medical works which profoundly enriched medieval medicine, and the thirteenth-century development of academic medicine and pharmacy in the new faculties of medicine (e.g. Paris, Montpellier, Padua, and Bologna).

FROM SALERNO TO MEDIEVAL UNIVERSITIES

Medieval Latin translations from the Arabic consistently modified the basis of contemporary *materia medica* by introducing new plants and substances, by amplifying the spectrum of therapeutical uses of those already known and employed in Western culture, and by paving the way for a complex theoretical debate on the correct acquisition of knowledge and use of medical remedies, as well as on the criteria of rational determination of their therapeutic properties and effects. Whereas the first two forms of renewal of phytopharmacy permeate medieval medicine in general, the theoretical *pharmacological* debate flourished particularly in the thirteenth to fifteenth centuries in universities.

The first among the most influential translators in the field of phythopharmacy and pharmacotherapeutics is Constantine the African. He translated several Arabic manuals of medicine. In the field of therapeutics, his most relevant works are his two translations: Ibn al-Jazzar's (898–980) *Liber de gradibus* (On the Degrees of Medicines), which is a collection of simple remedies structured, at least in its original version, according to the qualities (warm and cold, dry and humid) of the remedies; and a part of Book II of the *Practica* section of al-Majusi's *Pantegni*: *De probanda medicina* (On the Way of Acquiring Knowledge of Medicines), which set out the theoretical principles of pharmacology. Among the other translators of Arabic medical works, we may mention Stephen of Antioch (or Pisa; fl. *c.* 1127), who provided a new, and more precise, translation of al-Majusi's manual known as *Regalis dispositio*, and, more than any other, Gerard of Cremona, who was active in Toledo. Gerard produced a Latin translation of Avicenna's *Liber canonis*, the second book of which is devoted to remedies derived from natural substances. It is divided into two parts: the first comprises a six-chapter treatise on the theoretical principles of pharmacology, that is, the ways in which one may recognize the primary qualities of a natural object; and the second contains seven hundred entries on plants, minerals, and animal parts arranged alphabetically. Among the collections of simple remedies may also be mentioned Abraham of Tortosa's translation into Latin of *Kitab al-adwyia al-mufrada* (Book of Simple Medicines) by Ibn-Wafid (*c.* 997–1074), which was known as Pseudo-Serapion, *Aggregator de simplicibus medicinis* and contributed to the diffusion in western Europe of the Arabic medical compilations aimed at creating a concordance between Dioscorides, Galen, and other authorities.

If Avicenna and the Pseudo-Serapion were mostly read in university milieus, Constantine's translations deeply influenced the genesis and the development of a Salernitan phytopharmacy and therapeutics, the "golden age" of which was in the twelfth century with a vast literature comprising three types of works: *practicae medicinae* or clinical manuals like the *Practica* written by Platearius (see below) around 1150, where plants are mentioned as ingredients of medical recipes; *antidotaria* or collections of recipes, for example the *Antidotarium magnum salernitanum*, and the widely diffused *Antidotarium Nicolai*, both structured according to the alphabetical order of the compounds' names; and collections of *medicamina simplicia* describing the qualities and properties of simple remedies (hence single plants).

We know of three representative collections dealing with *medicamina simplicia* (simple medicines, that is, substances and products derived from plants, animals, and minerals, such as juices, powders, and fats) assembled in Salerno: the so-called *Glossae Platearii* (Glosses by Platearius), or, according to its opening words, *Liber iste* (This Book); the *Liber de simplicibus medicinis* (Book on Simple Medicines) attributed to Iohannes de Sancto Paulo;

FIGURE 5.3 Avicenna portraited as the king of medicine (manuscript Firenze, Biblioteca nazionale). Photo by Mondadori Portfolio. Courtesy of Getty Images.

and the *Liber de simplici medicina* (Book on the Simple Medicine) attributed to Matthaeus Platearius or, according to its opening words, *Circa instans* (On That [Matter]). The *Liber iste*, written during the twelfth century and preserved in *c.* eighty manuscripts with three different versions, is a collection of simple remedies aiming at providing information about the ingredients of compound remedies. It was probably meant to facilitate the understanding and use of the Salernitan *Antidotarium magnum* (Great Antidotary), a large collection of recipes assembled during the eleventh century. However, it was mostly employed as an independent collection or to accompany the *Antidotarium Nicolai*.

The *Liber de simplicibus medicinis* was written towards the end of the twelfth century, arguably by Iohannes de Sancto Paulo (d. *c.* 1214/15). Its connection with Salerno cannot be established with certainty, but there is no doubt that it heavily depends on Constantine the African's *Liber de gradibus*. The *Liber* deserves attention, even though it does not describe plants and natural substances, but regroups them in lists of items characterized by the same qualities and therapeutic properties. Rather than providing information, it could mostly be used (and, indeed, it probably was) to

FIGURE 5.4 A pictorial representation of the reputation of Salernitan medicine: Robert of Normandy (*c.* 1054–1134) is taken to Salerno to be healed (fifteenth-century Hebrew manuscript of Avicenna's *Liber canonis*, Bologna, Biblioteca universitaria, MS 2197, f. 210r). Photo by DeAgostini. Courtesy of Getty Images.

structure the natural world and to memorize and correctly employ substances sharing the same qualities and therapeutical features.

The most famous of the three is without doubt the compilation entitled *Liber de simplici medicina*, widely known by its opening words as *Circa instans* and once attributed to a certain Matthaeus Platearius (d. 1162). Today, we identify him simply as Platearius, in accordance with indications of authorship in some manuscripts. The compiler evidently had at his disposal a great variety of ancient and contemporary sources, such as Constantine the African's *Liber de gradibus*, and possibly the *Dioscorides alphabeticus*. He possibly also shared some data with the authors of the *Antidotarium Nicolai* and of the *Liber iste*. Not all his sources can be identified because of the lack of clear references and the considerable rearrangement of material. The work was soon reproduced in manuscript form after its composition. Over 250 copies are known, which bear witness to the transmission of the text in different versions. The success of the *Circa instans* is possibly due to its reader-friendly structure: plants and natural substances are listed in alphabetical order and described with respect to their main qualities (e.g. cold or warm), their medicinal properties, actions, and uses, together with a group of practical indications in the form of remedies for specific illnesses. The number of entries in the different versions of the text ranges from 250 to more than 800. None of these versions can be claimed to be the original. Nevertheless, these multiple versions testify to the success of the collection, its accessibility to different kinds of readers, and its capacity to absorb and incorporate new information. The *Circa instans* rapidly became an authoritative guide to individual drugs, as both its large Latin manuscript transmission and its many translations demonstrate, with versions in Anglo-Norman, French, Provençal, English, German, Dutch, Italian, Catalan, Hebrew, Serbian, and Gaelic, some of which expanded the text. Only the *Macer floridus* probably attained the same diffusion in the vernacular.

Within the manuscript tradition of the *Circa instans*, important is the strand represented, in Latin, by the *Tractatus de herbis* (the codex London, British Library, Egerton MS 747) and, in French, by the *Livre des simples medecines*, in which the text has a rich body of illustrations. The whole was probably meant to teach apothecaries about the origins of the substances they were using, and designed to be consulted while learning about *materia medica*, and not, as it has often been repeated, to be used when searching for plants in the wild, or to teach the medieval *medicus* about nature. Whatever its intended use, it is clear that the tradition of the *Tractatus* and the *Livre* helped determine the evolution of a tradition of plant illustrations in late medieval Europe with its realistic, observed, approach to nature. It is also of value for modern research dealing with the identification of the plants used in medicine across different cultures and the sometimes confusing transcultural nomenclature.

Later, *post-Salernitan* collections and herbals are characterized by an encyclopedic approach toward phytopharmacy and by attempts to amalgamate data and information derived from different sources into a single work. The compilers of such collections probably intended to achieve a double aim, namely to create exhaustive collections offering comprehensive information on any particular plant based on several sources (including the observations of the compilers), and to provide readers with texts that could obviate the consultation and reading of several sources. This approach is evident, for instance, in the *Tractatus de herbis*, the majority of which is derived from the *Circa instans* but was expanded using several sources (for example, the Pseudo-Apuleius and *Macer floridus*). This is also the case in the *Herbal of Rufinus* (*c.* 1287), whose author based the text on often-original descriptions on both personal observation and enquiry, as well as making reference to the practices of herbalists in Bologna and Naples. In the late fourteenth century the English Dominican friar Henry Daniel proceeded in the same way in his herbal known as *Aaron Danielis*, which is based on an equally wide, if not larger, range of sources.

The same tendency characterizes vernacular herbals, as evidenced by the second Old French translation of the *Circa instans* preserved in the thirteenth-century manuscript Garrett 131 in Princeton University Library. This verse text, which is in part translated from the *Macer floridus* into a southwestern French (from between Angers and Poitiers), is composed in largely correct octosyllabic couplets. It lists sixty-six herbs which are hot and dry and sixty-three cold ones in accordance with Galenic principles of qualities, together with their medicinal uses. Although the anonymous author claims to follow Macer "de mot en mot" (word for word), he added much other material. A modern edition, as well as a study of the manuscript tradition, is urgently needed. The explicit intention of its author was to make accessible material to readers who could not understand Latin.

FIGURE 5.5 Garden of medicinal plants (from a fifteenth-century copy of the *Livre des simples medecines*, the French translation of the Salernitan collection *Circa instans*, Paris Bibliothèque nationale de France, MS français 9136, f. 1v). Photo by Universal History Archive/Universal Images Group. Courtesy of Getty Images.

THE ENGLISH AND GERMANIC WORLDS

Translation was an important step in the gradual vernacularization of medicine and the diffusion of botanical knowledge in France, both topics that require further investigation. In contrast, Old English medical writings, the earliest vernacular texts to survive in western Europe, have undergone a process of rehabilitation, largely initiated over the past decades by medical historian Linda Ehrsam Voigts (b. 1950). Well before the Norman Conquest (1066), England had employed the vernacular in subjects that, on the Continent, were covered only in Latin. From about 900 there are a number of Old English medical compilations, some of which reflect popular traditions with a predominance of native plants, others representing translations from Latin in which those of the Mediterranean seem generally to have been well assimilated, though not always recognized. All of these compilations are full of recipes, some incorporating charms and magical elements, others being well organized on the basis of expert knowledge. The greatest focus on plants is found in a complex of texts known as the *Herbarium*, including a translation of the *Herbal* of pseudo-Apuleius. The translator's familiarity with Mediterranean plants was deficient, however, for he failed to attempt an English translation of the names of 41 of the 185 treated in the *Herbarium*.

Post-Conquest there were numerous writings in Middle English on botanical topics, largely after 1400. Indeed, the great herbals of sixteenth-century England by William Turner (1509/10–68), Henry Lyte (1529–1607), and John Gerarde (*c.* 1545–1612) were preceded by Middle English herbals, a number of which, of anonymous authorship, survive in manuscript form. Thomas Betson (d. 1517), librarian of Syon Abbey, transcribed an *Herbarium* in his notebook, that consisted of a list of seven hundred names of medicinal plants, animals, and minerals. The list was mostly copied from the *Synonyma* of John Bray (d. 1381), physician to king Edward III (1312–77), but the second part, consisting of about 450 remedies, was assembled by Betson himself and contains references to some 130 plants. The source of Betson's knowledge was exclusively books, which reminds us of the problem of endless copying throughout the Middle Ages.

Vernacularization played a determining role in the development of botanical and phyto-pharmaceutical knowledge in the German-speaking areas, too. Medieval German botanical and medical cultures were largely influenced by the botanical account provided by Thomas of Cantimpré (1201–70/1) in books X–XII of his encyclopedia *De natura rerum* (On the Nature of Things). In the same fashion as other encyclopedic compilers like Bartholomew the Englishman (fl. 1225–30, d. *c.* 1250) in his *De proprietatibus rerum* (On the Properties of Things) and Vincent of Beauvais (*c.* 1184/94–*c.* 1264) in his *Speculum naturale* (Mirror of Nature), Thomas of Cantimpré chronicled the multifaceted nature of the plant world with the help of works in different fields: lexicography, represented by the work of Isidore of Seville (560–636) and his *Etymologiae* (Etymologies), Papias (eleventh century) and his *Elementarium* (Elementary Dictionary), and Hugutio Pisanus (end of the twelfth–beginning of the thirteenth century) and his *Derivationes* (Derivations of Words); natural history, represented by Pliny the Elder; and medicine, represented by Dioscorides, the *Circa instans*, and *Macer floridus*. Thomas' work reached the German-speaking public through Conrad of Megenberg (1309–74) and his *Buch der Natur*, a vernacular encyclopedia largely based on *De natura rerum*, which became the encyclopedic authority for the knowledge of the natural world. *Macer floridus* was even more successful in the field of phytopharmacy; it was widely copied in Germany well into the fifteenth century, and was turned into more expansive prose and verse herbals written for the needs of middle-class readers (not necessarily specialized ones) by merging its content with other sources, such as Hildegard of Bingen's *Physica* and the *Circa instans*. It was transmitted by several manuscripts in different German versions.

By taking into account these two examples, it can be seen that late medieval German botanical and phytopharmaceutical literature was characterized by the wide use of the vernacular, attempts at encyclopedism, the intention of merging different sources into a small portable medico-botanical library in the form of a single collection, and the aim of reaching a larger audience via linguistic and structural simplification.

Does this literature reflect popular utility (many medico-botanical manuscripts reveal remarkably little annotation or signs of use) or, by contrast, a bookish desire to create a prestigious scientific literature, without much personal knowledge, as in the case of Isidore of Seville and also of Betson? There is no one answer. Some of the herbals are copies suited for a bibliophile. This is the case with the manuscripts of an Anglo-Norman herbal of *c.* 1160 based on pseudo-Apuleius' *Herbarium*, in the British Library, MS Harley 1585, London, and the closely related texts in the British Library, MS Sloane 1975 and Oxford, Bodleian, MS Ashmole 1462. The fourteenth-century physician and botanist Henry Daniel, quite exceptionally, combined a rich personal familiarity with plants (he kept a private botanical garden in Stepney, London, which contained 252 plants) with zealous, scholarly enquiry into an extensive range of scholarly sources, as in his herbal which features 170 of the plants he grew at Stepney. On the other hand, German herbals and medico-botanical collections are often reproduced in rather modest and carelessly copied manuscripts, with the important exception of the illustrated codices of the *Buch der Natur* (Book of Nature) by Conrad of Megenberg and of Johannes Hartlieb's (*c.* 1410–68) *Kräuterbuch* (*c.* 1450).

Whereas the production of vernacular herbals continued, almost without interruption, well into the fifteenth century, the redaction of Latin phytopharmaceutical collections lost momentum at the end of the thirteenth century, the last examples of them being the already mentioned, but unsuccessful, *Herbal* of Rufinus, and the ambitious, but, in fact, almost equally unsuccessful, *Liber pandectarum medicinae* completed around 1332 by the physician Matthaeus Sylvaticus. Matthaeus, probably born in Salerno, was not unaware of the developments of late medieval medicine: he drew material from Simon of Genoa's lexicon *Clavis sanationis*, compiled around 1290, and used the newly completed Greek–Latin translation of Galen's *De simplicium medicamentorum temperamentis et facultatibus* (On the Mixtures and Properties of Simple Medicines) made by Niccolò of Reggio (first half of the fourteenth century). He assembled a giant collection of *medicamina simplicia* consisting of hundreds of entries, some of which organize the data concerning the nature and therapeutic properties derived from several sources in a fashion imitating Avicenna's entries in the *Liber canonis*, while some others barely reproduce a glossary record mostly derived from Simon.

On the other hand, if the representation of the medicinal properties of plants did not encourage the production of new texts, the last centuries of the Middle Ages, especially the thirteenth and the fourteenth centuries, experienced a large *scholastic*, university discussion on the principles explaining nature and actions of medical remedies, their effects, and the action they perform within the human body, as well as the rules that help clarify the intensity of the power of the simple remedies and, therefore, contribute to establish their correct dosage in compound medicaments. This issue was displayed, for example, in the *Aphorismi de gradibus* (Aphorisms on the Degrees [of the Intensity of Effects of Medical Remedies]) of Arnald of Villanova (1240–1311). The complex discussion was furthered in the pharmacological sections of Avicenna's *Liber canonis* or Averroes' (1126–1198) *Colliget*. Although it was short-lived (ending by 1348, at the time of the Plague), the discussion was significant in that it shaped the theoretical background of pharmacology that lasted until the Early Modern Period.

A COMPLEX, YET PROMISING FIELD

The present review has attempted to trace the main tendencies of the medieval cultural perception of plants by taking into account both the historical development of that perception and the most representative texts and literary genres that influenced and conveyed it. It shows that the medieval cultural history of plants is by far and away a medical one. That does not mean, however, that other literary genres dealing with plants (natural history, philosophy, lexicography, agriculture) were not important, nor that they did not intersect with the medical tradition, as the perspective adopted by Albert the Great in his *De vegetabilibus* clearly shows.

Medical literature on plants can thus be divided into three different phases: late antiquity, a time characterized by the attempt to preserve the cultural heritage of Greco-Latin medicine; the era of monastic medicine; and the golden age of the Arabo-Latin medical literature, which played a significant role in widening the phytopharmaceutical background and in establishing a theoretical framework which became a subject of discussion. These three phases can be associated with some representative texts described above, albeit briefly.

One literary genre not discussed here in full needs to be mentioned, for it represents a research field that is still to be analyzed in its historical development, and in its role as an important source of information about the concrete therapeutic use of plants: recipe collections. In the Greco-Roman tradition recipes tend to be embedded in therapeutic treatises, but, with the rise of monastic medicine, they were sorted into two types of collections known as *receptaria* (collections of recipes) and *antidotaria* (collections of recipes against specific illnesses). The former, based largely on simples (mostly plant-based medicines), are loosely organized according to the disorder to be treated and their effects, and are of a fairly popular character. The *antidotaria* are more complex and more strictly pharmaceutical. They are arranged by medicinal form, often covering complex compound medicines, including *exotica* (exotic drugs, most often of Oriental origins), sometimes associated with famous physicians' or users' names, and frequently giving detailed instructions on preparation and measures. In both types, many unstated details, such as the part of the plant to be used, the site of application in the case of external use, dosage, and dietary factors, are left to the working competence of the healers (and perhaps dispensers or *pigmentarii*, the latter being sellers of unguents and pigments), who were presumed to possess considerable practical knowledge acquired through oral tradition. Collections of recipes could, however, also be structured in the form of therapeutic manuals arranged according to the illnesses affecting the body *a capite ad calcem* (from head to toe), as in the case of the immensely successful (more than eight hundred manuscripts preserved) *Thesaurus pauperum* (Treasury [of Health] for the Poor) written by Peter of Spain (*c.* 1240–50). Recipes also circulated singly and as "flyleaf literature" written in blank spaces, including the margins, of various manuscripts.

In England recipe collections were often composed in three languages (Latin, English, and Anglo-Norman), a process known as *code-switching* or *code-mixing*, as may be observed in many recipes contained in the major medical treatises, themselves written in Latin, of Gilbertus Anglicus (*c.* 1210–after 1250), John of Gaddesden (1280–1361), John de Greenborough (or Grandborough; d. after 1383), and John Mirfield (d. 1407). These recipes, which run to thousands, have yet to be studied properly, and would require systematic indexing first. Recipe collections in France and Italy are monolingual, or a mixture of Latin and French/Italian; they are in either prose or verse. One of the earliest-established prose-collections, the so-called *Lettre d'Hippocrate* (Hippocrates' Letter; thirteenth century) was also popular in England. No matter in which form they are preserved, collections of recipes deliver a great deal of information about the plants that were available and were used for "practical" therapeutical purposes, and they ought to be examined with respect to those purposes, and in a multidisciplinary way.

The preponderance of plants as ingredients in a vast number of recipes demonstrates the predominantly herbal basis of much medieval medicine, and palaeoclimatologists have greatly extended the list of plants that could have been cultivated in, say, southern England from *c.* 500–1200, where the climate was closer to that of continental Europe than it is today. Plant procurement is not so well documented. A set of forty-eight thirteenth-century line-drawings accompanying an Anglo-Norman translation of *Chirurgia* (Surgery) by Roger Frugard (before 1140–after 1195) in Trinity College, Cambridge, MS 0.1.20, depicts no fewer than thirteen scenes of the medieval dispensary with a full array of *materia medica* including herbs, storage jars, and medical instruments. Not until the fifteenth century is there the clear production of treatises *De collectione herbarum* (On Collecting Herbs), suggesting the best time of year and other conditions for the picking of herbs—their flowers, seeds, and leaves (though there is much useful material in Dioscorides and scattered elsewhere), and indicating that plants gathered from the wild would have stronger active constituents than those cultivated in gardens. It is clear that such vernacular productions on astrology, Christian liturgy, apotropaic rites, endless incantations, go back to antiquity, and the continuity of such treatises in manuscripts from the tenth to the eighteenth century is itself remarkable.

The understanding of medieval botany and the appreciation of its medicinal value today require the collaboration of scholars and phytopharmacists in identifying, isolating, and then synthesizing the active pharmaceutical ingredients—alkaloids, glycosides, essential oils, etc.—in a way that was not possible until last century. This has already led to much confirmation and verification of data from the past by contemporary pharmacists and ethnobotanists. The task is enormous. According to the preliminary—and yet unpublished—researches carried out by the research group *Klostermedizin* (see http://www.klostermedizin.de/index.php), of the 391,000 described plant species on the planet, it is thought that around 17,800 are used regularly for medicinal purposes, and many more may be waiting their turn. Medieval botany will continue to be part of future progress.

CHAPTER SIX

Plants in Culture

A Meaning-Laden Living World in Early South Asia

DIVYA KUMAR-DUMAS

Plants were used and treasured by the indigenous communities of South Asia well before the common era and by the groups who settled the subcontinent during the first and second millennia CE. These indistinct ancient origins have contributed to the prevalent association of plants with Indian religions and sacrality. However, various sources from the period between 500 and 1400 CE offer for exploration a more complex picture of Indic attitudes toward plants. It is hoped that, in time, more scholarship in this area will disaggregate investigations of sacred landscapes from quotidian experiences, although not because they did not intermingle. Rather, doing so would accommodate the documented practices of everyday life into present beliefs, which can have bearing on the available knowledge about what people in medieval South Asia thought and did. Although what is currently known about plants largely derives from sacred and courtly texts,[1] it is clear that plants informed many lived realities in a multitude of overlapping ways.

PLANTS IN CULTURE: ORIGINS BEFORE 500 CE

Cosmogonic myths from the earliest sacred texts abound in plant imagery. In one such Vedic narrative, primordial waters gave rise to a thousand-petaled lotus made of gold and radiant like the sun. This lotus was also the womb of origin gods within their pantheons (*Taittirīya Upanishad; Padma Purāna*).[2] In the beginning, when a great Nothing neither existed nor did not exist, dark shapelessness enclosed creation, which emerged when it was seeded by desire (*Rg Veda* 10.129, "Nāsadīya Sūkta"). From its golden womb, the demiurge brought forth all things (*Rg Veda* 10.121, "Hiranyagarbha Sūkta"), including wealth in the form of a mother goddess of fulfillment. Her epithets included Born-from-the-lotus, Standing-on-a-lotus, Dwelling-in-a-lotus, Lotus-colored, Lotus-loving, Lotus-footed, Lotus-handed, Lotus-faced, Lotus-eyed, Abounding-in-lotus (*Rg Veda* Mandala 5, "Shrī Sūkta"). Imperishable and beautiful, she was also the earth, who ensured abundant harvests, health, long life, and prosperity for the world in which certain early Indians dwelled.

From their continent of *jambudvīpa*, four oceans extended toward the four quarters of the universe from its lotus center. Its outer petals were foreign continents, and its staminal filaments were mountains and precious metals. From the mountains flowed waters understood as sacred rivers, elixirs of immortality, and subsequently places of pilgrimage (Dallapiccola 2003: 25; Eck 2012: 118–22). Aquatics were among a long list of plants ranging from the woody and herbaceous to herbs and grasses (Krishna and Amirthalingam 2014: 13). Although usually translated generically as "lotus," they were specified in the *Ṛg Veda* with distinctive names.[3] The lotus also became a meaning-laden metaphor for human ideals on earth. In literature, a beautiful woman is often lotus-eyed or lotus-faced and the compassion of enlightened beings, such as the Buddha and Bodhisattvas, reflect their lotus hearts (*Kāraṇḍavyūha Sūtra*; dated to the end of the fourth/beginning of the fifth century by Studhome 2002: 17).[4]

Apart from the lotus, origin myths also refer to a tree of life, emerging from the waters as source of creation. This tree, called *aśvattha*—translated as pipal (*Ficus religiosa*)—is a member of the mulberry family, which also includes the *nyagrodha*, or banyan (*F. benghalensis*). The banyan seems to reproduce itself infinitely like cosmic vastness, so that even the gods in heaven sit beneath its shade. Great ascetics, such as the Jain Tīrthankara (or "ford-crosser") Ādinatha and the Siddhārtha Gautama, the Buddha (or "awakened one"), attained enlightenment under such a tree, and both remain permanently associated with it. A respect (in fact a pantropical phenomenon) for fig trees, a species of *Ficus*, is also attested on several early steatite seals belonging to people of the mature phase of the Indus civilization (2600–1900 BCE) who enshrined fig trees and seemed to have revered figures dwelling within such trees.[5]

In later Hindu texts called *purāna*, which include chapters dated to the period covered by this volume (*c*. 200 to 1250 CE and beyond), other trees expanded the sacred plant repertoire. In the *Matsya Purāna*,[6] the *aśoka* tree (*Saraca asoca*) was planted by the divine wife of Shiva, pronounced Pārvatī. She likened the planting of a tree to bearing sons, thus linking trees to fertility. When asked why she took such care, with water and attention, to make this first sapling grow well, Parvati, who often assimilated other local goddesses and was understood to be a "mother" goddess, is said to have answered the gods and heavenly sages as follows:

> One who digs a well where there is little water lives in heaven for as many years as there are drops of water in it. One large reservoir of water is worth ten wells. One son is like ten reservoirs and *one tree is equal to ten sons* [my emphasis]. This is my standard and I will protect the universe and safeguard it with the *aśoka* tree.
>
> (*Matsya Purānam* Ch. 154: 506–12, cited in Narayanan 2001: 187)

In classical Tamil (Sangam) texts (100 BCE–300 CE) about warriors and chieftains, specific plant species, perhaps worn as totemic emblems, became group names to differentiate warring clans: "Your enemy is not the kind who wears the white leaf of the tall palmyra (Tam. *paṉai* [*Borassus flabellifer*]) / nor the kind who wears garlands from the black-branched neem trees (Tam. *vēmpu* [*Azadirachta indica*]). / Your chaplets are made of laburnum (Tam. *konnai* [*Cassia fistula*]) …" (*Puranāṇūṟu* 45, cited in Ramanujam 2011: 121n308).[7]

The deep respect for trees and plants evident in such diverse, early cosmogonic, sacred, and heroic narratives survived into the first centuries CE as a variegated Pan-Indic preoccupation spanning diverse ethnic and social groups. While the effects of social difference on these attitudes has yet to be teased out, it can be stated that the canon of cherished trees and other plants now included botanicals used for daily life as well. In everyday living spaces, these plants were food, medicine, decoration, and fragrance. They were also cut as offerings to the gods in the sacred spaces of the home and city. In fact, the widespread use of plants for offerings (*pūja*) to gods and greats, still practiced today, attests to a series of historical transformations to sacrificial cultures that have been associated with the rise of cities. Changes include: 1) the replacement of sacrificial animals and their products with plants for sacred rites (Inden 2006); 2) the transformation of a sacrificial culture into political societies based on great gifts (*dāna*), often of land (Thapar 1976); and 3) the rise of gardens, sponsored by urban elites, for the cultivation of flowers and fruits and as pleasure-grounds (Ali 2003).

The rise and spread of Sanskrit as the language of cosmopolitan culture across the varied urban geographies (Pollock 2009) affected plants. Botanical specificity became less important as Sanskrit poets used existing literary names for plants in new climatic zones based on their similar color, shape, form, and fragrance. Regional natives and their associated orally transmitted planting practices had naming conventions that only resurfaced in the primary sources after vernacular languages had been written down around the turn of the second millennium.[8] Although in the earliest sources ritual efficacy was tied to the species-specific qualities of plants, writing about plants across the transcultural "Sanskrit cosmopolis" (Pollock 1996) muddied the association of plant names with actual plants (e.g. Čejka 1999).

Given this context, plant names in Indian languages changed over time and could have varied meanings at once, complicating scholarly identification of the plants that are referenced in the sources. For instance, there has been a long-standing confusion in South Asia about the botanical association of an important plant, the *jambu* fruit tree. During the nineteenth century, those confusions were integrated into colonial knowledge systems by botanists. The *jambu* fruit tree was closely associated with South Asia's landform (*jambudvīpa*) in Hindu, Buddhist, and Jain texts, which was presumed to be an island (*dvīpa*) filled with *jambu* trees by those early Indians writing texts; today the *jambu* is most often understood as *jamun*, the Indian black plum (*Syzygium cumini*), or other *Syzygium* species. However, the architectural historian Michael Meister (2009) has suggested the need for significant rethinking of this identification based on a set of sculptural representations, dated to the third–second centuries BCE. First of all, the evergreen tropical flowering black plum is native in Southeast Asia, beyond the region of the "*jambudvīpa*" from early texts. More importantly, early descriptions of *jambudvīpa* had the fruits of its eponymous trees as "large as elephants" (*Viṣṇu Purāna* II.2, cited in Meister 2009: 110). Skeptical anglophone plant hunters and botanists, who understood "elephant-sized" to be a literary exaggeration, identified *jambu* as the small, blackish purple, olive-sized fruit from local people, who had themselves linked *jambu* with other fruits.[9] The crux of Meister's argument, however, is his discussion of a relief from Bharhut (third–second centuries BCE) with a label inscription in Pali/Prakrit that says "*jabū naḍode pavate*" (Skt. *jambū naḍode parvate*; En. "the Jambū tree on Mount Naḍoda") (see Figure 6.1). The relief features an image of a sanctified wish-fulfilling tree extending two "spirit" arms, one with a vessel of water and the other with a "bowl" of food that looks like a jackfruit (*Artocarpus heterophyllus*), cut in half, with its flesh piled up inside for consumption, according to Meister's convincing visual argument. When coupled with the idea of a tree as one that sustained sages, philosophers, and ascetics (Meister 2009: 111–12)[10] by both the first-century CE Roman encyclopedist Pliny the Elder[11] and in extant copies of *vṛkṣāyurvēda* (discussed below), the jackfruit's substantial size and nutritive value as well as the tree's natural occurrence in the forests of South Asia counters colonial ascriptions (and their sources) of *jambu* to *Syzygium* species. Dwelling in forest groves

FIGURE 6.1 Architectural relief depicting the jackfruit (*jambu*) tree on Mount Naḍoda, Bharhut Stupa. Indian Museum, Kolkata. Courtesy of Center for Art and Archaeology (CA&A) of the American Institute of Indian Studies (AIIS), 68613.

for eschatological pursuits would certainly have been better facilitated by the easy diet offered by a jackfruit tree than a *jamun*.

Beyond botanical confusion and misidentification, fruiting forest trees appear to have been worshipped in South Asia because they satisfied desires, from the most basic need for food to the most abstract need for spiritual awakening. Unfortunately, the wishes of historical forest dwellers, the largest group of people who benefited from tree gifts, are undocumented, although productive forestlands were themselves protected by kings, as indicated by early texts about kingship such as the *Arthaśāstra*, a text of uncertain date (possibly first century BCE with additions through the first century CE)[12] (2.2.5, cited in Nugteren 2005: 77). By partaking of a giving tree, a person could explore immortality, either through contemplation or through a productive, prosperous life, and sometimes both. Thus, "wish-fulfillment" metonymically reified the cosmogonic wish-fulfilling tree of life (*kalpa-vṛkṣa*). According to the classical Tamil (Sangam) literature (100 BCE–300 CE), trees embodied the divine, and the gods that dwelled within them had to be worshipped. A relief from the Bharhut stupa (third–second centuries BCE) depicts a tree, with compound opposite leaves, large flower clusters, and a decorative patterned band at the base of its trunk, that has been set apart in a constructed square altar. It is being propitiated with garlands, cut leaves and flowers, and worshipful revelry by figures with arms raised (see Figure 6.2). Depictions of worshipped trees were not limited to a single botanical type. Trees were treated like other deities and set apart or housed within temporary or permanent structures.[13] When cultivated and cared for, trees gave life to those populations who had become separated from the foraging habits of their forest-dwelling neighbors who continue to venerate trees even today (e.g. Fowler-Smith 2018; Nugteren 2005: ch. 5).

FIGURE 6.2 Architectural roundel relief depicting tree worship with cut flowers, leaves, garlands, and revelry including a label inscription translated as "The Bodhi tree of the holy Vipasi (*Vipaśyin*)" and identified here as the Aśoka tree (*CII* 2.2,1963: B13, 82–3), Bharhut Stupa. Indian Museum, Kolkata. Courtesy of Center for Art and Archaeology (CA&A) of the American Institute of Indian Studies (AIIS), 16368.

By 500 CE, ephemeral altars for plant sacrifice were increasingly being replaced by temples of stone. Plants continued to be used for worship in temples, and plant imagery decorated architectural surfaces, which also received a new type of epigraphic record. Stone inscriptions emerged in this period as an extensive primary historical source (in the range of one hundred thousand). These donor inscriptions point out varied cultural practices relating to plants in South Asia between 500 and 1400 CE. They document tree planting, tree protection, and garden construction (Dasgupta 2016); they also record plant lists (Wagoner 2012) but have not been exhaustively surveyed in this regard.[14] Uncovering the cultural histories of plants and gardens from these sources is a promising area for future research. However, it can already be seen that these inscriptions (along with literature from the period) both support and counter customary associations of plants with the sacred and the divine. Plants also figured in economic transactions, urban planning, romance, and recreation, and were used to make fragrances, medicines, and foodstuffs. Fruits were enjoyed in private and public, as metaphors and metonyms, or simply, as Amir Khusrau celebrates, delicious: "... among the juicy fruits ... we have Mango, Banana, and Sugarcane ... a number of fruits are found in India which are not grown anywhere else ... [Banana] has no parallel in the world" (*Nuh Sipihr*, 3:20–6). Full coverage of the topic will not be attempted here; rather a few examples from period sources will be highlighted, thus introducing a layered, multivalent depiction of how plants were typically imagined and how they did their work through culture in early South Asia.

TREES IN THE FIRST MILLENNIUM

Apart from use as economic resources and building materials, the tree (*vṛkṣa*) in South Asia carries varying cultural associations that might be distilled into two categories: first, as living beings associated with gods and spirits, trees can exert agency over human lives; second, when subjected to certain human interventions, trees bear witness to human actions upon them. In both cases, trees may appear semi-divine or more-than-natural. As beings that display unusual synergies with human desires and events, trees have been venerated since the earliest times (e.g. Jones and Cloke 2008).

Respectful attitudes toward trees can be observed in Late Antique and early medieval sources when trees become building materials. In technical treatises on architecture, immediately after choosing a building site, an architect would "take possession" of it by making ritual offerings and a spoken declaration: "That Spirits, Gods and Demons depart! That they leave this place and go elsewhere for I take possession" (*Mayamata*, ch. 4.1–3, 161–7).[15] Places inhabited by gods, whether natural or constructed, usually contained sources of water, gardens, groves, hills, and valleys, according to an oft-cited set of verses introducing encyclopedist Varāhamihira's chapter on temples and their sites (*Bṛhatsaṁhitā*, 56.1–8, cited in Asher 2012: 86).[16] Human dwelling sites (*vastu*) were not very different from temple sites because both were considered habitable places on earth, the first dwelling place for mortals and immortals alike. Many types of building sites were equally suitable for "gods on earth, the gods and the kings ..." (*Mayamata*, 2.15, 8–9).

When a building site was chosen, trees that were inappropriately positioned would be removed.[17] Trees for structural beams and pillars had specific qualities: "The chosen tree must be perfect, hard and vigorous; [trees] should neither be old nor should they be saplings; they should not be crooked and should be undamaged" (*Mayamata*, 15.62–3, 194–5). Suitable tree species for pillars included *puruṣa, khadira* (khair, *Senegalia catechu*), *sāla* (sal, *Shorea robusta*), *madhūka* (mahwa, *Madhuca longifolia*), *campaka* (champak, *Magnolia champaca*), *śiṃśapa* (shisham, *Dalbergia sisoo*), *arjuna* (arjun, *Terminalia arjuna*), *ajakarṇa* (white dammar, *Vateria indica*), *kṣīriṇī, padma* (sacred lotus, *Nelumbo nucifera*),[18] *candana* (sandalwood, *Santalum album*), *piśita, dhanvana* (possibly *Grewia tiliifolia*), *piṇḍī, siṃha, rājādana* (palu, *Manilkara hexandra*), *śami, tilaka, nimba* (neem, *Azadirachta indica*), *asana* (gamalu, *Pterocarpus marsupium*), *śirīṣa* (siris, *Albizia lebbeck*), *eka, kāla* (maybe *kālaśāka*—curry plant, *Bergera koenigii*), *katphala, timisa, likuca* (maybe *lakuca*—*Artocarpus lacucha*), *paṇasa* (*Artocarpus heterophyllus*), *saptaparṇaka, bhaumā(?)*,[19] and *gavākṣīn*.[20] When these types of trees were to be used as building materials, they were usually found in managed productive forests that were guarded against thieves (*Arthaśāstra* 2.1.19, cited in Nugteren 2005: 76) or at "a holy place, mountain, wood or river-bank" (*Mayamata*, 15.62–3, 194–5).

Although many of these trees cannot be matched with modern Latin scientific names, recorded practices surrounding the felling of trees named in Sanskrit are instructive in other ways. For example, on the eve of tree removal, tree spirits inhabiting the marked tree and any forest deities in the vicinity were first propitiated. Offerings included flowers, perfume, incense, meat, boiled rice with sesame, rice with water and fish, and other foods of all kinds (*Mayamata*, 15.82–3, 200–1). After these services, the architect spent the night east of the tree on a grass (*darbha*) mat with "his axe beside him to the right" (*Mayamata*, 15.87–8, 200–1).

> Then in the morning, after drinking pure water and facing to the west, the perfect architect, clad in white ... with axe in hand, pronounces this formula: ... "Let Spirits, Divinities and Demons disperse! On you O trees let Soma bestow power! May It be propitious for you O sons of Earth! Divinities and Demons I shall accomplish this act and you must change your dwelling place."
>
> (*Mayamata*, 15.87–90, 200–3)

Like building sites, trees served as dwellings for divinities and demons who were requested to move out before a tree could be felled. In fact, the destruction of trees, groves, and forests was prohibited on managed lands, with varying fines (*Arthaśāstra*, 2.17.1, cited in Nugteren 2005: 77). Although modern urbanization in South Asia seems to disregard these cultural attitudes toward trees, some traditional practices indicate respect even today: for example, ceremonial tree-planting might occur as a celebration around major life events; interplant marriage visible in public parks can include the close planting, enclosure, and horticultural training of the two plants together, as in, for example, neem and pipal; and religious festivals can re-enact the marriage of a plant to a deity, like the annual Hindu *Tulsi Vivah* between *tulasī* (*Ocimum tenuflorum*) and Vishnu.

Agency of trees upon humans

The notion of inhabited trees helps to explain the moments of tree agency found in early literature. Semi-divine beings formed a network of actors in human life. When they occupied trees, the tree became like a person and a protagonist. In one story, a semi-divine snake living in a tree intervened on behalf of a human, Vijjupaha (Skt. *Vidyutprabha*), the daughter of Aggisomma (Skt. *Agni Sharma*). Recognizing her unselfish nature and generosity toward other living beings as well as the obstacles she faced from a jealous stepmother, the semi-divine being visited her while she rested with the gift of a magical grove that surrounded her even as she moved from place to place so that she could always rest under his tree. This marvelous sign, along with continued protection from that semi-divine being, eventually led her to a secure position at the king's side as his wife and queen (Granoff 1998: ch. 28 "Ārāmasohā," 263–72). In a Santal folktale from Northeast India, the spirit of a woman sings from inside an instrument made out of the bamboo she once inhabited (Campbell 1891: 52–6); via its re-tellings in different parts of the subcontinent, the story morphed. The clump of tall bamboo could become a tree, and the plaintive instrument of an itinerant musician from an indigenous forest-dwelling tribe could become the *vīna* (Indian lute) of a goddess.[21]

Often called *yakṣa* (fem. *yakṣī* or *yakṣinī*),[22] these semi-divine beings were part of a set of divine beings with which they shared similarities. They also had certain capabilities, such as the ability to possess a woman's voice and ensure (or destroy) offspring. Like the water nymph (*apsara*), they were previously associated with watery places, like swampy forests, lakes, and rivers. Yaksha and yakshi enjoyed dancing, singing, and playing—especially the game of dice—and bestowing luck. They could be wise and beautiful, enjoying fragrance and music. When feathered or zoomorphic in form, their food included raw flesh including that of corpses as well as other meat and alcoholic beverages. For the humans with whom they cavorted, they could cause mental illness as well as sexual pleasure (Misra 1981: 2–4). Because these beings could be both benevolent and demonic, the term *yakṣa* seems to have compounded various sprites and spirits of Vedic attestation with local cult deities over time.

By the common era, such beings are described as delighting in more earthly locations, especially trees, and in medieval texts and imagery the female yakshi becomes more prominent (Misra 1981: 53–6), inhabiting, for example, the banyan and sacred *aśvattha* (*Ficus religiosa*) trees, or the *uḍumbara* (*F. racemosa*) and *plakṣa* trees (white fig, *F. virens*). In the *Mahābhārata*, the yaksha leader Kubera supplanted the Vedic god-king Indra as wealth-giver. Kubera and his retinue were also associated with powers of concealment, hidden treasure, and the intoxicating fragrant flowering *saugandhika*, found in a pond of Kubera's pleasure grove on the imagined Mount Gandhamādana (McHugh 2012: 94n11).[23] Over time the word *yakṣa* became interchangeable with the concept of a deity who lived inside a tree as a tree-spirit (*yakṣa devatā*) who granted wishes. The wish-fulfilling trees they inhabited (*kalpa vṛkṣa* and variant words) were eventually worshipped by people in similar ways across a shared first-millennium Indic world (Misra 1981: 4). Although cults of yaksha worship can be treated as relics of philosophical cosmogonies or indigenous animistic attitudes toward nature, ancestors, and the supernatural, references to *yakṣa* (Pali *yakkha*) are also found in texts associated with Buddhism, Jainism, and Hinduism that were redacted in the first millennium. In these texts, the worship of tree spirits represents a Late Antique and early medieval Indic attitude toward trees as beings—both hosts and actors in the human world.

Each of the twenty-four Jain Tirthankara (revered great men whose actions enabled them to successfully ford the waters of the interminable cycle of births and deaths and therefore point out a righteous path for others to follow) is associated with a tree under which he received special knowledge (*kevala*) as well as the yaksha spirit of the tree. For example, Munisuvrata, who descended from a heaven into the womb of the Queen Padmavati of Rajgir, is said to have attained omniscience under the *campaka* tree. The associate yaksha is Varuna and the yakshi might be Bahurūpā/Bahurūpini, Naradattā, or possibly Aparajitā depending on the Jain denomination (Misra 1981: 48–9; Shah 1987: 161–3). Some Hindus believe the *pārijāta* tree was the ultimate wish-fulling tree that emerged from the churning of cosmic waters and was removed from Indrani's heavenly garden by Krishna to his kingdom in Dwarka (*Harivaṃśa*, *Viṣṇu Parva*, chs. 65–76). Although this is commonly understood to be night-flowering jasmine (*Nyctanthes arbor-tristis*), some people identify it with the sub-Saharan baobab (*Adansonia digitata*),[24] furthering stories about movement across maritime networks. Later depictions of Dwarka, such as Kesu Kalan and Miskin, *Krishna and the Golden City of Dwarka* (*c.* 1585)[25] include botanically distinctive trees, and even later pilgrimage practices at Braj reinstate this important relationship of Krishna to groves (Shah 2007: 153–71).

Queen Maya (pronounced Māyādevī),[26] the Buddha's biological mother, was most often depicted giving birth to Gautama (the Buddha) in the grove of Lumbini while standing under a *sāla* tree (*Shorea robusta*) and reaching up to hold one of its branches for support. Previously, during her vision, a white elephant had walked all around her and entered her womb from the right side. This white elephant, according to Buddhist tradition,[27] was a bodhisattva (a Buddha-to-be) waiting for his final birth. Accordingly, in one panel of an important late-second–early-third-century Gandharan relief called *Scenes from the Life of the Buddha*,[28] the Buddha's birth is depicted as a figure exiting Maya's body from the right side, just like the scene described by the Sanskrit poet Aśvaghoṣa: "As the queen supported herself by a bough which hung laden with a weight of flowers, the Bodhisattva suddenly came forth, cleaving open her womb" (*Buddhacarita* 1894 [repr. 1997], 1.24, 5). In the *Nidānakanthā* version of the same scene, Maya entered an enchanted grove in Lumbini, profuse with flowers, fruits, vines, bees, and songbirds; "she was drawn to the most stately *sāla* tree and desired to take hold of one of its branches" and that tree bent down its limb (cited in Shaw 2015: 41).[29]

Considering textual and sculptural images together recreates the story in a more dynamic, multi-sensory fashion. Many things seem to happen at once in the image: first, a story is related about a royal mother-to-be who stopped with her entourage in a pleasant grove that seemed like a fine place to give birth. Second, like any good site for human habitation, the pleasant grove of Lumbini was teeming with spirits and divinities. One, in the form of a *sāla* tree, reaches out with a bough to help a laboring woman give birth to a divine being. Alternatively, a noble woman experiencing a difficult labor that would "cleave open her womb," held onto a winsome bough in Lumbini grove for support. In that instant, the desires of tree and woman acted in unison and upon the other: while Queen Maya added another divine being to the grove, the tree, also a divine being, supported the queen. Described as reciprocal and depicted as a union in the Gandharan relief (see note 28), her body stands in for the tree's trunk while its leafy branches emerge from Maya's head like a crown, one branch curving gently downward into the curled fingers that hold it: bough, limb, birth, tree, and figure become intertwined, cause and effect forgotten. According to religious studies scholar Miranda Shaw, the presumed site in southern Nepal of this miraculous birth continued to hold sway as a place where pregnant women prayed for solace, long after the site's association with the Buddha's nativity had been forgotten (Shaw 2015: 41).

The image of Maya holding a tree while giving birth corresponds visually with other early images of a yakshi inhabiting her tree. These representations of tree and female spirit are also called shalabhanjika (Skt. *sālabhanjika*) because of how the figure and sal tree[30] intertwine to produce a unified form of bent tree and figure. Such sculptural forms can tie female figures to other types of trees as well, such as the shalabhanjika at Sanchi's Great Stupa in Madhya Pradesh, India, in which a female figure is entwined with a heavily fruiting mango tree and positioned like an architectural bracket as if to support the eastern gateway to the stupa. Although the mango has been associated with Jain and Buddhist enlightenment, its depiction here suggests something of the fertile, material world (see Figure 6.3). In fact, the image of woman intertwined with tree became a very common sculptural motif during the first millennium: trees could include *aśoka*, *kadamba* (*Neolamarckia cadamba*), and *campaka*, among others (Krishna and Amirthalingam 2014: 38–40). Shalabhanjika[31] figures were also miniaturized as portable sculpture and exported or carried to places well beyond the subcontinent, along maritime routes, suggesting something of their potency. Two important examples of such "bhanjika" figures without their trees, found outside the subcontinent include an ivory from the excavations of Pompeii, Italy[32] and a bronze from the site of Khor Rori in Oman (see

FIGURE 6.3 Yakshi with mango tree in the form of shalabhanjika on the East gate to the Great Stupa, Sanchi, India. Photo by Eliot Elisofon, August 1949, The LIFE Picture Collection. Courtesy of Getty Images.

Figure 6.4). Comparing the posture of their nude bodies with the shalabhanjika from Sanchi (Figure 6.3), standing with legs crossed below the knee, curving upper body, hips thrust to one side, one arm bent at the elbow as if supporting something above, as well as similar physical features and ornaments, they are all a clear and therefore meaningful figural type, even when dissociated from the tree. By the fourth century, these already ancient images came to represent a Pan-Indic understanding of the yakshi or shalabhanjika as an important agent in the world. Just as images of Mayadevi at Lumbini continued to be venerated for over a millennium by local women with the wish for healthy children and as an aid in childbirth, shalabhanjika were also kept and carried as apotropaic and decorative symbols of plenitude and protection. The tree and its embodied inhabiting spirit were sometimes depicted separately while, at other times, one was a stand-in for the other.

Trees were venerated along with the spirits that dwelled within them, and in much the same way—for wish-fulfillment. Although specific trees have been cited as their dwellings, over time it was the composite concept of a wish-fulfilling tree (*kalpavṛkṣa*) that became absorbed by varied religious streams and courtly literature. Many tree

FIGURE 6.4 Miniature yakshi in the form of shalabhanjika. Bronze fragment excavated at Khor Rori, Oman. Height: approx. 8 cm. Courtesy of Arthur M. Sackler Gallery, Smithsonian Institution, Washington, DC. Gift of The American Foundation for the Study of Man (Wendell and Merilyn Phillips Collection), S2013.2.378.

species were thus incorporated into an ideology of trees as beings who, when pleased, could grant wishes, such that the popular, well-loved Hindu deity Ganesha could also be described as one.[33]

Agency of humans upon trees

The management of trees was a well-documented practice (*vṛkṣāyurveda*) comprising arboriculture, horticultural science, and forestry, in use well before the sixth century when Varāhamihira included a chapter on it in his *Bṛhatsaṁhitā* (2.54 or 55). This technical knowledge seems to have been developed to support the moral conduct of kings considered householders over their domains (Nugteren 2005: 77). Kings were responsible for settling people on undeveloped lands in the countryside to encourage productive fields, forests, and parks through irrigation works and infrastructure; land and water trade routes and ports; as well as sacred places and protected wilderness for ascetics (*Arthaśāstra*, 2.1.1, 2.1.19–21, 2.2.2–5, cited in Nugteren 2005: 77). *Vṛkṣāyurveda* was written into early treatises on statecraft and culture but was also embellished and developed via a long textual tradition of its own, continuing into the sixteenth century and beyond.[34] Although historical writing about methods of cultivation has proven significant for an understanding of garden history around the world (e.g. Clunas 1996; Subtelny 1993), *vṛkṣāyurveda* has not yet been explored for topics such as the detection of underground water for site selection, tree spacing, tree propagation, the nature of planting holes, seed treatments, plant nourishment, tree protection, and other aspects of planting and maintenance (Nene 2012: 424). These topics relate to a larger body of knowledge on agriculture, called *kṛṣiśāstra*, also compiled into texts during this period.[35]

One *vṛkṣāyurveda* text dated to the eleventh century was composed by Sūrapāla,[36] a medical practitioner at the Bengal court of Bhimapāla.[37] Surapala's text, considered the more exhaustive and systematic of these extant early works, expresses entrenched cultural values concerning human–plant relations in the first millennium. Although Indian literature often presents the city (or village) and forest in dialectical contrast (e.g. Stasik and Trynkowska 2010),[38] the plants and trees in both contexts were valued for similar reasons, and planting a tree was a meritorious act like having sons who could bring fame and fortune to one's name. Trees, however, returned to the planter the value of more than ten sons (*Vrikshayurveda*, 5–6:43).[39] In texts produced by city-dwellers, the act of planting also connected the tree as religious symbol of spritely fecundity with the civic practice of constructing outdoor public space. For example, planted trees along roads were valued more than trees in forests because they provided shade and therefore a public space for people to rest (*VrksA*, 4:43). This sentiment echoes one that was already present in a pillar edict by King Ashoka inscribed in the third century BCE: "Banyan trees have been caused to be planted by me on the roads, so that they will offer shade to beasts and men. Besides, mango groves have been caused to be planted by me for the same purpose" (Sircar 1957: Inscription 7 "Topra Pillar," 67).

Albertina Nugteren, a historian of Indian religion, has read the descriptions of tree planting in *vṛkṣāyurveda* in terms of their "karmic or heavenly rewards" (2005: 79). The planting of trees is seen as an effort to ensure indulgence from heavenly bodies and beings. In fact, each species was valued differently, and a tree's individual value was expressed by equating its planting to known ritual performances and concomitant afterlife benefits. For example, *tulasī* (*Ocimum tenuflorum*) multiplied one's expected years in Vaikuntha (heaven); *bilva* (bael, *Aegle marmelos*) ensured the presence of the goddess of wealth in one's family for successive generations; *aśvattha* (pipal) was a ticket to the realm of Hari; *dhātrī* (*Phyllanthus emblica*) was equivalent to performing several sacrifices, donating land, and practicing celibacy; a number of banyan trees ensured attendance by heavenly nymphs once in the abode of Shiva; neem and a righteous life led to an afterlife with the sun; four *plakṣa* were equivalent in merit to one *rājasūya* sacrifice; a grove of five to six mangoes ensured eternal life in the palace of Garuda; *palāśa* (palas, *Butea monosperma*) led to the dwelling of Brahma; eight *uḍumbara* led to the lunar world; *madhūka* equaled freedom from disease and the propitiation of Parvati; and *ksirini*, *dāḍima* (pomegranate, *Punica granatum*), *rambha* (also known as *kadalī*[40]—*Musa* x *paradisiaca*), *priyāla* (cheronjee, *Buchanania lanzan*), and *panasa* resulted in seven lives free from any affliction (*VrksA*, 9–20:43–4).[41]

However trees were valued for their afterlife benefits, tree-planting can also be read in terms of social benefits. For example, planting a jackfruit[42] ensured a certain type of social respect because the afterlife of someone with the "*dharma* of an ascetic (*yati*)" (Nugteren 2005: 81) was assumed to be a good one, not because the person who had planted a jackfruit already claimed that destiny. Rather, if one wanted to achieve the afterlife of an ascetic, one should plant a jackfruit tree: hence, "[h]e who has knowingly or unknowingly planted *jambu* is respected as a recluse even while staying in the house" (*VrksA*, 21:44). Planting other kinds of trees useful for fruits and flowers also resulted in rewards (*VrksA*, 22:44). There was even a prescription to avoid hell (*VrksA*, 23:44).[43] Certainly, the afterlife was written as more than an afterthought, but these prescriptions were also residential planting guidelines written within a culture of people who planted trees. Plant care was a respected vocation with significant social value. Like planting trees, writing such a text then became an assertion of the author's knowledge about plants and how humans could properly care for and control this socially significant good, as well as one's own destiny by doing so. Therefore, planting instructions, such as where best to plant certain trees, what the soil should be like, and how certain plants could be propagated (*VrksA*, 24–100:44–8) linked human horticultural mastery with esoteric concepts of merit, and not the other way round. Certain instructions may sound opaque, but that should not obscure the fact that the text attests significant human agency over trees and landscape through acts of planting and propagation.

It is specific about some ways human actions can alter plants, trees, and the social spaces they occupy, quite apart from a cultural desire for heavenly favor. For example, sacredness does not explain why a variety of species-specific supplements and fertilizers as well as treatments against common plant diseases is detailed at such great length (*VrksA*, 101–222: 48–55). Sixty-three verses propose various fertilizers and "nourishments" (101–64) and fifty-seven verses discuss plant diseases and their treatments (165–222). A skilled practice of planting design can also be inferred from a discussion of the spacing of trees and shrubs: if planted too closely, yields would be reduced; if planted too far apart, wind damage could become a problem. Customary planting plans, including a variety of named shapes and arrangements, also suggest an established aesthetics for this planting design practice (*VrksA*, 64–6:46 and 94–5:48). The tree-planting pit was another important subject of concern, including its measurements as well as its proper preparation with fertilizers (including cow bones and burnt cow dung), "good earth," and what could be called compost tea (*kunapa* water) (*VrksA*, 67–8:46–7). Surapala describes *kunapa* as "the excreta, marrow ..., and flesh, brain, and blood of a boar mixed with water and stored underground." When the compost was ready, certain parts of the *kunapa* were mixed with "water previously treated with fermenting seed for trees to produce flowers and fruit in abundance" (*VrksA*, 101:48; 113–14:49). Organic fertilizers are often used even today to promote tree flowering and fruiting. Transplanting instructions also reflected knowledge of woody-plant care. Small and large trees could be moved by preparing their roots with a transplanting agent before replanting; large trees were only to be transplanted during the evening with their roots well covered and after the recitation of an oath swearing protection to the tree with a lasting commitment of water (*VrksA*, 84–6:47). Watering programs for newly transplanted woody plants are still understood to be key to their root establishment and longevity in the garden. Seasonality was understood too: each species had a specific planting season (*VrksA*, 87–90:48). In conclusion, some of these planting instructions, when properly translated, might make sense to the modern horticulturalist, such as the recipe for *kunapa* (which could include the dung, bone marrow, flesh, brains, and/or blood of a boar, fish, goat, or other horned animal) or the sun protection and watering requirements for newly planted saplings (for a period of fifteen days until the ground is soaked).[44]

Other fertilizers may seem fanciful. Still, when read alongside recorded beliefs, like how certain trees were auspicious deterrents to disease and calamity when planted in specific arrangements and directions (*VrksA*, 91–3:48), even the far-fetched recipes and remedies reflected human attitudes towards the plants they attempted to use and influence and a widespread cultural ethos regarding the relationship humans were perceived to have had with trees. In a variety of ways trees remedy some of the difficulties humans faced in cities. In fact, the word encountered above for tree (*vṛkṣa*) was supplanted with a synonym (*taravah*; sing. *taru*) because "[t]rees alone on the earth, give happiness both here and in the hereafter. Since they 'save' from abject poverty they are named 'taravah' (the saviors)" (*VrksA*, 97:48).

Unusual human affect: The control over tree flowering

The literature makes clear that Indians wished to influence trees; a salient desire was to control bloom time. Flowering trees were highly prized. When trees are weighed down with flowers, their branches bend, making their ability to act in the world visible, as discussed in the description of Queen Maya's labor above. Bent down with flowers, boughs also become ornaments to their built environment. Profuse flowering caused bees to swarm and women to sing, inspiring repeated use of those clichés as literary tropes, as in this Gupta-period inscription introducing the ancient city of Daśapura.

> the trees weighed down with flowers ... the lakes, crowded with *karaṇḍava*-ducks, are beautiful—having the waters close to (*their*) shores made variegated with the many flowers that fall down from the trees growing on the banks, (*and*) being adorned with full-blown waterlilies. The lakes are beautiful (*in some places*) with the swans that are [trapped] in the pollen that falls from the waterlilies shaken by the tremulous waves; and in other places with the waterlilies bent down by the great burden of their filaments. Here the woods are adorned with lordly trees, that are bowed down by the weight of their flowers and are full of the sounds of the flights of bees that hum loudly through intoxication (*caused by the juices of the flowers that they suck*), and with the women from the city who are perpetually singing ....
>
> (*CII* 3, "Mandasor Inscription," 1888 (1970): 18, 84–5)[45]

Although Daśapura's buildings were multistory, "very white," and "extremely lofty," the city becomes beautiful due to its house gardens, banana groves, tree-lined streets, lakes, and surrounding wooded parks, all described as floriferous.

Among its many instructions, *vṛkṣāyurveda* offers species-specific prescriptions to favor flowering. Most of the concoctions are understandable, using plant, animal, and mineral materials in some combination. However, certain trees stand out because they require the presence of a beautiful woman to bloom.

> The *bakula* tree [*Mimusops elengi*] blossoms well when sprinkled with a mouthful of wine by a beautiful (young) lady.
>
> *Makanda* tree gets horripilation in the form of buds when it is scratched by a beautiful woman with the tips of her nails ...

> *Aśoka* tree yields rich blossom when kicked by a lady who on account of her beauty is like the banner of love-god, gracefully and gently with her lotus-like foot painted deep red with lac dye and adorned with lovely anklet jingling melodiously.
> The *kurabaka* and the *tilaka* trees yield rich blossom when embraced all around by a charming lady with her graceful creeper-like arms looking beautiful like the lotus stalks and adorned with dangling bracelet.
>
> (*VrksA*, 147–50:51)

Such prescriptions recall the intertwined, although now decidedly erotic, relationship of tree and woman as shalabhanjika. Their union, inspired by woodland divinities, became an ideal for elite urban women, and kicking the tree, a ritual of femininity and courtly culture; thus, the flowering of plants was especially responsive to the entreaties and touch of beautiful noblewomen. Besides attesting to a lady's perfection, control and influence over woody plants through perfected noblewomen like her could also confer royal authority—perhaps because a king ruled over his domains and their inhabitants as members of his household, broadly defined. Kings were supposed to control the seasons, time, and the flowering of plants; when a king lost control of bloom time, one could suspect moral failure or an illegitimate right to rule: quoting the king in a fourth–fifth-century play, "Do my sins stunt the flowering of the vines?" (Goody 1993: 326).

Kālidāsa[46] describes the courtly ritual of kicking the tree in his play *Mālavikāgnimitram* (dated to the fourth–fifth century CE by Western scholars). The king is newly enchanted by the beauty of the mysterious lady pronounced Mālavika who, although actually a princess who was promised without her knowledge to King Agnimitra, had entered the court in disguise after a skirmish in the surrounding woods. No one knew who she was, not even the king. The king's primary queen, Dharini, had taken Malavika in as a lady-in-waiting. Because she was young and beautiful, Malavika looked like she could fit in at court. Although normally the "ceremony of fulfilling the longing of the golden Aśoka" was the queen's own responsibility, Dharini had hurt her ankle. Just before Act III, Dharini decides to send Malavika in her stead to fulfill the desire of the tree, betting on a promise, "if within five nights from this time, [the tree] shows flowers, then will I bestow on you a favor which will gratify your desire" (*Mālavikāgnimitram*, III.5.7–11:60–3).[47] In the meantime, Malavika was developing a crush on King Agnimitra, who had also fallen for her looks, much to the dismay of his second queen Iravati who desired a romantic tryst with her husband. Iravati's amorous invitations were constantly thwarted by the king's budding romance with Malavika. In Act III, Malavika is in King Agnimitra's *pramadavana* (or "pleasure grove") preparing to kick the tree, while the king and his friend are hiding nearby and Iravati, drunk and depressed, was with her lady elsewhere in the garden.

Malavika, described as lovesick and like a *kunda* creeper (*Jasminum multiflorum*), barely dressed, "having its leaves ripened by spring and with only a few flowers," evokes the shalabhanjika figure once again: "Ample in the hips and thin in the waist, uplifted in the bosom and very broad in the eyes," she is a paragon of feminine beauty and grace. Bakulavalika, another of the queen's ladies, prepares Malavika for her encounter with the tree by painting her kicking foot with red lac dye and affixing the queen's anklet to it. The pedicure, once complete, is praised by the hiding king, whose words also introduce an ambiguity and correspondence between himself and his golden *aśoka* tree.

> **King:** Friend, look at this streak of wet paint laid upon her foot, which is like the first burgeoning of leaves on the tree of love consumed by Hara ... With the toes of

> her foot red-hued like young foliage, this girl deserves to smite two objects the Aśoka awaiting the fulfillment of its longing before it would blossom, and the lover who ... stands with bowed head.
>
> (*MaAg*, III.11–12:68–71)

Malavika admires her own foot as does Bakulavalika, who compares it with a red lotus before encouraging the encounter now with the king, confirming the introduced ambiguity: "May you, by all means, sit in the king's lap" (*MaAg*, III.23.3–5:76–7). The verbal exchange that follows excites Malavika to pursue the king's affection and an affair with the tree. At the end of this exchange, the conflation of king and tree recurs, but now from Malavika's point of view:

> **Bakulavalika:** Here before you stands one flushed (of increased redness, passion) and fit to enjoy.
>
> **Malavika:** (With joy) What? The king?
>
> **Bakulavalika:** (Smiling) Not the king, but a spray of leaves on a bough of the Aśoka. Make an ear ornament of it.
>
> (*MaAg*, III.14.16–21:80–1)

Malavika receives the gift of fresh leaves from the bough of the tree as a representation of the tree's ardor, and she arranges those leaves as an ear ornament before raising her own painted, ornamented foot toward the tree for her gift of a kick, reflecting an honored tradition of courtly love: the mutual exchange of similar gifts. The tree cannot help but do its part by flowering, especially since the exchange of gifts was witnessed and approved by the king who verbalized his understanding with an entreaty, "O Aśoka, if you are not immediately covered with blossoms, then in vain do you cherish the delicate longing ..." (*MaAg*, III.17:82–3). The scene concludes with the king revealing himself to Malavika, for an opportunity to touch her hand and offer her a few sweet words; they will meet again in a secret tryst in the following act. As the love between Malavika and King Agnimitra becomes accessible and realized, the *aśoka* tree comes into bloom. The play ends, after a series of plot twists, in their marriage.

FLOWERS AND LEAVES IN MEDIEVAL CEREMONY

Malavika demonstrates how trees might make gifts of leaves. Trees also offer flowers and fruit. In exchange, early Indians made similar gifts to them. These offerings fertilized the plant and could please inhabiting spirits. Gifts also expressed love between lovers and with lords. By the end of the first millennium, plant gifts became appropriate offerings of care not only for trees and lovers, but also great men, god-kings, and effigies of both. Cut flowers could be given in heaps along with leaves or offered as garlands as had long been the case (see Figure 6.2). Over time, specific flowers and leaves became associated with the particular gods who enjoyed them.[48] A courtly ethics surrounding the proper enjoyment of pleasures by lord and subject differs little from expressions of Hindu services to gods and the appropriate offerings for them (Ali 2004: 103–4). Therefore, sacred plant practices might be used to explain the aesthetics at medieval courts. However, as the scholarship increasingly makes clear for other aspects of courtly culture,[49] courtly ideas about plants seem to have been layered onto popular forms of worship with plants.

Thus, the sacred act of adorning a king-like god can be seen to be a relic of medieval court culture, when encountered in devotional settings. In both cases, the heady aroma of flowers and intoxication caused by plants became a signifier for refreshing sensations associated with ideal attitudes toward all sorts of deified beings—as well as love-laden encounters between humans or with the divine.

Garlands and plant gifts in religious and court life

Various types of symbolism might be read into floral forms and properties when considered by religious groups. Hindus adorn gods and men with garlands, while Buddhists use individual flowers whose ephemeral nature "stands for impermanence the central teaching of the Dharma." Although Islam generally "excludes flowers from worship and from iconography," flowers and incense were offered at the tombs of Muslim saints (*pirs*), perhaps because in the miracle literature the disappearance of a *pir* may leave "only flowers," and often "the odor of flowers" is described as "hanging over their tombs" (Goody 1993: 322, 330–2, 337).[50] Islamic jurist and physician Ibn Qayyim al-Jawziyyah (1292–1350) said "… there is a close relationship between [sweet] scent and the good spirit" (1998: 199, cited in Flatt 2016: 12).[51] Such descriptions suggest the valued qualities of flowers: fragrant, ephemeral, and components of garlands. In fact, pious notions and offerings often call for specific flowers and leaves that fit this description. For example, within the *purāna* literature which codified Hindu social rites and customs, one rite called the *vrata* stands out as identifying the plants used for worship (*pūja*).

A compelling picture of medieval plant worship (*pūja*) emerges when the description of a *vrata* rite, which can also be used to date its historical practice, is considered along with the list of named plants and flowers to be used within the rite. Different Hindu gods required specific varieties of cut flowers and leaves for their proper worship. According to Anne Pearson, scholar of South Asian religions, a *vrata* is a personal promise or pledge to undertake a set of observances regularly in order to achieve a particular set of objectives by following handed-down rules which may include "fasting, worship (*pūja*), the listening to or recitation of a narrative about the efficacy of the rite (*kathā*), and the giving of gifts (*dān*[*a*])[52] consisting of money and items of food and clothing to another person, often a Brahman." Although like other prayers they are instrumental in nature, a *vrata* takes the form, "I will undertake such and such a regimen for this period of time and may you, O God, pleased by my devotions, protect my family" not the more familiar bargaining pledge of making a pious action *after* receipt of a divine intervention (Pearson 1996: 2–3). When datable, the *purāna* literature can help scholars correlate modern plant lists with associated medieval plant worship, because plants were named as flowers and leaves within the verbalizations (*mantra*) to be said (and were therefore recorded) during the appropriate ritual *pūja*.

For example, in *Linga Purāna* (Ch. 84) dated to *c.* 800 to 1000 CE (Hazra 1975)[53] the Umā-Maheśvara *vrata* designed "for the welfare of men and women" involved preparing a sacrificial offering for a deity called "Bhava" at the full moon, new moon, as well as the eighth and fourteenth days of the lunar cycle. Its rules required women practitioners to observe certain restraints and have an image of Uma-Maheśvara cast in gold or silver installed for *pūja*. At the end of a year, brahmans were to be fed and offered gifts (cited in Pearson 1983: 83–4).[54] According to a contemporary account of the *vrata* by the same name, the appropriate *pūja* involves eight kinds of flower and five of leaf, to be "given up" to the image while pronouncing a list of titles for the god. Flowers are to be offered during *puṣpa* ("flower") *pūja* and leaves during *patra* ("leaf") *pūja* respectively. See Figure 6.5 for the plant list as recorded in a report by the Karnataka Forest Department and updated by the author and editors.[55] Although plant and flower offerings can overlap across different *vrata*, each rite has a distinct plant list. For example, the Mahasaraswati Vrata may correspond to the "Sarasvata vrata" documented in the *Matsya Purāna* (Ch. 66) dated to 550–650 CE, which required white flowers for the *pūja* (cited in Pearson 1983: 59–60). As expected, the contemporary plant list contains no leaves, and nine white flowers, two of which were in the previous example (*mallikā* and *karavīra*), but seven are new: *nandyavanda* (*Tabernaemontana alternifolia*), [*b*]*akula, punnāga* (*Calophyllum inophyllum*), *ketakī* (*Pandanus odoratissimus*), *jaji* (*Hibiscus rosa-sinensis*), *mandāram* (*Bauhinia purpurea*), *nilotpala* (*Nymphaea lotus*) (updated from Karnataka Forest Department, "Sacred Plants" 1988: 41–2, 67–145). Both gods and goddesses can have *vrata* plant lists, which can be taken as an authoritative list of the flowers and leaves deemed appropriate for a distinct deity that is usually worshipped in the form of an iconic image or an aniconic representation.

FIGURE 6.5 Plant list for "Shree Umāmahēshwara Vrata" updated from Karnataka Forest Department, *Sacred Plants* (1988), 43–4, 6–145 (see note 55). Courtesy of D. Kumar-Dumas.

Puṣpa Pūja		
Mahadevaya namaha	Mallika (*mallikā*) pushpam (*Jasminum* species)	Samarpayami
Bhavaya namaha	Malati (*mālatī*) pushpam (*Jasminum angustifolium*)	Samarpayami
Kshemankaraya namaha	Sevantika pushpam (*Chrysanthemum indicum*)	Samarpayami
Karunanidhaye namaha	Kamala (Tam. *kamalā*) pushpam (*Nelumbo nucifera**)	Samarpayami
Kumaraguruve namaha	Kurantaka pushpam (*Barleria prionitis*)	Samarpayami
Girijapataye namaha	Girikarnika (*girikarṇikā*) pushpam (*Clitoria ternatea*)	Samarpayami
Dhravyapradhaya namaha	Drona (*dronṇapuṣpī*) pushpam (*Leucas cephalotes*)	Samarpayami
Karicharmadharaya namaha	Karaveera (*karavīra*) pushpam (*Nerium oleander*)	Samarpayami
Patra Pūja		
Bhavaya namaha	Bilva patram (*Aegle marmelos*)	Samarpayami
Trilokeshaya namaha	Bhringaraja (*bhṛṅgarāja*) patram (*Eclipta prostrata*)	Samarpayami
Shivaya namaha	Tulasi (tulasī) patram (*Ocimum tenuiflorum*)	Samarpayami
Munisevitaya namaha	Machi (Tam. *mācipattiri*) patram (*Artemisia indica*)	Samarpayami
Akshayaya namaha	Ashwatha (*aśvattha*) patram (*Ficus religiosa*)	Samarpayami

Medieval Indian festival bronzes adorned with flowers are by now a familiar idea (Waghorne 1992). They have become widely associated with prayerful acts, called *darśan* by the devout, of seeing and being seen by their gods (Eck 1985). The procession of divine images during great Hindu festivals, such as Rathyatra of Puri, marries the ritual of *darśan* with social life. Such processions are still common in Tamil Nadu, particularly among Sri Vaiśnava communities that take images of their lord Vishnu through the countryside or for a refreshing dip in the sea (Eck 1985: 57–8). Flowers, other plant materials, unguents, and incense adorn processing deities when they go out, and to great visual effect (see Figure 6.6).

Perhaps less well known is the fact that "arraying and adorning an idol with rice and flowers" was among a list of sixty-four arts to be mastered by the medieval courtier according to the *Kāmasūtra* (dated to the early centuries of the common era). Eight of

FIGURE 6.6 Talasayana Perumal of Mahābalipuram during a procession of the festival bronze around villages in Chengalpattu district, Tamil Nadu, January 2016. Courtesy of D. Kumar-Dumas.

these prescribed proper etiquettes relate to fresh and artificial flowers—the number refers to its position in a longer list:

> 7 Arraying and adorning an idol with rice and flowers
> 8 Spreading and arranging beds or couches of flowers, or flowers upon the ground
> 15 Stringing of rosaries, necklaces, garlands and wreaths
> 16 Binding of turbans and chaplets, and making crests and top-knots of flowers
> 26 Making parrots, flowers, tufts, tassels, bunches, bosses, knobs, etc., out of yarn or tweed
> 40 Gardening; knowledge of treating the diseases of trees and plants, of nourishing them, and determining their ages
> 47 Art of making flower carriages
> 63 Making artificial flowers.
>
> (*Kāmasūtra* 1963, cited in Goody 1993: 323)

Flowers and plant materials were, in fact, central to the etiquette and ethics of a world of sensual pleasure at court, coined the "kama world" by the cultural historian Daud Ali who describes it as follows:

> Within this world of art and music, precious gems and metals, colorful flowers, luxurious fabrics, and enticing perfumes, the urbane, cosmopolitan sophisticate cultivated his or her physical, mental and affective capacities in order to succeed in experiences of "pleasure"—which included the creation and enjoyment of beauty, the experience of passion and attachment, and the pursuit of courtship and seduction.[56]
>
> (2011: 1–2)

This world of pleasures described in *kāmaśāstra*[57] texts can help explain how the aesthetics of court etiquette and protocol worked through pleasurable exchange (described below). Elites took enough sensual pleasure from aspects of their world to meticulously record its procurement in their how-to manuals. This preoccupation does not represent a decadent culture but a philosophical position in which pleasure was actually a moral technology of self-hood (Ali 2004: 179 after Foucault, *Technologies of the Self* [Martin 1988]) as explained below. It is the "kama world" and not the general symbolic meanings of flowers, flower garlands, and leaves that can contextualize plant offerings within patterns of elite cultural use during the first and early second millennia.

Aside from festival bronzes, other revered medieval bodies, like kings and saints, were often decked out with flowers and jewels (Ali 2004: 165; Branfoot 2017; Dehejia 2009). From medieval inscriptions and literature, this act of adorning the king with ornaments including flower garlands can also be understood as among a set of courtesies relating to lordship (Ali 2004: ch. 4). Three important concepts (*upacāra, alaṁkāra, pūja*) clarify how social actors used plants as gifts (*dāna*) to establish bonds of loyalty with their subordinates, in a cultural tradition peculiar to South Asia. The concept *pūja* has already been explained; below *alaṁkāra* and *upacāra* will be introduced. Not only could kings make their subjects eat leaves out of their hand, but they became decorated kings by receiving daily gifts of personal ornaments such as flower garlands from their subjects. These acts solidified bonds of loyalty through repeated actions that used plants as markers of the ruler-subject relationship. As adorned beings, kings were reified as lords by their subjects. Most importantly for our topic, plant cuttings were the primary witnesses to the intimacy of these political acts. This complexity is here explored via such ephemeral gifts of leaves and fragrant flowers.

Betel nut (*Areca catechu*)[58] wrapped in a betel leaf from the vine *Piper betle* was (and still is) enjoyed in a similar manner to chewing tobacco. When given as a gift, and because of the reciprocity implied by the "taking up" of such *pān* for voluntary consumption, the wad or "betel roll" also symbolized the desire to accept a command or new relationship through one's own free will (Curley 2003: 51). As *tāmbūla*, a preparation of betel leaf wrapped around betel nut and combined with slaked lime and other ingredients, the leaf figured prominently in traditions of hospitality. Traveler and historian Ibn Battuta (1304–69) recorded one feast arranged by the Sultan of Delhi Muhammad Tughluq during which such preparations of betel were offered to guests at the conclusion of the meal and received with solemnity: "It is their custom that the person to whom this is brought out takes the platter [containing betel] in his hand, places it upon his shoulder and then does homage with his other hand touching the ground" (*Travels* 1958: 3, 670–1; Curley 2003: 52). The giving and taking up of betel could also formalize a legal agreement and make it binding. At the turn of the tenth century, two inscriptions recorded the ceremonial exchange of *tāmbūla* which included a proclamation of the boundaries of a transfer of land from one party to another in the presence of village headmen (cited in Ali 2018: 538). In courtly settings, royal gifts of *pān* also served to mark the receiver as an honored subject. This practice was institutionalized during the tenth to twelfth centuries under Chālukya rule by the requirement of a titled courtier with the responsibility for carrying the betel bag at all times, whose presence was always required near the person of the king (Ali 2018: 538–40). The keeper of *tāmbūla* shared his proximity to the king only with the king's bodyguards, emphasizing his official importance. On one hand, he would prepare betel for the king's enjoyment, but, on the other, he was responsible for preparing *tāmbūla* whenever required in order to confer royal favor or solemnize an agreement between lord and subject. For example, when a lord (*mahāmaṇḍaleśvara*) named Tammiyarasa required military support in the siege of an enemy's fortress, he honored Sir Soḍḍiga (a *nāyaka*) with betel. Soḍḍiga agreed to fight by "'taking the betel from the palm' of his lord's hand" which was witnessed by the court as a public enactment of fealty and subservience (cited in Ali 2018: 542). Coercion to eat betel from the hand could have negative implications such as when a vanquished king was forced to do so after losing a battle.

Adornment as an ephemeral pleasure

Considering how the gift of *tāmbūla* would be received and enjoyed adds dimension to this political transaction and relies on the specific properties of the plant. First of all, not all betel wads were the same, and the preparation of a delectable wad was an important aspect of court pleasure. When used as a mouth freshener, the preparation of betel was considered part of the courtier's daily routine (*Kāmasūtra*, 1.4.16, 22, cited in Ali 2004: 167–8n75). A twelfth-century manual on court sumptuary written by the literatus-king Someśvara III (r. 1126–38)[59] includes an entire chapter about the preparation of *tāmbūla*, considered the highest pleasure (*uttamo bhogaḥ*) in a "kama world" of courtly enjoyments. Apart from the betel leaf and nut, additional ingredients could be rare luxuries acquired from afar, like oyster shell for the lime paste and spices like aloe-wood (*Aquilaria malaccensis*), ambergris, and camphor (*Cinnamomum camphora*). Such expensive aromatic products linked the fresh breath they produced to the prestige and social authority of those sweet-smelling mouths. At the turn of the second millennium, "possessing" materials associated with the "kama world" as well as "knowing about them" was essential to one's self-fashioning as a sophisticated urbanite or king. Most importantly, "now one needed to know how to consume them correctly" as "luxurious" or "intoxicating" substances for pleasure (McHugh 2011: 74). As the fifteenth-century central Indian cookbook called *Ni'matnāmah* (Book of Delights) corroborates, the betel wad was associated with intoxicating pleasure—and aphrodisiac properties: "it produces well-being and desire ... induces intelligent talk and produces an appetite ... The libido is strengthened, desire is awakened and erections are induced" (cited in Flatt 2016: 13). Contemporary science labels betel leaf an addictive stimulant and says the betel nut induces a feeling of well-being, heightened alertness, and greater capacity to work, with users describing the euphoria as similar to that produced by tobacco or cocaine.[60] Returning to how Soḍḍiga might have felt when he took the *tāmbūla* from his lord's hand helps link his sense of glory as a warrior to the pleasurable sensations he must have had with his king when he accepted the fight. His success in battle, therefore, was beside the point. Rather, it was Soḍḍiga's affirmation to the lord he served, in a brief moment of intimacy, that confirmed and sealed the deal and his fate. Although the court witnessed and recorded a ceremonial transaction of the gift of leaf, agreement was reached when Soḍḍiga said "yes" in his lord's presence. For Soḍḍiga, it was the intimacy of that liminal instant, alone with his king and through a plant, that bore witness to the contract—when vassal, momentarily just close enough to touch his lord's hand, took up the betel leaf into his mouth, and received an unusually refreshing panoply of sensations from the act. There is a record of this agreement only because the glorious Soḍḍiga died in that battle (*EC* 8 1904: Sagar 86, 291, cited in Ali 2018: 542).[61]

Kings and courtly elites also exchanged flowers, and, like betel leaf, the freshness and ephemerality of floral pleasures were key to their proper enjoyment. Yet flowers were worn as garlands and therefore were part of a wider practice of personal body culture at court: "Anything applied to the body—garments (*vastra*), perfumes (*gandha*), ornaments (*alaṁkāra*) or garlands (*mālā*)—was considered to 'adorn' it, and a large number of the skills or 'arts' (*kalā*) to be mastered by the *nāgaraka*[62] entailed expertise in various aspects of this elaborate regimen of self-beautification" (Ali 2004: 167). The notion of adornment or ornament (*alaṁkāra*) existed along a continuum. The decorated elite body was analogous to the beautified city and proper, moral action in the world. Physical appearance and personal ornament were thus linked to the many qualities of a highly decorated individual, those "virtuous, wise and powerful in the king's retinue ... considered to be the jewels or ornaments of his court" as well as ornaments to their families, lineages, and the earth (Ali 2004: 175–7). Elites should look adorned to be who they were in the world.

The lightness of floral decoration seems to have participated in this ethics of ornament because of the way flowers hang suggestively from bodies and buildings; due to their ephemerality, flowers played an important role in the complex world view of *alaṁkāra*. As ornaments, plants and flowers were equated metaphorically with the otherwise invisible qualities of persons. In one Gangadhar inscription (fifth century), architectural additions to the city including "pleasure-gardens of various kinds" are likened to the patron's "beloved wife with different sorts of ornaments" (*CII* 3, "Gangadhar Inscription," 1888 [1970]: 17, 77–8). The comparison enables a reader to see the city's more intangible, attractive qualities. Wearing flowers melds the visual impact of jewels and clothes with the invisible presence of perfume, which can hang over things and in spaces. Floral, especially lotus-like, fragrances were equated with the fresh breath of revered persons (McHugh 2012: 70–1). And fragrant flowers with the most admired scents, such as jasmine, lily, and lotus, were also most often used for garlands. By the fifth century, gifts of garlands had become one of the daily courtesies (*upacāra*) integrated into the royal routine (Ali 2004: 166).[63] These garlands made the body and face of the king recognizable as an ornamented and sweet-smelling part of the decorated world. Therefore, they were necessary accoutrements to the sartorial practices of lordship and excellent gifts of adornment from any subject.

Consider how the gift of a garland is unlike other gifts of scent derived from wood (like sandalwood or agarwood) or acquired as trade goods (like musk, saffron [*Crocus sativus*], and camphor for incense or perfume), all of which were treasured luxury goods. The fragrance of flowers, by contrast, cannot be stored without processing, making flowers inexpensive, terminal commodities (McHugh 2012: 92). Always procured locally, they are fragrant today and foul-smelling tomorrow. Yet, the literature equates a flower's beauty specifically with its fragrance, underlining the ephemeral nature of the floral gift. In fact, the fragrant garland was the principal means by which fragrance could be offered as a gift of adornment by every subject. Offering fragrance as the courtesy (*upacāra*) of a garland (*mālā*) established the donor not only as a dutiful servant but also as someone capable of giving the gift of adornment (*alaṁkāra*), regardless of his family's access to wealth. Furthermore, the ephemerality of flowers required a regular gift, likely to a set schedule. Fulfilling one's duty thus marked and regularized time. Because the king's garland would be seen by many, especially as a regularized gift, a donor's enactment of his ephemeral gift would be recognized. Like other daily courtesies of adornment presented as gifts by society, floral gifts could establish a donor's place within the social order at court.

Becoming an adorned being

The skill of flowers is how they please and "gladden the mind" (McHugh 2012: 227). In the *Anuśāsanaparvan* of the *Mahābhārata*,[64] Yudhishtira and Bhishma discuss why people offer flowers, incense, and lamps to the gods. They conclude that giving (*dāna*) in the form of fragrant offerings of flowers and incense pleases the gods who in turn reciprocate the pleasure to those who made gifts (McHugh 2012: 221–4). The reason, then, to give flower offerings was not only to see and be seen by a beautifully adorned king, but to please him and, if this was successful, be pleased by him. The fragrant garland was unmatched in its ability to deliver such a mutual exchange of visual and intangible pleasure. In fact, certain forms of marriage were ratified by the exchange of garlands, presumably for this reason. Offering a fragrant garland signified a reverence (*pūja*) that in turn triggered an exchange of mutual pleasure. The fragrance of flowers is cool, fresh, and heady acting as an ephemeral witness to the meeting of minds. When received as a garland, fragrant flowers transform the self by effecting pleasure. Also called *sumana*(*ḥ*) meaning "good minds," flowers cause *alaṁkāra* to happen as "a 'technology' of self-hood, the capacity to act upon oneself" (Ali 2004: 179). That garlanded self, in turn, becomes adorned by floral fragrance and associated virtues.

The act of decorating the king with a fragrant garland also transforms the actor.[65] In the moment of transaction, its fragrance, although destined for the king, wafts back toward the giver embracing their shared spatial intimacy with an olfactory suggestion of the king's imminent transformation. As flowers were accessible to everyone, the giver would know the pleasure of receiving fragrant flowers on the body and could therefore empathize with the king's experience when the gifted garland was placed around his neck. That the king's psychological disposition could momentarily be altered by the freshness of its fragrance empowers the flower garland with fleeting control and also, in the instant of giving, subverts the giver's status from subject to agent. Because of their ephemerality, flowers had to be carefully chosen to produce the intended effect. Thus, obtaining perfectly fresh flowers cultivated an attitude of reverence. The procurement of fresh flowers and preparation of a garland as a repeated courtesy, to be accomplished on schedule, disciplined and perfected this pious attitude with a regular reward: it was fragrance as pleasure that "hangs over" the moment of gifting and merges with one's pleasure of feeling empathy for the king's pleasure, the whole exchange resulting in an augmented pleasure for the giver. Bowing into the intimacy of that encounter with re-duplicating pleasure becomes an act of love. It opens one's heart to refreshing feelings that in turn mirror the refreshment of fragrance. Giving fragrant adornment cultivates an attitude in which the act of giving becomes enough for complete fulfillment. The term *alaṁkāra* literally means "to make [*alam*] sufficient," or, in other words, to magnify with just enough strength for something to become fully and completely itself. Later, via *upacāra*, *alaṁkāra* developed meanings relating to sophisticated decoration, suggestive ornament, and pleasurable fulfillment.

PLANTATIONS: GARDENS, GROVES, INSCRIPTIONS, ARCHAEOLOGY

The regularization of ephemeral plant gifts as courtesies raises questions about the gardens and fields necessary for their cultivation. In fact, land set aside for plantation was commonly recorded in medieval inscriptions, such as the Tiruvāḷangāḍu Plates[66] recording a donation of land by Chōla king Rājendra I (r. *c.* 1012–44). Previously read for how the gift was enacted through a ritualized land survey (in this case the village of Palaiyanūr, as a gift "to the god Siva residing in the temple at Tiruvāḷangāḍu" [Ali 2000: 172]), such grants also reified certain analogous relationships in medieval India: between subjects and their king on the one hand and devotees and their deity on the other. Great gifts of land to temple deities transferred a courtly understanding of the ephemeral gift of floral pleasure onto sacrificial values of a pious gift, associated with socio-political prestige through its ritual enactment and visibility. This particular grant required three provincial officials from Jayagondacolamandalam to go to the village with the order (originally written as a portable palm-leaf manuscript) "and accompany a local delegation in leading a female elephant around the circumference of the village Palaiyanur in order to mark out the boundaries of the donated land." Although it is not known how the elephant was dressed and ornamented, it is known that the representatives of several districts gathered in person for the arrival of the auspicious royal order from the (sweet-smelling?) "royal mouth" (*tirumukam*) and "respectfully received it and placed it on [their] heads" after which they accompanied the elephant, as if in a procession, around hamlets to delineate the royal order's land boundary. Only after this pomp and ceremony was the order permanently inscribed (Ali 2000: 172–4n21). Socio-political prestige and great gifts have been associated in Indic culture from the earliest times. However, the innovation in our period, based on the cornucopia of epigraphic sources, lies in how such enactments were revalued and made permanent, as written inscriptions.

The materiality of such enactment events in all their color likely included processional drums and melodic improvisation, crowds, fragrant floral ornament, the exchange of *tāmbula* and refreshment—the customary plant courtesies. When read as a place-making strategy (e.g. Mines 2008), a re-enactment of the original procession around received lands on another ritual occasion by the festival bronze of the resident god, seems to be one plausible explanation for the excitement surrounding religious processions and *darśan* which continue today. Ornamented thus, like a king, a festival bronze statue surveys its domains while enjoying the land, its people, plant courtesies, and pleasures like a beloved. This land was also the terrain upon which the ephemera of pleasures, leaves, and fragrances were cultivated for courtesies (*upacāra*) that could now be offered to him as the temple god—in order to please and be pleased.

As demonstrated above, plants supported significant cultural activities and human-plant relationships in medieval South Asia. Although religious and courtly prescriptive texts are the primary sources for current knowledge of this culture, actual plots of land would have been needed to support such an economy of ephemeral plant exchange. In fact, a largely under-explored set of inscriptions recording various meritorious acts by all sorts of people, including women, indicates that lands were, in fact, set aside for cultivation of flowers, leaves, and trees. Even without quantitative analysis, one can correlate the growing need for plant materials at court during the late first and early second millennium with records of gardens donated to temples, known from inscriptions of the tenth to twelfth centuries.[67] Noteworthy are those gifts of temple land slated for the production of flowers, presumably for fragrant garlands and cut plant offerings destined for the lord seated within. Lands were allocated to flower-sellers, along with other artisans associated with temple service,[68] or zoned for flower garden plots.[69] Some families contributed measured rations of betel nut and betel leaf along with gardens.[70] In certain instances, the garden was supported by additional lands for the domestic upkeep of its gardener.[71] Gardens with flowers and fruiting trees were also constructed as charitable acts for public enjoyment.[72] While it has been difficult to understand how medieval Indian gardens were laid out (Ali 2012: 46–7; Ali and Flatt, "Introduction" 2012: 3–4), endowed gardens and groves sometimes included plant lists.[73] Although gardens have not yet been studied as material facts,[74] it is possible to better understand how land was put to use for floriculture by looking especially at inscriptions detailing gifts of land both for various garden types and for the household income plots given to temple garden cultivators.

In preliminary conclusion, it can be said that the gardens and groves of early South Asia were largely defined by their plants and their architecture, both of which were valued for sacred and secular reasons. Stone inscriptions, often on the exterior and interior walls of standing *in situ* buildings, might also locate the actual medieval gardens to which they refer, now discernible via techniques of garden archaeology. Future work, employing such archaeological examples, should corroborate the current understanding of how a garden fitted into larger cultural patterns and processes with details from these actual medieval gardens indicated by inscriptions. For now, it can be concluded that the establishment of plantations, temple gardens, charitable public or medicinal gardens, and a variety of public, sacred, and scholarly groves, were considered meritorious acts which bought the donor social power and prestige as good works associated with a world of pleasures. While trees and plants might occur naturally in the forest, urban people with ethical and spiritual concerns brought a unique culture of plants to line their thoroughfares and ornament their public and semi-private spaces—gardens, groves, towns, and human bodies.

Plants in this culture, especially woody plants, were almost like persons with desires and minds of their own. Each plant had unique ways of being-in-the-world, and its plant-hood interacted with human beings who in return responded with care and gifts. This basic notion of interaction between humans and plants set the stage for a highly codified courtly culture based on mutuality and gift exchange—in which ornamented bodies saw and were seen, gave and received, pleased and were pleased. Plant products became the currency for these types of exchanges, whose values rested in one's ability to constantly procure pleasure and refreshment especially through the most ephemeral gifts. Elites then valued the means of production—that is, the productive gardened landscape. Ultimately, especially in the south, the greatest gift would eventually become the grant of land.

This relationship between plants and humans was complex and can be evasive to historians who often represent it as a tense dialectic between the urban wealthy whose texts survive and forest dwellers who lived in a natural, sacred world. However, when plants are understood as beings, agents, and active participants in human stories, they can shed light on other discernible relationships between the city and the forest. Further work on plants and their cultural landscapes can also illuminate the historical facts recorded in South Asian sources.

ACKNOWLEDGEMENTS

I thank Annette Giesecke for inviting my contribution; my daughter, Mahi Céleste Dumas, for her astute suggestions on which good stories to include; two teachers: my father, Narendra Kumar, for sharing his perspectives on Indic sources throughout my life, and Michael Meister, for training me to share mine through scholarship. Among the mentors, colleagues, and friends who read drafts, three deserve special mention: Deven Patel for his suggestions, corrections, and unwavering presence; Rory Turner for his supportive comments around sticky issues of culture; and Anastasia Amrhein for her careful editorial feedback and thoughts on early landscape. I am also grateful to Daud Ali for his ongoing contributions to the relevant literature and well-founded critiques of this essay; Darshan Shankar, S. N. Venugopalan Nair, M. A. Alwar, and Varghese Thomas for their initial approach to this vast and untrammeled subject; and David Mabberley for sharing his unfaltering botanical knowledge. All mistakes remain my own.

CHAPTER SEVEN

Plants as Natural Ornaments

Pre-modern Iranian and Eastern Islamic Lands

YVES PORTER

Vegetal ornament in Islamic art is often grouped with non-naturalistic ornament, both being referred to as "arabesque," literally "in the Arab style," and defined as "a form of artistic decoration based on rhythmic linear patterns of scrolling and interlacing foliage, tendrils, or plain lines" (Fleming and Honour 1977). As we will see below, this term is not nuanced enough, effacing as it does the various and variously derived forms that Islamic ornament could take (Komaroff 2010; Kühnel 1949). The fact that visual idioms developed in the immense geographic expanse of the lands of Islam, and through millennia, allowed for great variety of expression (Bernus-Taylor 1989: 9–10). Vegetal forms were employed in numerous decorative contexts, ranging from pottery to architecture, and in an equally wide assortment of media (including stone, clay, wood, ivory) in the Middle East since protohistoric times (see also Chapters 7 and 8 of Volume 1 in this series). In the course of thousands of years, these motifs experienced periods of simplification, sometimes merging into abstraction. Indeed, flowers and plants were seldom treated in a naturalistic way before the Modern Period (*c.* 1500–1800), when some trends coming from the European visual arts were adopted, mainly through printed herbarium-plates.

Our purpose here is neither to rival Owen Jones' *Grammar of Ornament* (1856) or Alois Riegl's magisterial *Stilfragen* (1893), but rather to offer a brief, new overview of the ways vegetal and plants motifs evolved and were used in the arts of Eastern Islamic lands as well as to suggest how such ornament can inform our understanding of long-lost plantings, "living" ornaments, in early Islamic gardens. The question of the meaning or symbolism of vegetal ornaments, be it for instance the all-covering palmette-scroll, the cypress, the palm tree, or the so-called "*waq-waq*" motif, not to mention the elusive "tree of life," has generated an important literature (Grabar 1992). The "tree of life" issue is a particularly thorny one; if some symbols, especially in Christian iconography, are recognized/recognizable, most of the ancient "oriental" examples do not necessarily correspond to a codified "symbol of life." On the other hand, the Zoroastrian religious texts mention several trees (see below), but never a specific tree of life. Unlike the "symbolist" approach, other methods of analysis, such as the "formal occasion," which studies subjects from the point of view of their location (on a wall for instance), and not of their meaning, have rarely been applied to the Islamic arts (the "formal occasion" method has notably been studied, in the frame of European art, by H. Focillon, L. Réau, and O. Pächt). These two approaches oppose thus the view of the patron, with its consequent *Kunstwollen* (or "artistic volition"), and that of the craftsman in the practice of his art.

In this vast domain we will focus here on the Iranian world and its neighboring areas, such as Turkey and Central Asia, with a brief excursus into Northern India. After a survey of pre-Islamic Iran, we will proceed through the first centuries of the Caliphates, before examining proper Islamic Iranian examples in a chronological sequence. Some surveys on specific forms or vegetal species (the palm tree, the *waq-waq* motif, the "Chinese" lotus, for instance), will allow for highlighting their development in a longer time-span and over an expanded area.

THE PRE-ISLAMIC IRANIAN SUBSTRATA

Islamic art is frequently said to offer a distanced or remote vision of Nature so as not to imitate divine Creation. However, naturalistic depictions of flora certainly did exist in Iran well before the advent of Islam. Moreover, the pre-Islamic decorative repertoire yielded important references for the arts of subsequent periods. Before entering the domain of Islamic art, it is thus necessary to survey some aspects of the previous eras. Before Islam, flowers, trees, and other vegetal forms were ubiquitous in Iranian mythology, poetry, and art. When we look at the beautifully carved reliefs of Persepolis (sixth century BCE), we might be amazed to see, amidst the guards, ambassadors, and princes, rows of cypresses (*Cupressus sempervirens*) bordering the monumental staircases of the Apadana audience-hall representing a kind of eternal garden (see Figure 7.1). Yet, we often view the lands of

FIGURE 7.1 Persepolis (sixth century BCE), staircase of the Apadana. © Yves Porter.

the Iranian world as an arid area, dominated by the desert. In fact, five million years ago, an inner sea covered an important part of what is now the Iranian plateau (Ghirshman 1951: 24). Rivers flowed down from the mountains and reached this sea. The first human settlements were initially in the upper valleys, but they later descended towards the lower parts of these rivers, following the progressive drying of the sea. Nowadays only the Salted Lake remains as a trace of this sea, and many of the rivers have dried. The protohistoric sites of Sialk, Giyân, Hurvin, or Hissar, some dating back well before the third millennium BCE, were situated on the sea's ancient shore.

The Iranians probably came to the plateau, in a succession of waves of the so-called Indo-Iranian migrations, starting during the second millennium BCE. They brought with them their mythologies and religions, which included numerous references to fertility, plants, trees, and miraculous essences, as some bronze or silver artifacts from Luristan (eighth–seventh century BCE) attest (Ghirshman 1963: 52, 56–7). It is thus inappropriate to define the civilization of these first Iranians as deeply conditioned by the desert-like features of their final location. After the drying of the rivers, the inhabitants of the plateau developed an ingenious way of catching and collecting groundwaters. Originally, the miners realized that while digging galleries (underground passages), water tended to accumulate and had to be evacuated. This observation led to the invention of the underground water-channel system known as *qanat* (or *kariz*) (De Planhol 2011), which provided irrigation for agriculture but also for the development of ornamental gardens.

Little is known about the arts of the Medes, an ancient Iranian people whose empire arose during the seventh century BCE, but the so-called *Oxus Treasure*, now in the British Museum, assuredly sheds light on this period (Ghirshman 1963: 90–4). Among the artifacts constituting this Treasure are engraved gold plaques depicting men in what has been identified as Median clothes holding bunches of *barsom* twigs, a typically Iranian item. *Barsom* twigs, coming from the *haoma* tree (sometimes identified as the pomegranate), were essential in the religious rites mentioned in the *Avesta*, the sacred text of the Zoroastrians (Kanga 1988; Taillieu 2003). From a a slightly earlier period the helmet of the Urartian King Sardur (r. 764–735 BCE), found at modern Karmir Blur, Armenia (which became part of the Persian Empire), displays motifs of winged divinities flanking a kind of tree, very reminiscent of Mesopotamian figures (Ghirshman 1963: 91). Indeed, the art of the pre-Achaemenid peoples of Iran was as a blend of older, Middle Eastern elements together with new trends brought by newcomers to Iranian lands. These characteristics also appeared during the Achaemenid period (sixth–fourth centuries BCE), which was issued in by Achaemenid King Cyrus the Great, founder of the vast Persian Empire that extended from the Mediterranean to the Indus River.

The art of the Persian dynasty known as the Achaemenids (*c.* 700–330 BCE) drew its manifold features from as many sources: besides its own endogenic patterns, some others came from the Mesopotamian Elamite civilization, together with that of Babylonia (the so-called "tree of life"), to which Greek Hellenistic elements were later added (the Greek palmette, for instance). Besides the already noted cypress, vegetal motifs notably included lotus flowers (*Nymphaea* spp.) and buds, as we can see on the representation of the *Shahinshah* ("King of kings"), or of Persian dignitaries holding a lotus flower sculpted at Persepolis, which became the ceremonial capital of the Achaemenid Empire (*c.* 550–330 BCE) lying about 60 km (40 miles) northeast of modern Shiraz. These and other lotus decorative motifs are likely derived from the Egyptian artistic repertoire, but other vegetal motifs were surely inspired more directly by nature. The Achaemenid kings famously had expansive walled enclosures, planted with a multitude of trees, where they liked hunting. Greek authors named them *paradeisos*, using a term derived from the Avestan *pairi-daeza*, and originally meaning a "hunting-enclosure"—or simply "enclosed space," as the term extended also to ornamental gardens with flower- and fruit-tree-filled planting beds such those at Pasargadae, the capital of Cyrus the Great, Persepolis' predecessor. Our modern word "Paradise" derived from this old-Iranian name. In biblical Hebraic, the word passed as *pardes*, while in Arabic, it became *firdaus*, and was used in the Quran for the description of Paradise (Porter 2007: 638).

The invasion of Alexander the Great (356–323 BCE), who spread Greek culture through the many lands he conquered, signified the end of the Achaemenids; however, the following period (known as Seleucid, 305–188 BCE in Iran and Mesopotamia), acknowledged a certain Achaemenid continuity intermixed with the Hellenistic Greek presence. This, and the following dynasty (known as the Arsacid Parthians, 247 BCE–224 CE), are still considered to be obscure periods. The Sasanian period (224–651 CE), acknowledged as the second flowering of the Persian Empire, overshadowed its predecessors. There is no doubt that this period was decisive in the formation of the later Iranian and subsequent Islamic art. As in the previous periods, the visual arts continued using an idiom far removed from naturalistic depiction. The representations of plants in Sasanian art followed this trend. Decorative motifs inspired by vegetal forms can be seen on the pediments of Taq-e Bustan's arch (near Kermanshah, northwestern Iran, early seventh century): they form a symmetrically axial composition of stems, leaves reminiscent of the acanthus, and unidentifiable hybrid blooms (Degeorge 2010: 39). At the same time, inside this rock-excavated and fully carved monument, we also find hunting scenes, a favorite theme of Sasanian iconography. Here, for instance, appears the King Khosrow II (590–628 CE) shooting boar in the marshes, surrounded by what are unmistakably water-reeds. However, in the same way that we will find in much later depictions (like in the so-called miniature painting), these seemingly naturalistic scenes are devoid of depth, not to mention perspective. Nature is rendered in a quite two-dimensional, synthetic, and frequently repetitive way, with vegetal elements abstracted, often allowing them to assume symbolic values.[1]

Vine and grape-harvest motifs appear in various Sasanian metal vessels, offering yet another sort of vegetal motif; a gilded silver pitcher in the British Museum shows vine-scrolls with naked grape-harvesters, together with a variety of animals and birds. This theme is naturally evocative of the Dionysian iconographies inherited from the Hellenistic period and is thus ultimately of classical Greek derivation. The theme was used as a symbol of fertility and eternal life. But the Dionysian contents were transformed and adapted here to the needs of the Mazdean pantheon, which centered on the worship of Ahura Mazda, supreme god in ancient Iranian religion, especially Zoroastrianism. Indeed, images of Dionysus were largely diffused from the Mediterranean coasts up to Yemen and Afghanistan; they tended, however, to mix with proper Mazdean deities, often mingling their personalities, Dionysus appearing as a variant of the Iranian god Verethragna, god of victory in war. These motifs were also exported eastwards, to China, where we find them in Tang metal mirrors (Demange 2006: 100–1). Vine motifs were also largely used in the early Islamic art (Degeorge 2010: 142).

The *Spring of Khosrow* (first half of seventh century), a huge, late Sasanian royal square carpet about 30 m (approximately 100 ft) wide, is only known through textual sources but is of great importance in so far as it sheds light both on the design of—and plants in—Sasanian ornamental gardens and on representations of gardens and their plantings as ornament for indoor spaces, an eternally blooming garden. According to these literary descriptions, it featured plant-lined allées and streams of water embroidered with gems against a ground of gold. Streams and walking paths intersected with pavilions and delimited rectangular planting beds filled with flowers, grasses, and fruit-bearing trees. Expanses of grass were rendered with emeralds, and colored gems formed flowers having gold, silver, and silk foliage and stems (Morony 1988). This masterpiece of Sasanian art was torn to pieces following the fall of Ctesiphon (637), the Sasanian capital, when the Arab army invaded the empire, under the caliphate of Omar. The idea of depicting a garden on a carpet re-emerged in later periods; unfortunately, the earliest preserved examples are not older than the sixteenth century, during the Safavid period (1501–1732). However, the poetic metaphor comparing a garden or a meadow to a flourished carpet as a literary *topos* in early medieval Persian poetry appeared from the ninth century CE onwards (Porter 2004). As for the garden depicted on the Spring of Khosrow carpet, its design ancestry can certainly be traced to the paradise garden of Cyrus the Great at Pasargadae, and its general layout has been confirmed as representative of Sasanian royal gardens on the basis of archaeological finds at Emarat-i Khosrow, a seventh-century garden built by Khosrow II Parviz (591–628) in what is now Kermanshah Province.

In sum, the succession of pre-Islamic Iranian empires, of which several were particularly brilliant, led to a rich and composite repertoire of vegetal motifs merging their different sources. All these were available when the first Islamic dynasties came to power.

FROM THE LATE ANTIQUE HERITAGE TO THE FIRST ISLAMIC EMPIRES (SEVENTH–TENTH CENTURIES CE)

In 651, the murder of the Sasanian emperor, Yazdegerd III (r. 632–51), signified the end of the dynasty and the empire. Despite the Arab presence, no Islamic art can be fully recognized for several generations, Arabic writing being the most identifiable trace of this. A small pottery cup, now in the Louvre, with a square saucer decorated with vine scrolls found at Susa would easily be considered Sasanian if it did not have an epigraphic band in Arabic.[2] Metal vessels adorned with motifs derived from the Sasanian repertoire continued being made up to the tenth century.

As far as vegetal motifs are concerned, the Late Antique arts of the Mediterranean regions continued using decorative motifs such as acanthus, vine, olive-tree, or laurel leaves, which derived from the Greek and Roman decorative arts. In the early arts of Islam, however, these widely used motifs progressively tended towards simplified forms. We find several references to trees (*ashjar,* this word also referring to some kinds of shrubs; Ballanfat 2007: 78–9) in the Quran. However, since the sacred text was necessarily not illustrated, the vegetal species described in it did not have a specific iconography. Moreover, the very first decades of Islam, that is, from 622 to the establishment of the Umayyad caliphate (661–750), were a period of a certain artistic austerity. The Prophet Mohammad himself was probably reluctant to display any sign of vanity, and the contents of some chapters (or *Surah*) of the Quran make clear that ornament is intrinsically a signifier of pride (Porter 2007a: 126–7).

The Umayyads (661–750 CE)

In contrast to the first decades of Islam, the Umayyad empire represented a disruption in the political and economic bases of the caliphate. The importance of agriculture in the newly conquered lands placed it at the core of the empire's wealth. The immensity of a territory including heterogenous populations both required a new form of organization to control this territory and generated a cultural porosity between the different communities residing in it. These facts gave birth to a process of land-marking through the construction of secular and religious monuments. These buildings were erected and embellished by local craftsmen, who were naturally inclined to use their own techniques and decorative repertoires. Mixing typical local with exogenous elements allowed for the renewal of the artistic and decorative references, without necessarily being linked to any definite Umayyad *Kunstwollen*.

By the end of the seventh century we can observe a definitive tendency in this regard. The Dome of the Rock, in Jerusalem, built in 691 during the caliphate of Abd al-Malik (685–705), represents the first real achievement in Islamic architecture. The mosaics covering its inside upper parts, made according to a Byzantine-based technique of gilded or colored glass cubes, gave a large role to vegetal ornaments (see Figure 7.2). These fall into two categories: some motifs consist of relatively "naturalistic" representations of palm trees (*Phoenix dactylifera*), vine scrolls, or acanthus leaves, while others figure non-natural species, in the shape of hybrid vegetal creations, reminiscent of those found in Sasanian art.[3] Sasanian textiles were probably the favorite means of transmission for such patterns. Iranian silk brocades were later imitated in the Byzantine area; from there, Sasanian motifs found their way up to the westernmost medieval Europe, where we still can see their trace in twelfth-century Catalan Romanesque decorative arts, for instance.

Much has been said regarding the "symbolic" meaning of these vegetable motifs, the prefiguration of Paradise being among the proposed hypotheses (for a synthesis, see Grabar 2006). Indeed, other approaches, such as the "formal occasion"-approach

FIGURE 7.2 Jerusalem, Dome of the Rock (691 CE), mosaic of the inner drum. © Yves Porter.

mentioned at the beginning of this chapter, have seldom been applied to the study of these motifs. It is striking that the tree-form fits remarkably in the almost triangular space of the spandrels between the arches, or between the high windows of the cupola's drum. These spaces to fill were ideal opportunities for experimenting with patterns deriving from a variety of repertoires, thus allowing for a cultural porosity which further enriched early Islamic art. However, it is not necessary to postulate that the motifs retained any of their original meanings in this transfer process. This implies that the motifs were not semantically determined. This process translated into an extreme liberty in the use of varied patterns, among which vegetal ornament was a favorite.

In contrast with the vegetation of the Dome of the Rock, we find more naturalistic plants and trees in the slightly later mosaics of the Great Mosque of Damascus (started *c.* 705) (Degeorge 2010) (see Figure 7.3). Here, too, trees occupy several of the spandrels in the building. The façade of the prayer-room's axial nave displays a landscape with buildings; the upper corners of the wall are occupied by trees depicted to a much larger scale than the buildings so as to fill the available space, with the bent leaves and branches thus following the architectural lines. The western portico of this mosque, which opens onto the vast courtyard, has its back wall entirely covered with an over 34-meter-long (100 ft) panel featuring a landscape on the shore of a river. This section is sometimes named the Barada Panel, after the name of the river passing by

FIGURE 7.3 Detail of mosaics, West portico, Great Mosque of Damascus, Syria. The mosque was originally constructed by the Umayyad caliph al-Walid from 705–15 CE. The mosaics, partially restored, are original Umayyad designs, showing trees and buildings. This is part of a portion that is also called the Barada panel as some medieval commentators saw in it the river which runs through Damascus. Photo by Bernard O'Kane/Alamy Stock Photo.

Damascus. The composition alternates architectural ensembles with oversized shrubs and trees; their foliage is rendered in different hues of green suggesting volume, but the trunks are generally rendered with bands of colors in a much less naturalistic approach.[4]

The façade of Mshatta's palace (present-day Jordan), now in the Berlin Museum of Islamic Art, is also representative of this transition period of late antiquity (Enderlein and Meinecke 1993; Grabar and Ettinghausen 1987: 67–71). Built at the end of the Umayyad Caliphate, the façade was planned to be covered with vegetal decoration, although some parts were left unfinished. Amidst the vine-scrolls, some animals are represented, either real (lion, bull) or fantastic, among which are winged horses, *senmurv* (mythological hybrid birds, half a lion and half bird-of-prey), and griffins. To the right of the door is a panel without animals, the central figure of which is a kind of palm tree. Its silhouette detaches itself from a background of non-symmetrical vegetal scrolls among which we can easily recognize grapes and vine-leaves. But other adjacent plant forms are unnatural, such as the finial with a pair of wings-motif, originally a Sasanian symbol (the royal crown) that had, over time, accrued a meaningless decorative function through its "vegetalization." The Mshatta "palm tree" itself has a checkered trunk; the "palms fronds" are actually a combination of palmettes and double-wings. This kind of hybrid vegetation already figured on the mosaics of the Dome of the Rock; however, in the Jerusalem monument, the double-wings were used both in the representation of crowns or trophies, and of fantastic trees.

The Palm Tree

Up to the present day, the "palm tree" motif has been perceived as emblematic of the Islamic lands; it figures, for example, on the flag of Saudi Arabia. Orientalist painters and theater set-makers have used it extensively, in an almost mnemonic or metonymic way, to evoke Middle Eastern countries. Almost-naturalistic examples can be found at the Dome of the Rock, or at the Damascus Great Mosque, particularly in the so-called Treasury. In later periods, this motif was widely used but mainly in a stylized form. Some ninth-century carved wood plaques in the *minbar* of Kairouan's Great Mosque figure a "tree" derived from this pattern.[5] A very simplified rendering can also be observed on a blue and white tin-glazed bowl from ninth-century Iraq, now in Los Angeles (see Figure 7.4).[6] On the *Pyxis of al-Mughira* (Spain, dated 968), we can see a couple of horsemen plucking dates on both sides of the palm tree trunk.[7] This kind of pattern (i.e. a pair of horsemen, or a pair of lions, on both sides of a tree), was already used on late Sasanian and Byzantine textiles, although it was rendered in the stricter symmetry commanded by the weaving technique. It continued to appear on tenth-century Iranian textiles.[8] Later derivations of the palm tree pattern could include only the upper part of the tree, forming a fleuron-like design, appearing in diverse media: book illuminations, textiles, ceramic, and carved stone.[9]

The Abbasids, Samarra, and the So-called "Arabesque"

The Abbasids seized power in 750, bringing the Umayyad caliphate to an end. Their dynasty lasted up to the Mongol invasion in 1258. Little remains of the first city of Baghdad, founded in *c.* 762, and not much more from the decorative arts of this period. By contrast, their "second" capital, Samarra, situated about 130 km (80 miles) north of Baghdad, has preserved numerous vestiges of its occupation during the ninth century.

FIGURE 7.4 Palm tree motif. Earthenware bowl, Iraq, ninth century CE, inglaze-painted. Height: 6.66 cm (2⅝ in.); diameter: 23.49 cm (9¼ in.). The Madina Collection of Islamic Art, gift of Camilla Chandler Frost (M.2002.1.34), LACMA. Photo courtesy of LACMA.

Archaeological explorations started in the first decade of the twentieth century, mainly under the authority of Ernst Herzfeld (1879–1948). The five volumes of his monumental *Ausgrabungen von Samarra* were published, with contributions by F. Sarre and C. J. Lamm, between 1923 and 1930. Notably, Samarra was discovered to have been the site not only of a series of palaces but also of these complexes' expansive ornamental gardens. Ninth-century geographer al-Ya'qubi noted that all the land had been converted to palaces and well-watered gardens, including manicured fields for playing polo, for the enjoyment of the upper classes. Among these palaces was the Balkuwara Palace, built in the mid-ninth century. Now desertified and devoid of vegetation, its planting schemes un-recovered, this palace assuredly boasted luxurious plantings of native and imported plant species. Traces of the gardens' hardscape—sequences of planted courtyards, fountains, and water channels—intimate their former splendor (Hobhouse 2004: 74; Ruggles 2008: 26–7).

Among the many artifacts coming from the exploration of the palaces are some fragments of mural paintings, together with pieces of stucco decoration that scholars have sorted in three different styles (see Herzfeld 1923; for a synthesis on this question, see also Grabar and Ettinghausen 1987: 102–5). Most of the patterns draw their inspiration from vegetal forms; however, due to their over-simplification, they definitively tend to abstraction. In the early twentieth century, Herzfeld (1913) proposed that these stylized vegetable forms were the origin of the arabesque. This interpretation was challenged later (Komaroff 2010). In general terms, we could admit that the endless scroll-motif, or at least the repetition of a pattern, with a predilection for symmetric, all-covering, compositions, are among the principles of the so-called "arabesque," in spite of its non-naturalistic tendency. However, the variety of motifs used at Samarra makes it difficult to identify Samarra's decoration solely with the term "arabesque."

The vine-leaf enclosed by an almost circular stem is representative of the so-called Samarra Style A. This design is clearly inherited from the preceding period, although it is now simplified, and a renewed attention is given to the rhythm of the general composition.[10] The way of carving is reminiscent of the drilled technique that can be observed on Byzantine capitals, for instance. The Samarra Style B is smoother than the former, and even less naturalistic, since the vegetal components are not recognizable anymore; palmettes, trefoils, double-wings, and a kind of yin-yang motif, are among its most popular patterns. The Style C is also named "beveled" because of the way of carving without sharp edges; its formal characteristics are abstract patterns of leaves and palmettes encompassed in all-over, repetitive, and symmetric compositions (see Figures 7.5a, b and c). These stuccos from Samarra had a tremendous impact in a large

FIGURE 7.5a Samarra "Style A" bas-relief, from Samarra Palace (Iraq), ninth century CE. Museo Arqueológico Nacional-Colección, Madrid, Spain. Photo by Album/Alamy Stock Photo.

FIGURE 7.5b Samarra stucco "Style B," Samarra, Iraq, ninth century CE. Museum of Islamic Art, Pergamon Museum, Berlin. Photo by Richard Mortel/Flickr. Courtesy of Creative Commons.

part of the Islamic world, as they were reproduced not only on their original material (mural painting and stucco decoration), but also on wood or ivory, as well as on ceramics or metals.[11] In sum, the development of the Abbasid decorative repertoire attests to a trend of growing abstraction in the treatment of the motifs, enhancing their rhythmic decorative effect without a view to naturalistic depiction. This kind of normative formality coincided with the development of the standardization of the Arabic "six styles" of calligraphy.

The Advent of Autonomous Aesthetics in Iran

The architecture of the first centuries of the Arabo-Muslim occupation in Iran has, for the most part, fully disappeared. Very little is known of these lands under the Umayyad and early Abbasid Caliphates. From the ninth century CE onwards, some buildings have preserved parts of their decoration. This is the case of the Noh-Gumbad mosque, in Balkh (present Afghanistan). This now half-ruined brick structure was decorated with stucco ornament having a variety of vegetal-inspired motifs reminiscent of the patterns used in Samarra's A and B styles. Slightly later, the Friday Mosque of Na'in displays some stucco decoration that can be dated to the tenth century CE. In addition to epigraphic bands, this includes multilobed roundels and geometric compositions filled with a simplified version of the vine-leaf with circular stem. The spandrels of the arch in front of the main mihrab display a kind of fleuron also reminiscent of Samarra.[12] The *mihrab*, for its part, has a spectacular tympanum carved in high-relief with pomegranate-like fleuron and palmette.

During the ninth and tenth centuries, local dynasties started exerting a degree of autonomous power. The reappearance of new, specifically Iranian kingdoms favored the advent of original creations in the fields of both artistic and literary production. This period

FIGURE 7.5c Samarra stucco "Style C," the flat cut "bevelled" or "slant" style of carving with stylized ornaments repeated in endless series, Samarra, Iraq, ninth century CE. Museum of Islamic Art, Pergamon Museum, Berlin. Photo by World History Archive/Alamy Stock Photo.

is known, mainly in the domain of poetry, as the Iranian Renaissance. The Samanids (819–1005), reigning over eastern Iran and Transoxiana (roughly corresponding to present-day Uzbekistan and parts of Turkmenistan, Tajikistan, and Kazakhstan), were particularly representative of this phase. Not much is known of their monuments, since most of them were destroyed during the Mongol invasions, in the first part of the thirteenth century. Yet, the excavations at Nishapur, under the supervision of The Metropolitan Museum in New York, yielded some examples of mural decoration in both stucco and painting. Some motifs are definitively of vegetal inspiration, mixed, however, with curious designs resembling human hands and eyes.[13] Vegetal motifs in the form of leaves, palmettes, and florettes are quite ubiquitous in the crafts of this period, on ceramics, metals, or textiles; their general feature, however, is to be heavily stylized towards an almost abstract simplification (Bernus-Taylor 1989: nos. 6–8).

Only a few illuminated manuscripts of this period are preserved; they are primarily copies of the Quran. One of them, written on parchment and dated to *c.* 950, displays a double-page frontispiece mainly decorated with geometric interlaced patterns (intersecting circles, tresses, knots). However, the margins show some fleuron formed by palmette motifs and double-wings derived from older patterns. Half a century later, another Quran was copied on paper, at Baghdad, by the renowned calligrapher Ibn al-Bawwab. Its frontispieces exhibit elegant palmette-scrolls from a renewed repertoire at the center of the composition, while the side fleuron retains the archaic features of the preceding example.[14]

Although there are only a few known remains of the visual arts of this period, contemporary Persian poetry provides many examples of plants, flowers, and trees mainly used as metaphors and surely reflecting plants found in contemporary ornamental gardens. The cypress undulating with the wind is compared with an elegant gait or with a slender silhouette. Red roses, like tulips, pomegranate flowers, and fruits, as well as many other reddish plants, are metaphors for the lips and/or cheeks of the beloved. Yellow roses, on the other hand, might signify the love-sick lover, while narcissus evoke the eyes. Ebony is used to qualify the color of the hair or of the eyes; hair is also compared with the willow branches. As we will see later when discussing manuscript painting, the poets most usually describe Nature during spring, devoid of shadows or deepness (Fouchécour 1969: 251).

THE GREAT EASTERN ISLAMIC EMPIRES: A GATE TO NEW INSPIRATIONS (ELEVENTH–FIFTEENTH CENTURIES CE)

From Anatolia to the Indian subcontinent, and passing through Iran and Central Asia, the expansion of new empires yielded a renewal of visual references to ornamental motifs from other cultures, either through invasions or commercial and other contacts.

The Turkish Dynasties and the Development of Palmette-Scrolls

The rise of the Turkish dynasties (Ghaznavids, 977–1186 CE; Seljuks, 1026–1194 CE) brought with it new trends in the arts of the Middle East. Vegetal ornament concentrated on the further development or evolution of the fully unnaturalistic palmette-scrolls which henceforth dominated the Islamic decoration. This was later known as *eslimi* in the Eastern Islamic lands, and *tawriq* (or *ataurique*, in Spanish) in the Arab countries, the Maghreb, and the Iberian Peninsula.

The earliest standing monuments of the Seljuks are relatively severe in their decoration, often limited to brick-based geometric designs, as on the cupola of Nizam al-Molk in Isfahan's Friday Mosque, dated 1072/92. Vegetal ornament developed at a later stage; several outstanding examples of Seljuk stucco can still be observed in situ. The Ardestan Friday Mosque (Central Iran, twelfth century) displays an elegant mihrab decorated with stucco; its tympanum has an axially symmetric composition of high-relief palmettes and half-palmette scrolls intersecting at various levels, creating the impression of an in-depth movement.[15] A monumental epigraphic band runs on the upper part of the square domed prayer room's wall; the inscription, written in an elegant *thuluth* style, detaches itself against a vegetal ground formed by endless palmette scrolls (see Figure 7.6). The Gonbad-e Alavian, in Hamadan (western Iran, twelfth century) is another remarkable example of Seljuk stucco: all the interior of the room in this tomb is covered from floor level to the octagonal transition zone with striking palmette-based stucco compositions (Shani 1994).

FIGURE 7.6 Ardestan (Iran), Friday Mosque (twelfth century CE): entry to the prayer hall. © Yves Porter.

The very high-relief stuccos found in Seljuk Iran find an echo in their Anatolian counterparts, which are carved in stone, however. One of the most amazing examples of this kind of sculpture are the portals of the mosque-cum-hospital at Divriği (modern Turkey, 1228–9). Among the huge protruding fleuron and floral arabesques in the decor of this monument, we also find curious combinations of vegetal and animal forms, which call to mind some Celtic illuminated manuscripts (Grabar and Ettinghausen 1987: 325).

The Ghaznavid Dynasty had its origin in the Turkish slave-officers who served the Samanids. It was mainly under the reign of Sultan Mahmud (998–1030) that this lineage reached its climax, especially thanks to a series of expeditions towards the Indian subcontinent. The architectural decoration of the Ghaznavids is mainly known through the ruins of their capital palace, Ghazni, in present-day Afghanistan. Among the most spectacular remains are a series of carved marble panels coming from the palace of Mas'ud III (r. 1099–1115); some of these are decorated with animated scenes, while others display only vegetal and geometric decorations in addition to the first monumental inscription in Persian (Ruggiadi 2010).

The Waq-Waq *Tree*

A marble panel in the Ghazni Museum is probably one the first examples of vegetal scrolls with human and animal finials, mixed here with griffins, harpies, and sphinxes. Unfortunately, most of the artifacts preserved at Ghazni Museum are now lost, due to a Taliban blasting in 2014; moreover, information on the present state of these collections is almost non-existent. However, the IsMEO (Istituto Italiano per il Medio ed Estremo

Oriente) has some photographs of items in the collection in their archives (Ruggiadi 2010). A slightly later example (late twelfth or early thirteenth century) of a similar motif can be seen in a stone-paste, pear-shaped, and cock-headed pitcher in Paris; it displays a scroll-pattern painted in black against a turquoise ground; its finials are also human and animal heads.[16]

This kind of animated scroll with animal or human head finials continued to be used in later periods. It is often misleadingly labeled the "*waq-waq* motif" (Bernus-Taylor 2001: 165–75). This name refers to the Waq-Waq Islands, described in several versions of the *'Ajâ'ib al-makhluqât* or *Marvels of Creation* (see Figure 7.7). The most famous version is Zakariya al-Qazvini's; the earliest illustrated manuscript is at Bayerische Staatsbibliothek, Munich.[17] Among many other marvels, these texts describe an island inhabited by trees the fruits of which are human heads (or full bodies, according to the sources) called "*waq-waq*." However, the relation between these trees and the development of this decorative motif is far from clear (Porter 1992: 115).

FIGURE 7.7 The Waq-Waq Tree, *'Ajaib al-makhluqat* (The Marvels of Creation), Zakariya ibn Muhammad ibn Mahmud Abu Yahya al-Qazwini (1203–83 CE), sixteenth century CE manuscript. Photo by CPA Media Pte Ltd/Alamy Stock Photo.

Later examples, dating to the last quarter of the thirteenth century, include some luster tiles with verses from Firdausi's *Shahnama* (Book of Kings) in an arched form, its spandrels being occupied by this kind of animated *waq-waq* scroll. As suggested about the Dome of the Rock's trees, the scroll motif is well suited to the spandrel form. However, the motifs on these specific tiles have been identified as evoking the Talking Tree appearing in the *Story of Alexander* included in Firdausi's *Shahnama* (Mélikian-Chirvani 1984: 305). While searching for the Source of Life, the Conqueror came across a Talking Tree which told him his fate.[18] Firdausi did not describe the tree as having head-finials of any sort, however, meaning that the iconography of the Talking Tree was borrowed here from that of the *Marvels of Creation*. This motif appears on other luster tiles as well, but also on book-bindings, carpets, and metal objects.

By the second half of twelfth century, the production of ceramics, mainly at Kashan (Central Iran), exhibits two significative improvements: the first is the so-called *mina'i* polychrome, low-fired, over-glaze painting; the second is luster decoration created using metallic oxides. Both these techniques provide us with a great number of objects—vessels and tiles—on which images abound. The ubiquitous checkered cypress is characteristic of this period. It is used in scenes depicting episodes from the *Shahnamah* epic, but also for more standard compositions such as rows of galloping horsemen on a large plate dated 1191 now in the Chicago Art Institute.[19] We also find in these ceramics the palmette-scrolls, often inhabited by leaf-like birds, as on a large dish dated 1211 now in Philadelphia.[20]

The oldest illustrated Persian manuscript which has come down to us, and is now in Istanbul, is a romance known as *Varqe & Golshah*.[21] The manuscript was probably copied at Konya, in central Anatolia, by the very end of twelfth century. Its illustrations display characters against a monochrome ground, sometimes adorned with a schematic vegetal item; the scenes are devoid of a landscape setting without the illusion of depth. The stories of the jackals *Kalila wa Dimna* come from an original Sanskrit version of fables (*Pancatantra*), translated first into Middle Persian during the Sasanian period, and subsequently into Arabic by the end of the eighth century. Unfortunately, the oldest illustrated manuscript of these fables, preserved at the Bibliothèque nationale de France in Paris, dates only to the early thirteenth century; earlier illustrated copies are known only through literary mentions. Probably copied and illustrated in present-day Syria, the Paris manuscript shows numerous outdoor scenes in which the animals feature.[22] The vegetation in these illustrations is very schematic. Its role is more geared towards balancing the pictorial scheme than faithfully reproducing the natural environment. Its presence might simply be related to the principles of the formal occasion, the choice of a motif fitting a space being determined by the general balance of the composition. This implies that the botanical species can hardly be identified. Curiously enough, this schematic treatment also appears in scientific illustrated manuscripts. The *Book of the Theriac*, copied in 1199 CE, offers botanical plates with different plants that are identified in the inscriptions or labels.[23] However, the rendering of the plants is so graphically simplified that it would prove difficult to recognize any of them.

The Mongol Flowering

From 1220 to 1258 CE, the eastern lands of the Caliphate experienced a disaster: the Mongol invasions, which culminated with the sack of Baghdad and the massacre of the Abbasid Caliph's family. The ensuing period, dominated by the Il-Khanid Dynasty, is sometimes qualified as the "Pax Mongolica." The three successors of Genghis Khan, represented by the Yuan in China, the Golden Horde in Central Asia and Siberia, and the Il-Khans in Iran, formed a kind of family-based alliance which made it possible, for instance, for the renowned Marco Polo to travel safely from Anatolia to Khanbaliq.

The impact of the Mongol period on Iranian art was enormous, particularly as the visual arts are concerned. It is during this period that landscape made its appearance in Persian painting. Preserved illustrated Persian manuscripts are very rare before this time. Among the oldest surviving Il-Khanid manuscripts, a copy of Ibn Bakhtishu's *Manâfi' al-Hayawân* (Benefit of the Animals) made towards the very end of the thirteenth century at Maragha (Iranian Azerbaijan), depicted the subjects in a landscape of rocks, shrubs, and trees which is reminiscent of Chinese painting (Schmitz 1997: 9–24) (see Figure 7.8). Tree trunks resemble the skin of a dragon, and strange mushroom-like rocks are scattered through the landscape. Persian artists quickly adapted the Far Eastern conception of landscape to their own world-vision. If some forms are reminiscent of those seen on Chinese artifacts, others—like the rendering of misty atmospheres in a tri-dimensional space, characteristic of Chinese landscape painting—are completely absent. Moreover, landscape as a painting genre by itself never really made its way into Persian painting, possibly apart from the "funerary landscape" discussed below. Additionally, if the rocky

FIGURE 7.8 "Pair of Eagles," folio from *Manafi' al-hayawan* (On the Usefulness of Animals) of Ibn Bakhtishu (d. 1058), Iran, *c.* 1300 CE. Ink, opaque watercolor, and gold on paper. Rogers Fund, 1918. Accession Number: 18.26.2, New York, The Metropolitan Museum of Art. Photo courtesy of The Met Museum.

landscapes of the late thirteenth–early fourteenth centuries demonstrate obvious Far Eastern influences, they were rapidly adapted in such a way that the settings appearing in the paintings of the second half of the fourteenth century completely integrated a truly "Persian" painting idiom.

The "Chinese" Lotus

In the decorative arts and crafts, the contacts with the Far East allowed for the adoption of patterns such as the lotus flower, together with its characteristic leaves. The Chinese lotus motif was probably first borrowed from Far Eastern silks, since these goods were a commodity very much sought-after in the courts of the Islamic lands (Crowe 2000). Blue and white porcelain developed somewhat later in China; its success in the Islamic lands was enormous. From the borrowed elements such as lotus flowers and leaves, marsh-weeds, and peonies (to which we could add the dragon and phoenix, but also the "Chinese cloud," and the rendering of waves, for instance), a new repertoire emerged, mostly composed of naturalistic flowers and plants, which were later known in Persian as *khatâ'ī*-style, that is the "style from Cathay" (Rawson 1984) (see Figure 7.9).

FIGURE 7.9 Tile with image of phoenix and Chinese-inspired floral ornament; from Iran, probably Takht-i Sulaiman, thirteenth century CE. Stonepaste; modeled, underglaze painted in blue and turquoise, luster-painted on opaque white ground. Rogers Fund, 1912. Accession Number: 12.49.4, New York, The Metropolitan Museum of Art. Photo courtesy of The Met Museum.

This differed from the "Islamic" or *eslimi*-style (for a discussion on the *eslimi-khata'i* styles, see Porter 1992: 112–13; Porter 2000: 113–14). The geography of the dispersion of such an artistic idiom in the lands of Islam during the fourteenth century spread over an immense area, from territories around the Mediterranean Sea (Turkey, Syria, Egypt) to the sultanate of Bengal (Porter 2016). The first representations of the "Chinese" lotus in Iran probably appeared at the Takht-e Soleyman palace (northwestern Iran), built *c.* 1275.[24]

The Timurids: The Complex Deepening of a Decorative Repertoire

Timur (or Tamerlane, 1336–1405) rose to power after the disintegration of the Mongol empires, in the last quarter of the fourteenth century. The dynasty that he founded—i.e. the Timurids (1370–1506)—lasted up to the first years of the sixteenth century in Central Asia. A branch of this family stayed in power much longer in India: the Moghuls (1526–1857). The Timurid artistic production has often been compared to that of the European Renaissance. However, biased as this comparison might be, we must recognize that the achievements of Timurid arts are overwhelming, not only in monumental architecture, but also in manuscripts and objets d'art.

The Timurid princes proved remarkable patrons for the creation of books. Their ateliers, first based in Samarkand and Herat, produced many illuminated and illustrated manuscripts. These princely workshops (*karkhana,* or *kitabkhana*) were organized with an almost military hierarchy; they produced designs that were later used for book illumination or book-bindings, as well as for the decor of buildings, carpets, or other objects (Lentz and Lowry 1989: 159–232). This served to promote the dissemination of motifs in various artistic media, also their diffusion from their original homeland, from Ottoman Turkey to the Indian subcontinent.

Timurid manuscript paintings (or "miniatures"; this word is ambiguous, since it today describes "something small," whereas its etymology comes from "minium" [lead oxide], thus quite distinct from a question of size) are often considered as the Persian painting par excellence. Vegetal species appear abundantly on these pages, either as elements of the landscape where an action takes place (the setting), or in the marginal decorations and on book-bindings. As noted earlier, plants are seldom fully recognizable, however. Notable exceptions are the cypress, and the chinar or Oriental plane tree (*Platanus orientalis*), for instance; other species might be more loosely identified, such as junipers, almond-trees, or poplars. The painters show meadows in full-light, in a kind of eternal spring, without shadows, where cherry and almond trees are always in bloom. Time is as if negated, since there are no variations of luminosity, of seasons, or of day-hours. Flowers are often difficult to identify with accuracy, although hollyhocks, lilies, iris, or peonies can sometimes be more easily detected (Lentz and Lowry 1989: 117).[25]

In architecture, many of the monuments of Tamerlane's period distinguish themselves by their large size. This is the case of the so-called Bibi-Khanum mosque in Samarkand, or the Gur-e Mir tomb, in the same city. They are mainly decorated with colored tiles with vegetal ornamentation—mostly in an abstract form (see Figure 7.10).

However, smaller buildings, such as the series of tombs in the Shah-e Zende necropolis, also in Samarkand, likewise display treasures of refinement in their ornament. The spandrels of the entrance arch in the façade of Shad-i Mulk Agha's tomb, dated to *c.* 1371, show an extraordinary composition based on lotus flowers intermixed with foliated scrolls, under-glaze painted in white against a cobalt-blue ground. Other tombs in the same ensemble offer examples of ceramic panels adorned with floral compositions

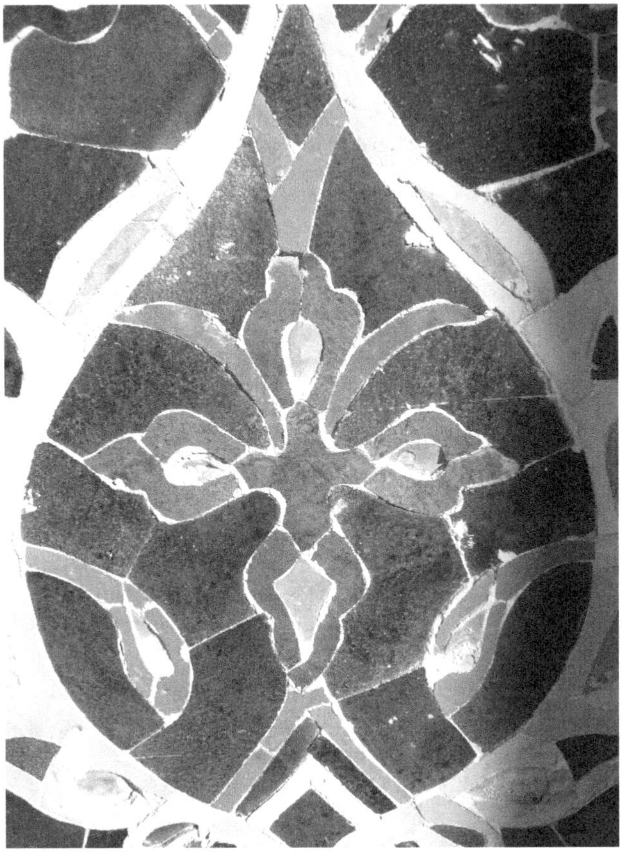

FIGURE 7.10 Samarkand (Uzbekistan), Madrasa of Ulugh Beg (early fifteenth century), tile mosaic. © Yves Porter.

also based on the lotus-flower motif (Soustiel and Porter 2003: 89–95, 130), which were later imitated in fifteenth-century Ottoman architectural decoration (Porter and Degeorge 2001: 195).

A specific vegetal-ornament theme developed by the end of the fourteenth century CE in Central Asia: the so-called "funerary landscape" painted on tomb walls (Golombek 1993; Soustiel and Porter 2003: 238–40). Although many of these frescoes have faded since the early twentieth century after they were photographed, we can still see trees (cypress, chinar) painted in blue against a white ground and animated with flying magpies and other birds. Such landscapes in funerary contexts clearly evoke the promised Paradise. In the first decades of the fifteenth century, we find in Syria and Ottoman Turkey some examples of blue and white painted tiles showing trees, birds, and plants, which are reminiscent of these mural paintings (Porter and Degeorge 2001: 188–93, 197).

Timurid gardens were, like miniature paintings, highly illustrative of the aesthetics of this period. Very little actually remains of them, however, in Iran, Afghanistan, or Central Asia. Their best representative descendants are certainly the Moghul gardens, built by their successors in Afghanistan and northern India. Their later Iranian counterpart is represented by the few remaining, and greatly transformed, Safavid gardens (Porter 2016b).

THE INDIAN SULTANATES: ADAPTING NEW NEEDS TO VERNACULAR FUNDAMENTALS

The Sultanate of Delhi was founded at the very end of the twelfth century, following the raids of the Ghurids (*c.* 1000–1215 CE), a dynasty coming from present Afghanistan. The Delhi sultanate lasted up to the devastation caused by Tamerlane's invasion in 1398. Its territory was divided between several independent sultanates, each developing its own artistic personality.

The art of the Indian sultanates has often been considered as a mere offshoot of the Turco-Iranian culture. This is of course an over-simplified interpretation. While the newcomers tried to impose Islam as the dominant religion, they also relied on local craftsmen and artists to realize their architectural and artistic ambitions. We see the local Indian decorative repertoire intermixed with Arabic inscriptions, along with the re-use of pre-Islamic elements, on the first important monuments built at Delhi. Among the vegetal-inspired motifs, the lotus plays an important part, either as the full-blooming flower or as its bud. It appears, among others, on the screen built as a façade for the prayer-hall of Delhi's Qubbat al-Islam mosque, started in 1199. Later examples, such as the gate known as 'Ala'i Darwaza (1305), also included in the Qubbat al-Islam precincts, display elegant friezes of lotus buds in rows of spear-like intrados, or in delicately stone-carved reliefs. These lotus-motifs continued to be used, in both their forms, for a very long period. Other, external influences, either coming directly from China or passing through Il-Khanid Iran, reached as far as the sultanate of Bengal (1336–1576). This is the case of the extraordinary lotus decoration found at Pandua's Adina Masjid, completed in 1375. The decorative repertoires of the Indian sultanates were nourished by a complex sum of elements allowing for an extreme diversity of expressions (Porter 2016a: 175, 176, 184).

The Trees from Gujarat

One of the most amazing contributions to the Islamic sultanate architecture in India certainly was that of Gujarat. This topic is particularly of interest as far as vegetal designs are concerned. The mosques of Gujarat display a series of decorative motifs based on representations of trees that are unique in Indian art. A series of marble-carved tombs made at Cambay in the 1330s display rows of repetitive, yet easily recognizable banana trees, obviously inspired by the illustrated contemporaneous Jain manuscripts, made in the same area (Lambourn 2004: 99–133) (see Figure 7.11). By contrast, the Great Mosque of Ahmedabad, built in 1424 CE, is decorated with remarkable representations of trees on its prayer-hall façade, each one being different. Placed in architectural niches, some of these trees grow in asymmetrical form from an undulating trunk, still retaining something of a "naturalistic" aspect. Other examples, by contrast, show an exuberant foliage made of flame-like whirling shapes. Later examples, such as on the pediments of the Nagina Masjid Tomb (*c.* 1525) in Champaner, exhibit an extraordinary series of variations based on the pattern of a creeper-like undulating stem. These Gujarati vegetal ornaments culminated with the pair of *jali*-screens adorning the *qibla*-wall (back wall of the prayer hall) of the Sidi Sayyid Mosque, in Ahmedabad, dated to 1571. The expansiveness of the layout is unparalleled here: both *jali*-s display a variation on the theme of two different species of trees embracing each other: one has a single palm tree against an all-covering undulating foliated tree, while the other shows four palm trees alternated with three undulating trees.

FIGURE 7.11 Cambay (Gujarat, India), Jami' Masjid (Friday Mosque, *c.* 1330), carved funerary monument. © Yves Porter.

CONCLUDING REMARKS: LOOKING BACKWARD AND FORWARD

Although vegetal ornaments took very varied forms in the lands of Eastern Islam, as we have shown, most of them are admittedly represented in a non-naturalistic way. The progressive contact with other artistic idioms, such as Far Eastern aesthetics after the Mongol invasions, and later those coming from the expanding countries of Europe, contributed to progressively transform this vision. In the last quarter of the sixteenth century, for instance, we find numerous examples of fully recognizable species of flowers such as tulip, rose, carnation, hyacinth, and cypress, together with other non-natural plants in the Iznik under-glaze painted pottery. By the same period, naturalistic miniature paintings started being produced in Moghul India under a clear influence of European engravings and herbarium-plates. By the early seventeenth century, the design of plants inspired by these herbarium-plates flourished in architectural decor, as well as on other media such as textiles. This process can also be observed in Safavid Iran, though at a slightly later date. We must stress, however, that if the forms of the plants are quite accurately rendered, the projected shadows, for instance, are never represented. This somewhat two-dimensional, fully-lighted space vision, characteristic of the arts of the Middle East, persists to the present day. We might add that the palmette-scroll (or arabesque), which probably emerged during the reign of the Seljuqs, also continues to exerting its fascination in the works of modern *ornemanists,* as many contemporary monuments demonstrate.

CHAPTER EIGHT

The Representation of Plants

Mediators of Body and Soul

SARAH R. KYLE

A curious early fifteenth-century "found" object reflects the plurality of roles plant imagery played in medieval culture. It reveals visually the vital roles of plants both in spiritual life and in bodily health and well-being. The *Root Christ* or *Christ of the Vine* (*Cristo de la Cepa*; see Figure 8.1), now housed in the Museo de la Catedral, in Valladolid, Spain, was considered a kind of *acheiropoieton*-object ("not made by human hand"). Donated to the Benedictine monastery of San Benito in Valladolid, the *Root Christ* consists of a crude figuration of Christ-crucified as shaped from the root and rootlets of a grapevine. It was found, or so the legend goes, by a wealthy Jewish man from Toledo who, while out tending his vines, unearthed the figure from a patch of weeds (Robinson 2006: 415n60). The man fell to the ground weeping and immediately converted to Christianity. From the vantage point of the Christians who tell the tale, the man's soul was saved, and his conversion became an exemplum to other non-Christians in Castile (Robinson 2006: 415–18). In the *Root Christ* we find an example of how plants and their representations possessed an ability to cure the spirit in a manner similar to the healing miracles of Jesus and the saints. Yet implicit in the very form of the *Root Christ*, as a figure emerging from the roots of a medicinal plant, we also find a set of remedies for the body.

The *Root Christ*, with its associations to both medicine and faith, illustrates the hermeneutical flexibility and wide-ranging appeal of many medieval representations of plants. Today it is customary to look at an image of a plant—especially one found in a botanical book or nature guide—as a key to the identification of the plant's genus and species, following the binominal nomenclature espoused by Carl Linnaeus (1707–87) and his followers in the eighteenth century. But in the pre-modern periods such an approach to plant imagery is anachronistic (Jones 2006: 1). Visual representations of plants did not solely serve to aid in the plants' identification; they were not, simply by their very fact, depictions of actual, natural objects.

Rather, plant images served many functions, ranging from tools for identification, classification, or indexing of plant medicines, to metaphors for spiritual edification, articulation of personal characteristics, and allusions to moral and physical health—sometimes serving several of these functions at once. This flexibility in the generation

FIGURE 8.1 *Root Christ* (*Cristo de la Cepa*), Valladolid, Spain, Museo de la Catedral, approx. 30 cm (12 in.) height, vine roots, Toledo, Spain, early fifteenth century. Photo © Julio Puente Mateo. https://www.luzyartes.com.

of meaning made plant imagery a valuable didactic tool. Instead of inviting only identification of the object represented, such imagery invited connections with and between other categories of meaning. Plants could help, heal, or harm the human body and they also could delight the mind and the senses. They are justly studied for their roles as foods, medicines, and commodities. Plant imagery becomes a place of intersection for these aspects of cultural history. And artists working in a variety of media—from wood, ceramics, textiles, and stone to ink or paint on panel, paper, vellum, and silk—navigated these intersections, developing techniques to convey the multiplicity of roles played by plants and by the representations of them.

In its many manifestations, medieval plant imagery often reveals a complex and interdependent relationship between physical, spiritual or emotional, and moral wellness: plants and their imagery ministered both to the body and the soul. In this sense, plant imagery follows the pluralistic understanding of health that governed medieval conceptions of medicine and healing in which the bodily and the spiritual intertwined (d'Andiran 2010; Oslan 2003; Ziegler 1998). Partly to convey plants' dual healing nature and their broad applicability in human life, artists turned to a diversity of styles, media, and compositional techniques. They portrayed plants as rigidly schematic, overtly or covertly symbolic, and even as anthropomorphic. Yet they also portrayed plants in highly detailed and realistic fashions.

Historically, scholars have interpreted the variety of medieval plant imagery, especially that produced in Europe, the wider Mediterranean, and the Islamic world, by connecting it either to the cultural and religious significance of nature and of gardens—that is, to the soul and to the immaterial world (Corazzo 1996; MacDougall 1986; Rhodes and Davidson 1994)—or to developments in medicine, pharmaco-therapeutics, natural history, and the rise of empirical observation of the natural world—that is, to the physical or to the material world (Ackerman 1985; Collins 2000; Givens et al. 2006; Pächt 1950; Smith 2006). As a consequence, although perhaps an unintentional one, scholars have separated the spiritual, moral, and bodily influences of plants expressed in such imagery. This chapter, however, argues that these threads of interpretation are not mutually exclusive; rather, they weave together, overlapping and enhancing one another, and reveal the centrality of plant imagery—often through its very ambiguity and inconsistency—in narrating the depth of an ensouled humanity. While this tendency is visible in settings sacred and secular (Givens 2005, 2006; Horowitz 1998) and across diverse cultural and geographical areas, it is perhaps most strongly and consistently visualized in illustrated books on plant medicine, better known as "herbals," and it is on examples from these books that this chapter focuses.

MEDIEVAL HERBAL ILLUSTRATIVE TRADITIONS

Broadly speaking, the illustrated herbals produced in the medieval world, especially in the Pan-Mediterranean region and across Europe, contain textual and visual compilations of information about medicinal substances—predominantly, but not exclusively, plants. This information derived from Greek and Arabic works, particularly those of Dioscorides (first century CE), author of the influential pharmacopeia *De materia medica* (On the Natural Substances to be Used for the Preparation of Medicines) (Nutton 2013: 178–82; Riddle 1986); Galen (129–after 216 CE), physician to Roman Emperor Marcus Aurelius (121–80 CE) and systematizer of the humoral theory of medicine (Arikha 2007; Nutton 2004, 2013: 222–53); and the Persian philosopher and physician, Ibn Sīnā (980–1037 CE), best known in the West as Avicenna, whose *al-Qānūn fi'l-ṭibb* (*Liber Canonis*; Canon of Medicine) became a central text in medieval medicine (Grendler 2002: 30; Siraisi 1987a: 43–76). Medieval herbals also incorporate information from Late Antique Latin sources, such as the fourth-century *Medicina Plinii* (Pliny's Medicine) and the sixth-century Pseudo Dioscorides' *Ex herbis femininis* (From Feminine Herbs), which borrowed from earlier Greek and Latin sources and, in some cases, included material drawn from myth and from folk medical practices (Riddle 1981: 46–7; Sigerist and Howald 1927: xix).

The *Herbarium Apuleii Platonici* (The Herbal of Apuleius Platonicus) was—together with the works of Galen—the principal treatise on medicinal substances used in the Latin West from the sixth through the thirteenth centuries (Collins 2000: 165–6). Figural or anecdotal scenes most often appear in the illustrated traditions of *Herbarium* manuscripts, suggesting the "immaterial" and usually apotropaic values of the *materia medica* through visualization of ritualistic harvesting practices, traditional folkloric uses, or origin myths. However, from the eleventh century onward, and due in large part to the translation and transmission efforts of erudite physicians and others, Greek and Arabic medical texts increasingly gained ground in both university curricula and in book collections, especially in Italy (Touwaide 2011). The merging of these separate traditions in part accounts for a convergence of folkloric and more academic medical theories and practices in the fourteenth-century illustrated *Tractatus de herbis* tradition (Baumann 1974; Segre Rutz 2002; Touwaide 2013: 53).

The illustrative traditions that accompany these and earlier treatises on medicinal plants have long and varied histories of production and collection and are wildly heterogeneous. Scholars have charted the complex familial relationships between the illustrative and textual programs across the centuries in which these books were created and collected (for instance, Baumann 1974; Collins 2000; Touwaide 2013). Within the diversity of illustrative traditions—in their variety of representative styles, in their use of anecdotal imagery, and in the very fact of their presence in books of *materia medica*—we can recognize the multivalent roles plants played within the human community as agents that connect narratives of bodily and spiritual health. Artists used different stylistic cues to portray plants at the crossroads of the physical and spiritual worlds; they revealed plants as tools capable of healing, restoring, or even harming the body and the soul. Not all plants represented necessarily possessed these multiple characteristics or "virtutes" (powers). Nevertheless, they were understood as part of a greater sense of divine order and creation. Health practitioners—from physicians to apothecaries, root-gatherers, and folk healers—considered plants as part of God's arsenal for aiding a frail and fallen humanity (Park 2013: 614).

To explore this dynamic further, artists sometimes accompanied images of certain plants with narrative vignettes, clarifying the plants' apotropaic, moral, or spiritual function(s). Alternatively, artists portrayed plants in a range of styles—from schematic and unrealistic, to strikingly life-like and to hybrids of the real and the imaginary (sometimes in the same manuscript). Like the addition of anecdotal scenes, the shifting between styles, especially when encountered within a single manuscript, was a visual cue, signaling to readers and viewers the plants' use(s) as both bodily and spiritual medicines. Such scenes and stylistic plurality suggest that, while the images may have been used in select cases to help identify plants—as much of the scholarship on these books has alleged (Baumann 1974; Collins 2000; Pächt 1950), they served other functions for the books' readers as well (Givens 2006; Touwaide 2013).

The stylistically varied representations of plants in medieval illustrated herbals assume new—and more flexible—significance when considered in light of the multiple roles played by the *Root Christ*, as both a sacred image and tool for spiritual salvation and as a physical plant with diverse healing properties. The *Root Christ*'s power manifests in part through its interplay between the very reality of the vine-root and the unreal, schematic figuration of Christ emerging from it. The vine-root is a literal remedy for many sufferings of the human body, as revealed in the plant's roles recounted by a range of medical authorities (Lev and Amar 2008: 177–9). Consider, for instance, the entries on grapes, grapevines, and wines in Dioscorides' *De materia medica* (5.3.1–2 and 5.6.1–2 [ed. Wellmann 1914: 3.2–3, 3.5–6; trans. Beck 2005: 331, 335]) or in the *Aqrābādhīn* (The Medical Formulary) by the Arab physician Al-Kindi (d. *c.* 865) (§121 [ed. and trans. Levey 1966: 134]). At the same time, the schematic figure of Christ that emerges from the root makes it a heuristic device—an object that, through its material agency and metaphorical associations, enables beholders to understand or perceive their spiritual health.

How might this dynamic apply to medieval plant imagery more broadly? Like the fruitful plurality of functions and meaning emerging from the *Root Christ*'s combination of representative styles, the illustrations in medieval herbals reflect the complex and nuanced relationship between human bodies and souls and the plants that can heal or harm both. In the case of the *Root Christ*, the plant's literal realness illustrates its power as a medicine for the body while the metaphorical, unrealistic Christ image illustrates its power as a touchstone for the soul. This dual role reveals how the blending of plants' realistic and metaphorical attributes can navigate the correlativity of bodily and spiritual health. Precisely because it is ambiguous and reflects the material and the immaterial at once, the *Root Christ*—like the combination of schema, realism, and narrative scenes in illustrated herbals—demonstrates how plants and their representations served to cultivate and protect physical and spiritual wellness.

A SYNCHRONICITY WITH LITERARY TRADITIONS

Visual representations of plants were not the sole sites in which medieval people encountered this dynamic understanding of nature and humanity's relationship with it. Interest in the natural world—from its plants to its animals, minerals, winds, and climates—crossed social and educational lines in the Middle Ages. Not only did highly literate physicians or scholars of natural philosophy pursue this knowledge, popular preachers and teachers brought knowledge of the natural world to bear in their exegeses, sermons, or lessons, often using the plants and animals as mirrors for the virtues and vices of humanity. As theologian and poet Alan of Lille (*c.* 1128–1202) described in *De incarnatione Christi rhythmus perelegans* (On the Incarnation of Christ in Rhyme): "All the world's creatures are as a book and a picture / And a mirror for us" (ed. Migne 1855: col. 579; trans. Reeds and Kinukawa 2013: 581n56). Mystics and saints celebrated the plants and animals found in nature, while poets used imagery drawn from the natural world to illustrate moral truths or satirize courtly love, and schoolteachers used it to help students learn their grammar (Orme 1982: 314n9; Reeds and Kinukawa 2013: 580). Vernacular translations increasingly made knowledge of the natural world—from contemporaneous and classical sources—available to growing readership. Furthermore, educational expectations among gentry and ruling classes began to demand a facility with such knowledge, which accounted, in part, for the increase in collection of herbals, "books of secrets," and "natural wonders" by wealthy collectors and bibliophiles (Eamon 1994; Findlen 1994; Park and Daston 2001; Williams 2003).

In diverse literary settings—from poetry to religious, medical, and agricultural treatises, prayer books to homilies—medieval writers and orators mobilized sets of metaphors and allegories to explore the associations between plants and nature and the human body and soul. For instance, in the allegorical poem *Hortulus* (Little Garden), Walahfrid Strabo (*c.* 808–49), the abbot of the Benedictine monastery in Reichenau, Germany, and tutor of Charles the Bald (823–77), future Holy Roman Emperor, recounts the literal medicinal virtues of the plants in his garden, connecting them through use of allegory and metaphor to Christian virtues and the communal life of the monastery. The garden becomes a metaphor for the soul (*Hortulus*, §2, ll. 41–6 [ed. and trans. Payne 1966: 27–9]). At the outset of the poem, the poet's garden is choked with nettles, their "barbs / Tipped with a smear of tingling poison;" but, after the waning of winter and through his hard labor, the poet-gardener makes his "little patch" ready for the seeds of beneficial, healing plants (*Hortulus*, §2, ll. 32–6 [ed. and trans. Payne 1966: 26–7]). As the reader progresses through the abbot's didactic poem, the role of the plants and their associated imagery as both physical remedies and allegories for the soul and its readiness for the "medicine" of a spiritual life becomes increasingly apparent (Wallis 2010: 98).

Strabo clarifies his message in the closing section on the virtues (medicinal and spiritual) of the rose (§26). He lauds not only the rose's extraordinary medicinal value—"No man can say, / No man remember, how many uses there are / For Oil of Roses as a cure for mankind's ailments" (*Hortulus*, §26, ll. 402–4 [ed. and trans. Payne 1966: 60–1])—but also praises its metaphorically rich status as a perennial (it dies back each fall and grows anew—resurrects—each spring) (§26, ll. 398–9 [ed. and trans. Payne 1966: 60–1]). In this section, the abbot also returns to the lily, which he had discussed at the mid-point of his poem (§15) and which grows "over against" the rose. Making clear the spiritual medicine residing in both plants, he writes "For in this flower [lily], / Shines Chastity, strong in her sacred honor" (§26, ll. 410–11 [ed. and trans. Payne 1966: 62–3]), and further notes that:

> These two flowers, [the lily and the rose,] so loved and widely honored,
> Have throughout the ages stood as symbols
> Of the Church's greatest treasures; for it plucks the rose
> In token of the blood shed by the Blessed Martyrs;
> The lily it wears as a shining sign of faith.
>
> (*Hortulus*, §26, ll. 415–18 [ed. and trans. Payne 1966: 62–3])

To cinch his allegory, in the closing lines of his verse on the rose, Strabo offers a prayer to the Virgin, "Our refuge and our friend forever," and reminds the reader that to Mary "came a flower of the royal stem of Jesse, / A single Son to restore the ancient line;" and to Jesus, this flower, the abbot attributes the medicinal and spiritual values of the lily and the rose: "By his holy word and life he sanctified / The pleasant lily; dying, / He gave its color to the rose" (*Hortulus*, §26, ll. 419–26 [ed. and trans. Payne 1966: 62–3]).

Through the application of allegory, symbolic association, and metaphor to the healing virtues of medicinal plants, the abbot's poem revealed the deep connection between the body, soul, and plants. Using rhetorical devices, writers and preachers could illuminate the connection between plants and a more multivalent and inclusive view of health that includes spiritual care. Yet artists likewise echoed and visualized these "unseen" or "invisible" associations, in part by exploring through the stylistic diversity of their imagery the complementary relationships between plants, the representations of them, and the bodies and souls of the people who cultivate, harvest, and minister with them.

VISIONS AND VISUALIZATIONS OF PLANTS

To explore this dynamic, let us turn to two well-known plants with lengthy histories of medicinal uses and manifold representations across the illustrated herbal traditions: artemisia and mandrake. Ancient and medieval sources describe these plants as physically and spiritually therapeutic. Artists across the illustrative traditions have represented them schematically and realistically, and they are included in illustrated herbals that blend realistic and fantastical representations of plants, as well. Furthermore, both plants are accompanied by anecdotal scenes in certain illustrative traditions, elucidating their plurality of functions and meanings.

Artemisia

Today *Artemisia* is the name of a genus of bitter shrubs and herbaceous plants that comprises over five hundred species in the north temperate zone to southern Africa and South America (Mabberley 2017: 75–6); contemporary uses include the ornamental (Springer

FIGURE 8.2 "Artemisia leptafillos" (*A. campestris*, field wormwood) with Chiron the Centaur and the Goddess Artemis, from *Herbarium Apuleii Platonici* (London, British Library, Harley MS 5294, f. 14r), 25.5 × 15 cm, southern Italy, twelfth century. © British Library Board, Harley 5294.

and Proctor 2000: 72), the culinary, and the medical—such as vermifuges, stimulants, and treatment for malaria (Riddle 2010: 89). For medieval physicians and their Greco-Arabic predecessors, however, "artemisia" referred to only a small number of *Artemisia* species. Principally the plants considered artemisias are those now commonly known as wormwood (*Artemisia absinthium*), mugwort (*A. vulgaris*), and southernwort (*A. abrotanum*) (Riddle 2010: 81). The ancient medical authorities valued artemisia in part because of its affinity for what Pliny the Elder (23–79 CE) termed "the troubles of women" (*Naturalis Historia*, XXV.36.73 [ed. and trans. Jones 1956: 7.188–9]; Riddle 2010: 83). Pliny's description of the plant's uses aligns with its legendary origins as a gift to humanity from the Greek goddess Artemis, the Roman Diana (see also Chapter 5 of Volume 1 in this series).

Yet artemisia's therapeutic uses were not limited to women's health. The so-called "mother herb" was a critical component in the treatment of other maladies and mental illnesses, as well. For instance, Dioscorides praised artemisia as an emmenagogue and anti-inflammatory, especially for uterine disorders (*De materia medica*, 3.23.5 and 3.113.2 [ed. Wellmann 1906: 2.32, 2.126; trans. Beck 2005: 189, 233]). Galen prescribed it to treat intestinal parasites and fevers, noting the dangers of overdosing (*De simplicium medicamentorum temperamentis et facultatibus* [On the Mixtures and Properties of Simple Medicines], 6.1.1 [ed. Kühn 1821–33: 11.798–807 and 839–40]). Al-Kindi discussed artemisia as a tonic for the mouth (*Aqrābādhīn*, §177 [ed. and trans. Levey 1966: 296]), while Avicenna recorded the use of several types of artemisia for prompting appetite, preventing intestinal worms, and regulating menstruation (*Liber Canonis*, Bk. II, tract. 2, cap. 2, ff. 88v–89r [Venice, 1507]).

The anonymous author of the *Herbarium Apuleii Platonici* describes the plant and its mythology, noting: "Diana [i.e. Artemis] is said to have found three types of artemisia, the virtues of which she gave to Chiron the Centaur who initially instituted their medical uses. For this reason the herb derives from Diana" (*Herbarium*, XII.2, ll. 12–15 [eds. Sigerist and Howald 1927: 45; trans. Riddle 2010: 83]). The goddess of the hunt, daughter of Leto and Zeus, and twin sister of Apollo, Artemis was venerated as the protector of women, especially as they labor in childbirth (Riddle 2010: 80). Her helper Chiron was an apt recipient for Artemis' gift. Teacher of heroes and of gods, the Centaur (half man, half horse) is celebrated by ancient Greek and Roman sources alike. Homer lauds Chiron as "the most righteous of the Centaurs" (*Iliad* 11.832 [ed. and trans. Murray 1924: 542–3]); Statius (*c.* 45–96 CE) praises him as the loving foster-father to the young Achilles (*Achilleid* I [ed. and trans. Mozely 1928: 508–82]); Ovid (43 BCE–*c.* 17/18 CE) remembers him as the wounded healer who, tragically unable to heal himself after one of Hercules' poisoned arrows falls upon his foot, transforms into the constellation Centaurus (*Fasti* 5.379–414 [ed. and trans. Frazer 1931: 448–9]). It is Pliny, though, who describes the Centaur as the inventor of the disciplines of botany and medicine (*Naturalis Historia*, 7.197 [ed. and trans. Rackham 1942: 2.638–9]).

In the *Herbarium*'s illustrative tradition, the images that accompany entries for artemisia often include a schematic—or abstracted—representation of the plant accompanied by a narrative vignette of this legendary beginning. For instance, in the entry for "Artemisia leptafillos" (perhaps *A. campestris*, field wormwood; see Figure 8.2) from a twelfth-century copy of the *Herbarium*, likely made in southern Italy and now housed at the British Library, London (shelf-mark Harley 5294), the artist included a scene that shows Artemis giving two varieties of the plant that bears her name to the Centaur. This scene accompanies a general, sketchy pen drawing of the plant itself, which shows only vaguely its identifying features, particularly its crowning spherical inflorescences.

In its imprecision, the image of the artemisia plant in the Harley *Herbarium* shares with many of the representations of plants in this and other illustrative traditions a tendency to emphasize not the individual, particular details of the plant that would lend it illusionistic reality, but rather its overall shape and its significant parts (Hoeniger 2006: 67). On the one hand, this deliberate stylistic ambiguity cultivates a connection to an artistic tradition. Art historian Minta Collins (2000: 199) called the imagery in this *Herbarium* "consciously archaising" and a "renewal of a respected text," drawing attention to the similarities the illustrations bear to a ninth-century *Herbarium* (now Florence, Biblioteca

Medicea Laurenziana, plut. 73. 41). She considered the artist's copying of the ninth-century imagery a deliberate effort both to show respect for his model and to improve it (Collins 2000: 200). On the other hand, and especially when considered in conjunction with the vignette of the plant's legendary origins, the artist's representation of the plant invokes its many uses as both a physical and a metaphysical medicine.

Artists connoted artemisia's history of use in cultivating physical, spiritual, and emotional health both through their renditions of the origin-myth vignette (which associates divine figures with a medicinal plant) and through the more ambiguous styles in which they represented the plant itself. These artistic practices consciously visualize long-standing aspects of the textual traditions and folkloric practices associated with the plant. For example, in the description of artemisia which accompanies the illustration of the plant and its legend, the *Herbarium*'s author noted the plant's habitat, the many names by which it was known in other languages, and briefly enumerated both artemisia's physical and talismanic or spiritual virtues:

> For travel: If one takes wormwood [*artemisia*] along on a trip, and holds it, [s]he will not feel tired from the trip. It chases away the demons and, indoors, it neutralizes toxins. It treats illnesses of the eye. Foot aches: Wormwood triturated with oil and applied as an ointment on feet, eliminates pain. Stomach aches: Wormwood, triturated or ground and reduced to powder form, mixed with water and taken as a drink, calms intestinal pain.
>
> (*Herbarium,* X.1–3 [eds. Sigerist and Howald 1927: 42; trans. Touwaide 2013: 58])

The *Herbarium* points not only to the plant's value to the body (in what might be called internal medicine), noting its anodyne and digestive usefulness in plaster, ointment, and infusion preparations, but also its value to the spirit. Its author describes the virtues of simply holding the plant: it possessed apotropaic powers, protecting the spirit from demons and the body from the fatigue of long travels.

Like many of her medical predecessors, Hildegard von Bingen (1098–1179), German Benedictine nun, mystic, healer, and polymath, recounted the physical usefulness of artemisia (*A. absinthium*, "Wermuda" in German), especially in treating the ailments of women. Interestingly, she also noted its virtues in the treatment of *melancholia*, a malady of body and soul that today is associated, in part, with what we call depression (*Physica*, I "De plantis," cap. 109 "Wermut" [eds. Migne and Hamman 1855: cols. 1172b–1173b]). Hildegard described *melancholia* in a manner that reveals her understanding of Galenic medicine. According to Galen's humoral theory, *melancholia* is a state of imbalance of the four fluids, the "humors," within the body: blood, yellow bile, black bile, and phlegm. An excess of black bile defines the temperament, complexion, or disposition—what today we might call one's "personal physiological constitution"—of a melancholic person, and physicians devised complex treatment plans (including medicinal treatments and health regimens) to offset this physical imbalance.

Broadly speaking, for Galen and his followers the humoral balance of the body determined the state of one's physical health; yet, because the body and the soul intertwine, one's personal characteristics were understood, in part, to be a manifestation of humoral balance or imbalance, as well. Hence, imbalances of the humors led both to physical and mental illness (when acute) and to characteristic dispositions or temperaments (when chronic). These temperaments correspond with their dominant humor: sanguine (blood), choleric (yellow bile), melancholic (black bile), and phlegmatic (phlegm). In this framework of humors and personal characteristics, plants—alongside diet, hygiene, and lifestyle—served as mediators and instruments to rebalance and to harmonize the health of peoples' bodies and souls.

The Cordovan physician, Ibn Sarābīyun, known in the Latin West as Serapion the Younger (fl. mid-thirteenth century) and author of the *Liber Serapionis aggregatus in medicinis simplicibus* (Serapion's Book of Aggregated Simple Medicines), according to the usual title of its Latin translation, shared Hildegard's view of artemisia's usefulness. In the *Carrara Herbal*, a fourteenth-century manuscript produced in Padua that contains an Italian translation of Serapion the Younger's treatise on medicinal plants, the anonymous artist alluded to artemisia's ability to help return a melancholic to a state of healthy humoral balance in the dynamic way he chose to represent this versatile plant (see Figure 8.3). Throughout the manuscript, the artist represented plants in many different styles, which range from the highly abstracted (lacking naturalism) to the highly particular (articulating exact, individuating details of specific plants), and to fantastic hybrids of these styles. In the entry for "absenço" (*A. absinthium*, f. 12r/v

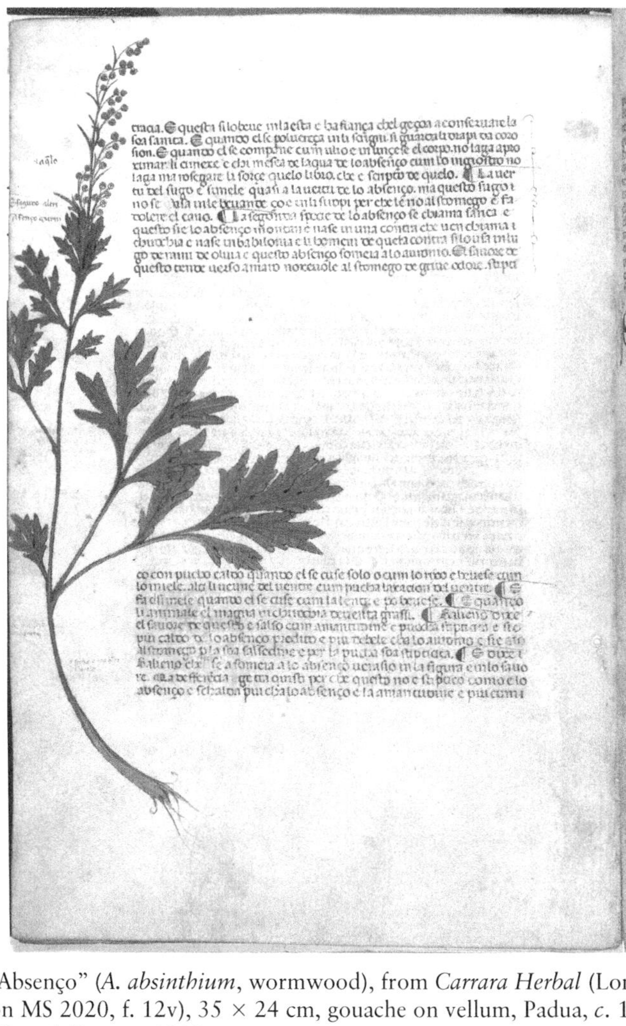

FIGURE 8.3 "Absenço" (*A. absinthium*, wormwood), from *Carrara Herbal* (London, British Library, Egerton MS 2020, f. 12v), 35 × 24 cm, gouache on vellum, Padua, *c*. 1390–1400. © British Library Board, Egerton 2020.

[ed. Ineichen 1962–6: 17–20]), the artist included an image of a cutting of the plant, showing a woody stem with several of the characteristic bi- and tripinnate, craggy leaves and topped with a whorl of small, globular heads of yellow flowers. The cutting appears in striking, life-like detail, looking as though it had been pressed into the margins of the very book for study and preservation.

The artist complicated this illusion, however, by layering yet another illusion on top of it; he portrayed one of the deeply incised, silver-green leaves, the one closest to the text-block, as though it has slipped behind the neat row of regular gothic minuscule script. This playful illusionism not only points to the fundamental connection between the text and imagery in this manuscript, it also stops the readers' progress through the manuscript, delighting them and distracting them from the didactic details of Serapion's treatise. Yet this disruption immediately relates to the medicinal value of the plant in the treatment of *melancholia*, in particular. Delightful and joyous activities—including reading, viewing beautiful things, or walking in a garden—were prescribed by physicians as part of health regimens to those suffering from an excess of black bile (Bertiz 2003: 208; Olson 1982: 57). For instance, in the plan of care devised for a patient suffering from stomach problems, the Paduan-trained physician Gentile da Foligno (d. 1348) cautioned the patient to counter-balance his *melancholia* with healthy pleasures in order to ensure a complete recovery from his ailments (*Consilia Cermisoni. Consilia Gentilis*, f. 63v [Venice, 1495? ed. and trans. Olson 1982: 59n33]). In the *Carrara Herbal*, the artist made the process of viewing the illustrations—an act central to the process of reading the book—integral to the construction of meaning from the book's content, and so created an experience that itself exemplifies the book's role in cultivating health and well-being and in mitigating the potential pitfalls of humoral imbalance, such as those described by Hildegard (Kyle 2017: 67–87).

Unlike the more secular approach to Galen's humoral medicine found in Serapion's text, humoral imbalance for the devout Hildegard reflected the fallen state of humanity—a post-lapsarian reality with which humans must contend (Radden 2000: 80). In her *Liber compositae medicinae* (Book of Holistic Healing), Hildegard describes the black bile of melancholy as:

> black, bitter, and [able to release] every evil, sometimes even a brain sickness. It causes the veins in the heart to overflow; it causes depression and doubt in every consolation so that the person can find no joy in heavenly life and no consolation in his earthly existence. This melancholy is due to the first attack by the devil on the nature of man since man disobeyed God's command by eating the apple.
>
> (II, § "De melancoliae morbo" [ed. Kaiser 1903: 38; trans. Pawlik and Madigan 1994: 35])

The author of the *Herbarium* may also have suggested this connection between artemisia as a remedy for melancholy, and melancholy's connection to the devil and to fallen humanity, when he described the ability of artemisia to "[chase] away demons" (*Herbarium*, X.1–3 [eds. Sigerist and Howald 1927: 42; trans. Touwaide 2013: 58]), in this case, the "demon" of *melancholia*.

In a similar way, common folkloric ritual practices associated with artemisia suggest the plant's ability to offset the bitterness of melancholy and other "demons." For instance, one of the names for mugwort (*Artemisia vulgaris*) in medieval German hints at the ritualistic life of this plant as rooted in popular religious practices. The plant was called *St. Johannsgürte*—St. John's girdle. On the feast day of St. John the Baptist, June

24, people wove garlands of artemisia and wore them as girdles while they performed "undisclosed rituals and spells," to borrow historian John Riddle's description (2010: 88). Singing and dancing accompanied these rituals, which culminated in the casting of the girdle into a bonfire, an act believed to offset or ward against personal adversity (Grimm 1854: 2.1161–2 [trans. Stallybrass 1976: 3.1211]). In his poem *Hortulus*, Walahfrid Strabo perhaps alluded to these folkloric practices and integrated them with medicinal ones in his description of wormwood (*A. absinthium*) and the healing practices associated with it. He praised the plant, noting that to make an effective remedy for headaches not only must the head be bathed in an infusion of wormwood, but ("Do not forget this") it must also be wreathed in a garland of its leaves, lest the remedy be ineffective (*Hortulus*, §9, ll. 189–94 [ed. and trans. Payne 1966: 40–1]).

The connection between artemisia and suffering of the body and the soul becomes increasingly clear when we consider how the plant functioned metaphorically in literary contexts, especially in the Hebrew Bible and New Testament (*la'anah* is the Hebrew word for wormwood; ἀψίνθιον *apsinthion* or ἄψινθος *apsinthos* in Greek). Artemisia is referred to several times in the Hebrew Bible (Prov. 5:4, Lam. 3:15 and 3:19; Deut. 29:18; Job 30:4; Jer. 9:15 and 23:15, among others) and once in the New Testament (Rev. 8:10–11). In each case, it is used as a "stand-in" for extreme bitterness—not of taste, for the plant is certainly extraordinarily bitter, but of situation: the bitterness of sorrow, adversity, remorse, or personal suffering (Moldenke and Moldenke 1952: 48–50; Riddle 2010: 82; Zohary 1982: 184). For instance, Lamentations 3:19 reads: "Remember my affliction and my bitterness, the wormwood and the gall." In this and other instances in which artemisia is mentioned, the plant serves a metaphorical role, but, in a manner akin to the representation of artemisia in the *Carrara Herbal*, this role is one that speaks directly to its many uses in medicine and to the association between bitter suffering and the diverse healing properties of this bitter plant.

Artemisia is given a place of prominence on the frontispiece of the herbal allegedly written and illustrated by the physician known as Manfredus de Monte Imperiali (see Figure 8.4). Produced in the early fourteenth century in southern Italy, the *Tractatus liber de herbis et plantis* (Paris, Bibliothèque nationale de France, Lat. 6823) was copied widely and became very influential, especially to the production of illustrated herbals created in the latter part of the fourteenth century in northern Italy. The herbal contains a compilation of texts and imagery drawn from other herbals, especially from the *Tractatus de herbis* known as Egerton 747 and housed at the British Library (Collins 2000: 268–73; Touwaide 2013: 46–8).

The frontispiece shows Manfredus seated with a book nestled in the crook of his arm. He speaks to a group of young students, two of whom are holding out plants for their teacher's attention—one is a kind of artemisia, perhaps "Artemisia leptifillos,"[1] and the other, betony (*Betonica officinalis*). The artist's choice of artemisia as one of the two plants being shown to Manfredus for discussion and analysis points to its central importance in textual and illustrative herbal traditions and in the teaching and practice of medieval medicine. The context of the scene illuminates this point. Manfredus is portrayed speaking: words stream from his mouth toward his students. He says "Moderation, by means of *body and soul*, is the first and greatest medicine" (*Prima et ultima medicina propter* corpus et animam *est abstinentia* [my emphasis]). This phrase, which emphasizes the interdependence of the body and spirit in the maintenance of good health, mirrors the many virtues of artemisia.

Compositionally, the words of Manfredus' speech mirror those of another phrase, which streams across the page from the upper right corner. Accompanying the hand of God, which

FIGURE 8.4 Lippo Vanni or Robert d'Oderisio, author portrait of Manfredus de Monte Imperiali, from *Tractatus liber de herbis et plantis* (Paris, Bibliothèque nationale de France, MS Latin 6823, f. 1r), 34.5 × 24.7 cm, southern Italy, *c.* 1330–40. Source: Bibliothèque nationale de France, Département des manuscrits, Latin 6823.

is shown descending from a series of concentric spheres and giving a gesture of blessing to Manfredus and his students, the text reads: "Investigate all things, holding fast to what is good" (*Omnia probate, quod bonum est tenete*). This phrase, drawn from the first letter of Paul to the Thessalonians, 5:21, denotes Manfredus' teaching and medical practices as divinely ordained and assisted. But it may allude to the interpretation of the biblical passage found in the medieval *Glossa Ordinaria* tradition, as well (Kyle 2017: 50). This twelfth-century "ordinary interpretation" of the Latin Bible is attributed to the theologian Anselm of Laon (d. 1117) and his school. The *Glossa Ordinaria*'s exegetical commentary on this biblical verse and those that immediately precede it focus on the necessity of learning from elders who have "tested" their knowledge and determined what is good. It reads:

> Extinguish not the spirit because it is the will of God to do all these things, you older people who—through the Holy Spirit—have the gift of intelligence, do not conceal it, because if you do that you too may lose it. Do not spurn their prophecies [of older,

intelligent people] and substitute yourselves, but [on the other hand] do not accept all things [i.e. all prophecies] indiscriminately, but test [probate] them, i.e., break them down with reason, and hold on to whatever is found to be good.

(*Biblia Latina cum glossa ordinaria*, 1. Thess. 5.19–21 [eds. Migne and Hamman 1852: col. 620, unpublished trans. M. Musgrove])

If the presence of the biblical verse on the frontispiece of Manfredus' herbal meant to invoke the *Glossa*'s interpretation as well, then the text connotes Manfredus' vision of himself as a member of a genealogy of physicians and teachers. Together with the plant imagery and portraits of Manfredus' students, however, the phrase and its exegetical associations uphold the role of the plants in the continuum of medical learning and practice. "Tested" and found to be good, *artemisia* has pride of place as an invaluable medicinal material for the body and the spirit. *Artemisia* played this role in the past, plays it in the present, and—as the reader intuits from the inscription and the placement of the plant in the hands of the next generation of physicians—will play it in the future.

Mandrake (Mandragora officinarum)

In a manner similar to artemisia's appearance on the frontispiece of Manfredus' herbal, the anonymous artist of the series of six full-page illustrative scenes that preface a sixth-century illustrated copy of Dioscorides' *De materia medica* alludes to the history and healing properties of mandrake. Known as the *Vienna Dioscorides* (Vienna, Österreichische Nationalbibliothek, *medicus graecus* 1), this Greek copy contains a compilation of information drawn from Dioscorides' full text and then ordered alphabetically (Touwaide 2006: 35). It was dedicated to, and further belonged to, Anicia Juliana (*c.* 462–528), the daughter of the Byzantine Emperor Flavius Anicius Olybrius (d. 472). An inscription on the dedication page (f. 6v), which also contains a portrait of Juliana enthroned, describes the book as a gift for Juliana from the citizens of Honorata, a district in Constantinople (Killerich 2001: 171; Premerstein 1903: 110). In the two prefatory illustrative scenes immediately preceding this dedication page, the artist featured mandrake. The portrayal of the plant in these scenes is significant: it is the only plant pictured in the illustrative frontmatter of the manuscript, which itself is otherwise illustrated with over three hundred images of plants. In conjunction with the remaining prefatory illustrations, the mandrake's presence in these two scenes both denotes its cultural significance in the early Byzantine period, especially, and teaches the book's readers how to interpret the textual and illustrative contents of the manuscript.

Related to deadly nightshade (*Atropa belladonna*), mandrake is a powerful narcotic with a long history of medicinal and folkloric uses (Harrison 1956: 87; Zarcone 2005). It has almost as long a history of visual representation. A common plant native to southern Europe and the eastern Mediterranean, mandrake has rough, large dark-green leaves and bell-shaped flowers that become round reddish-yellow fruits resembling small apples in the autumn. Famously, it has a long, parsnip-like root that tends to bifurcate, crudely resembling the arms and legs of a human. Due to this characteristic form, the plant accrued legends and rituals, which a plenitude of medical, literary, and artistic sources document. As they did for images of the illustrious artemisia, artists working across the medieval illustrated herbal traditions represented mandrake in diverse ways and settings to convey its history of medical use, metaphorical and legendary associations, and contemporary relevance to health and well-being. We encounter one such popular way of representing mandrake and its powers in the series of opening illustrative scenes in the *Vienna Dioscorides*.

In the two scenes that feature mandrake (f. 4v and f. 5v), the plant is shown anthropomorphically—as a small brown homunculus with dark green leaves sprouting from its head. In the first (see Figure 8.5), the artist portrayed a female figure—identified by an inscription as *Heuresis* (the personification of Discovery)—giving the mandrake to Dioscorides, who is seated in a chair beneath an inscription of his name. The artist depicted the mandrake, with its characteristic root "limbs" dangling, at the exact center of the scene. From the right, the arm of *Heuresis* connects the mandrake to the female personification, who is dressed in a flowing white robe with a red stole draped across her shoulders in the manner of a goddess. From the left, the arm of Dioscorides reaches out to take the plant. Between the two figures and beneath the mandrake the artist portrayed a contorted dog, reminding the viewer of the legends surrounding the harvest of this powerful plant, legends discussed below.

FIGURE 8.5 Scene of *Heuresis* giving Dioscorides the Mandrake, from Dioscorides, *De materia medica* (Vienna, Österreichische Nationalbibliothek, *medicus graecus* MS 1, f. 4v), 37.6 × 31.2 cm, traditionally dated to the early sixth century, probably Constantinople. Source: Österreichische National Bibliothek.

In the second illustrative scene featuring mandrake (see Figure 8.6), Dioscorides sits at a desk and writes in a codex, while another female personification, *Epinoia* (Knowledge), displays the plant to an artist who sits at an easel and paints its image. Not only does the presence of the artist suggest the centrality of plant imagery to the herbal tradition, but, in showing the artist actively collaborating with Dioscorides and contributing his "Knowledge" to the physician's work, the scene also reveals the centrality of art to conceptions of health and its active role in therapeutic practice. As discussed above, illustrations could serve not only to identify the plant but also to emphasize the traditions, legends, and rituals surrounding it and to give insight into the very role that seeing such an image could play in the cultivation of good health. The artist with his representation of mandrake becomes an active agent in creating an experience of well-being.

Centrally placed in both scenes, the mandrake image becomes a heuristic device—standing in for the histories, traditions, uses, and representations of the other plants contained in the volume. Mandrake was an appropriate choice for the artist to feature in this role because of its extensive history of use and its many associated legends: from

FIGURE 8.6 Scene of Dioscorides, *Epinoia*, and artist painting mandrake image, from Dioscorides, *De materia medica* (Vienna, Österreichische Nationalbibliothek, *medicus graecus* MS 1, f. 5v), 37.6 × 31.2 cm, traditionally dated to the early sixth century, probably Constantinople. Source: Österreichische National Bibliothek.

Homer and the Hebrew Bible to the ancient Greek and Latin medical authorities and their medieval compilers and commentators, mandrake has been both loved and feared for its powerful medicinal and aphrodisiacal qualities. Circe may have enchanted Odysseus and his men with mandrake ("moly") (Homer, *Odyssey*, 10.302–6).[2] Mandrake cures Leah's barrenness in Genesis (30:14–18), and the Lover gathers the plant's fragrant fruits as gifts for the Beloved in the Song of Solomon (7:13). While uniformly advising judicious and cautious administration of it, classical and medieval medical authorities enumerate the plant's many medicinal virtues (Riddle 2010: 67–9; Scarborough 2006: 17–18).[3] In a tale Pliny repeated (*Naturalis Historia* XXV.94 [ed. and trans. Jones 1956: 7.147–8]), Theophrastus (*c.* 372–287 BCE), Aristotle's student and author of the *Historia plantarum* (Enquiry into Plants), recounted mandrake's curious harvest-ritual, writing:

> Thus it is said that one should draw three circles round mandrake with a sword, and cut it with one's face towards the west; and at the cutting of the second piece one should dance round the plant and say as many things as possible about the mysteries of love.
> (*Enquiry into Plants*, 9.8.7–8 [ed. and trans. Hort 1916–26: 2.258–9])

Like artemisia, mandrake possessed supernatural healing and apotropaic qualities. The *Herbarium* attributed to Apuleius notes "For witlessness, that is devil-sickness or demoniacal possession, take from the body of this said [plant] mandrake [...] administer to drink in warm water [and][...] soon he will be healed" (*Herbarium*, CXXXI.4, ll. 37–41 [eds. Sigerist and Howald 1927: 224; trans. Grieve 1931: 511]). A treatise from what is currently known as the *Hippocratic corpus* described perhaps a less frightening application of the plant in helping to heal a broken heart or disturbed soul. *De locis in homine* (On Places in Man), probably written *c.* 440–330 BCE, advises the use of mandrake to ease melancholy or madness and the administration of it particularly to those suffering from suicidal thoughts (§39.1 [ed. and trans. Craik 1998, 75]; see also Peduto 2001: 758; Stannard 1962). Mandrake, then, as attested by both ancient and medieval sources, had physical, mental, and spiritual applications as a medicine.

To allude to the plant's illustrious history, uses, and its appearance in essential medical sources, illustrations of mandrake often refer to the plant's legendary powers and to the curious rituals that surrounded its collection. In his chronicle *Bellum Judaicum* (The Jewish War, 78 CE), historian Flavius Josephus (37–after 100 CE) described the mandrake as a temperamental, animated homunculus that could flee if provoked. He further noted the danger of uprooting it and recounted the critical—and sacrificial—role of a dog in harvesting the plant, a role perhaps denoted by the artist of the *Vienna Dioscorides*' frontispieces. Josephus wrote:

> Flame-coloured and towards evening emitting a brilliant light, [mandrake] eludes the grasp of persons who approach with the intention of picking it, as it shrinks up and can only be made to stand still by pouring upon it certain secretions of the human body. Yet even to touch it is fatal [...] [An] innocuous mode of capturing it is as follows. [Rhizomists] dig all round it, leaving but a minute portion of the root covered; they then tie a dog to it, and the animal rushing to follow the person who tied him easily pulls it up, but instantly dies—a vicarious victim, as it were, for him who intended to remove the plant, since after this none need fear to handle it. With all these attendant risks, it possesses one virtue for which it is prized; for the so-called demons—in other words, the spirits of wicked men which enter the living and kill them unless aid is forthcoming—are promptly expelled by this root, if merely applied to the patients.
> (*The Jewish War*, 7.3, ll. 180–5 [ed. and trans. Thackeray 1927, 3.556–9])

Historians of medicine Anne van Arsdall (2009: 67) and John Riddle (2010: 62) have suggested that the frightening tales of improper harvesting leading to death may have served root-cutters well: people would be too afraid to seek out this useful, and so perhaps expensive, plant themselves. Regardless of the possible economic benefits of sowing rumors about the plant's potentially lethal harvest, artists rendered the ritual in their portrayals of mandrake across several illustrative traditions of herbals, including that of the *Herbarium Apuleii Platonici*, the *Tractatus de herbis*—Manfredus de Monte Imperiali's much-copied fourteenth-century version depicts this scene—and in the fourteenth- and fifteenth-century *Tacuinum sanitatis* (Tables of Health) illustrative tradition.

In a fifteenth-century representation of mandrake found in a luxury *Tacuinum sanitatis* manuscript produced in Rhineland, Germany, from imagery likely derived from a northern Italian original (König 2009: esp. 89–94), the artist shows a "sleeping" mandrake, with the top of its root "body" exposed and a cord tied from around its neck to that of a dog, which rushes toward a bowl of food while a young man flees in the opposite direction with his red hat pulled down tightly over his ears (see Figure 8.7). The representation of the plant is fanciful—in addition to the facial features drawn onto the crest of the mandrake's root, the leaves are nondescript, linear with smooth, entire margins, no shading or detailed venation, and the fruits hang suspended above the plant in schematic, berry-like clusters. Despite the plant depiction's lacking accuracy, those of both the young man and the dog show fine attention to detail—the man's clothes are tailored to his proportionate frame and the wrinkled fabric of his coat bunches along his forearms as he raises his hands to cover his ears; the dog's gaunt ribs protrude from its sides, and dark shadows hollow its body, indicating its hunger and malnourishment. So, while capable of portraying objects more realistically, the artist chose to represent mandrake more abstractly. Why? Partly because the plant's anthropomorphic qualities are so strongly represented in historical literary and medical sources and partly because of the role these scenes play in the *Tacuinum* itself.

The *Tacuinum sanitatis* tradition differs from the other textual and illustrative herbal traditions that I have discussed in that it includes information beyond the therapeutics of medicinal materials found in other herbals (Bertiz 2003; Hoeniger 2006; Sarton and Thorndike 1928; Segre Rutz 2000, 2002: 123–70; Touwaide et al. 2009). While the *Tacuinum sanitatis* includes information about medical substances, it also includes prescriptions that address the so-called six "non-naturals." The "non-naturals" are factors drawn from environment and lifestyle that were believed to influence health: quality of air, food and drink, exercise and rest, sleeping and waking, repletion and excretion, and emotions and sexual activity. The treatise begins with medicinal substances—plants and the foods made from them, animals, animal products, perfumes, and wines. The remainder of the treatise turns to the "non-naturals"—from music to emotions, climates, bathing, clothing, and leisure activities (Hoeniger 2006: 53). Each medicinal substance and non-natural is classified according to Galenic medical theory, which identifies its medicinal use(s) and humoral and elemental "complexion" (i.e. hot, cold, dry, or wet, which are qualities associated with the body's four humors—blood, black bile, yellow bile, and phlegm—which are, in turn, associated with the four humoral temperaments—sanguine, melancholic, choleric, and phlegmatic; see Nutton 2013: 242, 246–8). The textual contents of the *Tacuinum* derive from the *Taqwīm al-Ṣiḥḥa*, an Arabic treatise written by Ibn Buṭlān (d. *c.* 1063), a Christian physician working in Baghdad, and translated into Latin in the mid-thirteenth century (Elkhadem 1990). Although the identity of the translator is debated, it seems likely that the translation emerged from the courts of Charles of Anjou and Manfred in southern Italy (Bertiz 2003: 67).

FIGURE 8.7 Scene of Mandrake Harvest ("Fructus mandragore"), from *Tacuinum sanitatis* (Paris, Bibliothèque nationale de France, MS Latin 9333, f. 37r), 33.5 × 22.5 cm, Rhineland, Germany, mid-fifteenth century. Source: Bibliothèque nationale de France, Département des manuscrits, Latin 9333.

As we have seen in other contexts in which plants are represented in many styles, the artists illustrating the *Tacuinum* manuscripts participated in an ethos in which health was understood holistically, as a state not bound to the body alone, but to the mind and soul that, together with the body, constitute the human being. Art—like plants—can minister to the ensouled body and help to negotiate health and well-being. In the representation of mandrake and its associated mise en scène in the *Tacuinum*, the artist creates a viewing experience that complements aspects of the readers'—especially elite readers'—health requirements. Not only are the readers learning about the natural world and the *materia medica* it provides, an activity itself connected to moral virtue and bodily health, they are also captivated by the complex and multivalent visualizations of these substances, legends, and legacies. This captivation, in turn, was one component of nurturing a healthy mind and body through a temperate joyfulness (Kyle 2017: 72–80), which the *Tacuinum* both prescribed and embodied. Among its lifestyle, hygiene, and diet recommendations, the *Tacuinum* advocated walking, horseback-riding, singing, hunting, storytelling, and conversing with friends in a garden—all activities thought, through their enjoyment, to ward off disease (Bertiz 2003: 205–12). These enjoyments, in turn, are envisioned and recorded by the artist in ways equally delightful—a double therapeutic dose of medicine for the reader.

CONCLUDING REMARKS

The roles of plant imagery in herbals were changing by the end of the medieval era. In manuscript and printed illustrated herbals produced in the fifteenth and sixteenth centuries in Europe, particularly, we see a growing tension between the dynamic explored in this chapter—in which representations of plants in diverse styles alluded to the plants' many physical and metaphysical uses and legends and also themselves possessed healing agency—and a new current in plant illustration that was gaining momentum. By this time, physicians and naturalists were increasingly seeking to identify definitively the particular plants described by the ancient sources (for instance, to what exact kind of artemisia did Dioscorides or Galen refer?). Humanists contributed to this project through their efforts to translate Greek texts and transmit them in Latin and vernacular translations to an educated class increasingly fascinated by the literature, art, and ideas of antiquity. For different—but complementary—reasons, physicians and apothecaries, humanists and philologists, and natural historians and botanists shared in these translation efforts, resulting in the creation of disciplinary hybrids like medical humanism (Nutton 1997; Ogilvie 2006: 11–12, 121; Reeds 1976; Siraisi 1987b, 2007: 63–105) and medical philology (Nutton 1985, 1988).

In this context, while some artists continued to portray plants in both stylized and realistic ways and to accompany their imagery with vignettes narrating the plants' legendary and folkloric associations, as in the mid-fifteenth-century *Tacuinum sanitatis* illustrations discussed above, others took plant illustration in new directions. Through careful observation and recording of nature's "particulars," its identifying details, these artists sought to provide recognizable, classifiable plant images that could serve, in part, as a kind of visual key, "unlocking" and organizing a growing lexicon. That is, artists created images of plants—often, but not always, quite naturalistic ones—to accompany lists of synonyms for plant names in different languages. This compositional strategy is visible, for instance, in the *Tractatus de herbis* (*c.* 1440) now London, British Library, Sloane 4016 (Touwaide 2013) and in the *Herbarius* (*c.* 1445) written by the Venetian physician Nicolò Roccabonella (1386–1457) and illustrated by artist Andrea Amadio (fl. mid-fifteenth century) now Venice, Biblioteca Marciana, Lat. VI, 59 (coll. 2548). The sixteenth-century developments of the so-called *herbarium siccus* (collection of dried plants) and collections of living plants in newly founded botanical gardens complemented these efforts (Baumann 1974: 91; Findlen 1994: 166). By collecting more naturalistic images of plants alongside actual living or dried specimens and accumulating and synthesizing the assorted names by which these plants were known, artists, physicians, humanists, and botanists together were seeking consensus on plant identities. They were beginning to build a unified, cross-cultural body of knowledge about medicinal substances, while continuing to document in text and image the complex relationship between people and plants (Grafton and Siraisi 1999; Kusukawa 2012; Egmond 2017).

ACKNOWLEDGMENTS

This work was supported by the Obermann Center for Advanced Studies at the University of Iowa. I am grateful to the 2017–18 Obermann Center Fellows for their valuable suggestions and encouragement.

NOTES

Chapter 6

1 For Indic practices, the primary and secondary sources used in this essay weigh the discussion of plants in culture toward Hindu and Buddhist (over Jain) communities. Early forest-dwelling and later Muslim communities are only represented in a cursory fashion. Jewish and Christian communities are not represented in secondary literature here, although their presence in South Asia is attested.

2 Michel Angot believes Taittirīya was composed around the sixth century BCE (see Angot 2007). *Taittiriya-Upanisad avec le commentaire de Samkara*, 7; *Padma Purāna* is a much later text, although the dates of the composition have not been pinned down and range from the fourth through the fifteenth centuries CE. See Rocher (1986: 206–14).

3 The authors translate three aquatic plants referred to in the *Rg Veda* into Latin, as follows: *pundarika*—white lotus (*Nelumbo nucifera*), *pushkara*—blue lotus (*Nymphaea caerulea*), and *kiyambu*—waterlily (*Nymphaea alba*) (Krishna and Amirthalingam 2014: 13). Although I have offered Latin names for plants in this essay, the practice of ascribing Latin names to Indic words is fraught with uncertainty. The veracity of any botanical identification here and elsewhere must, therefore, be treated with circumspection. When secondary sources have not offered Latin names, I have followed the Pandanus Database of Plants (http://iu.ff.cuni.cz/pandanus/database), which is a useful source collecting textual occurrences of plants in early Indian languages and relating them to proposed Latin names. The database requires Indian plant names to be entered in their accepted diacritical format, which is often not the way they are spelled in Indian scholarship and translations, including works I have cited. Plant names will therefore be presented in this essay with scholarly diacritics appropriate for the language in question to facilitate research. For cross-reference or when a plant could not be found in the Pandanus Database, I have used the table given in Krishna and Amirthalingam (2014), labeled "Plant Names in Indian Languages," 281–4, which lists plant names across four columns (English, Sanskrit, Hindi, and Tamil).

4 The more famous *Lotus Sutra* is a Mahāyāna Buddhist text that was translated into Chinese between the third and fifth centuries. It is considered the final teaching of the Buddha by many Asian Buddhists, because of its extensive instruction in "skillful means" as a path to Buddha-hood. See Reeves (2008) for a poetic English translation or Teiser and Stone (2009) for excerpts and commentary.

5 Two steatite seals as well as other visual evidence from Indus Valley sites collected by various scholars corroborate this: https://www.harappa.com/search/site/pipal (accessed March 19, 2021); see also Krishna and Amirthalingam (2014: 6–7).

6 The dating of *purāna* literature is hotly debated by scholars and those for whom it forms part of religious liturgy. On dating, Ludo Rocher (*The Purāṇas* 1986: 199) writes: "According to one opinion the Matsya developed from the fourth century BCE up to the third century CE according to another opinion the development spread from the fourth century BCE until 1250 CE. After having assigned the Matsya to a date not later than the sixth

century CE, Kane more specifically placed it between 200 and 400 CE. Hazra concluded that 'no one date is sufficient for it' [dating different chapters to different periods.]"

7 Plant emblems correspond to the Tamil kingdoms of Chera (*paṉai*), Pandya (*vēmpu*), and Chola (*konnai*), see Ramanujam (2011: note for 121, 308).

8 The great exception in South Asia to this history of vernacular languages and plants is the case of Tamil. With a formalized landscape culture (called *tinai*) documented in extant Sangam texts, Tamil plant names emerged independently of Sanskrit sources. See Sekar (2012) for an evocative introductory summary. Later medieval writings in Kannada, Telugu, and other vernacular languages also point to prior botanical nomenclatures.

9 Colonial ascriptions were probably based on already altered ideas about *jambu*. By the late fifth–early sixth century *Amarakośa* (a "dictionary" for writers in the form of a long list of words with a chapter called *Vanauṣadhivarga* devoted to plants) *jambu* is separated from other words for jackfruit (*lakuca, likuca, dahuh, panasah, kaṇṭakiphalah*) that are similar to words used in some Indian languages today, suggesting that by the fifth century the word "*jambu*" may have taken on a significance that separated it from the common jackfruit. Nicholas Roth, who treats plant knowledge in Mughal India, translates the word *kaṭahala* from a sixteenth-century text as jackfruit; he does not translate the word *jamun* and *jambu* does not appear. See Roth (2018: 56–7).

10 The idea that sculptures at Bharhut may have represented jackfruit was first raised in *Hobson-Jobson*, a nineteenth-century compendium about botany in India, and was supported by a discussion of Pliny the Elder. See Yule and Burnell, *Hobson Jobson* (1903 [rev. 2013]: 440–3). I will translate *jamun* as jackfruit in the following discussion about *vṛkṣāyurvēda*.

11 Pliny, *Naturalis Historia*, first century CE, 12:24 is a reference to "the tree is called the pala, and the fruit ariena" describing a tree in India "the fruit is larger and superior in flavour; the sages of India live on it." Pliny, *Natural History*, ed. Rackham (1945): 16–9. "Pala" might refer to the Sanskrit *phala*, which simply means "fruit," or the compound "*jambūphala*" referring to the *jambu* tree's fruit. Alternatively, it might refer to the Tamil word for jackfruit (*pala*) or alternative Sanskrit words (*atibṛhatphala* or *phalavṛkṣaka*). In classical scholarship, this tree is usually identified as the banana tree.

12 McHugh (2012: 65n10) cites Scharfe (1993: 292) and points out the contested dating. Olivelle (2013) dates parts of the text to the second–third century CE.

13 See Meister (2007: fig. 3) for a sculptural depiction of a multistory hypaethral tree shrine from the stupa at Kanganhalli (alternate spelling, *Kanaganahalli*).

14 Surveys of plants called out in inscriptions are generally written in one of several regional South Asian languages, based on where the inscriptions were found. See, for example, Swamy (1975). Reading inscriptions requires knowledge of multiple languages for proper documentation and decipherment.

15 The *Mayamata* (see *Mayamatam*, Dagens, trans. 1994, repr. 2000) is an authoritative example of a comprehensive text about architecture and city planning from South India, composed in Sanskrit and compiled into its present form around the eleventh to twelfth century, with possible later interpolations. It is one example within the first-millennium genre of writing about architecture called *vastuśāstra* ("treatise on dwelling"). Other similar well-known architectural texts include: *Bṛhatsaṁhitā*, *Agnipurāṇa*, *Viṣṇudharmottara-purāṇa*, *Mānasollāsa*, and *Aparājitapṛcchā* [all not in bibliography]. A text such as this one, although often translated as a "treatise," derives its prescriptions from observing practices that were in use rather than through an abstracted theoretical method. In fact, it can be difficult to understand the context for certain statements without knowledge of architectural practice, which the text's author may not have had.

Architectural knowledge was hereditary and passed down within communities in the building professions. Some texts, like this one, present their knowledge of architectural practice more systematically than others, although none are frozen texts; each belongs to a stream or recension.

16 The chapter called (*prasāda lakṣaṇa*) in the *Bṛhatsaṁhitā*, a Sanskrit compilation meaning "Long Compendium" attributed to *c.* the sixth century CE, discusses locations in which "deities reside with pleasure," as ideal sites for temples.

17 The text does not say that such on-site trees would be used for structural components of the building, and certain trees were avoided for this purpose including those "growing in a place inhabited by spirits …" (*Mayamata*, 15.71–6, 198–9).

18 Here, although translated as the sacred lotus, *padma* seems to refer to something sturdier than *Nelumbo nucifera*.

19 The question mark was present in Dagens' transliteration, perhaps due to a confusion about the meaning of the word.

20 I have corrected spellings from Dagens' transliteration (198) based on his text in *nāgari* script (197) to facilitate searchability within the Pandanus Database of Plants; we updated any incorrect scientific names from the Pandanus Database following Mabberley (2017). I have not, for the most part, considered the suitability of named trees for their intended structural purpose, a process that could change our understanding of their Latin identification.

21 I heard a rendering of this tale from my father, who grew up in South India, before finding the short story in Jacobs (ed.), *Indian Fairy Tales* (1892; repr. 1969) as a young girl. The retelling, from versions my father had heard, had lost its original context as folklore passed down within an Indigenous, former forest-dwelling "Scheduled Tribe." Santal reservation land is in Jharkhand, a state in Northeast India.

22 I will use the English form "yaksha" and "yakshi" without italics to refer to these creatures for ease of reading and in figure captions, unless I am specifically referring to the Sanskrit term, in which case I will use diacritics and italics. The English terms are widely used for museum labels to describe sculptural representations. A Sanskrit spelling may figure in referenced titles.

23 For this image, McHugh cites a scene in the *Āraṅyakaparvan* (Forest Book) of the *Mahābhārata*. The importance of the idea of the *saugandhika* pleasure grove of Kubera in the medieval period, is attested by the reproduction of this scene in two plays: "*Kalyānasaugandhikavyāyoga* ascribed to a Nīlakaṇṭha, of uncertain date, between the tenth and fifteenth centuries CE. Also, the Saugandhikāharaṇa of Viśvanāthakavi, probably from the early fourteenth century. Also, Krishnamachariar mentions references to the performance of a Buddhist dramatic version of this story. See Krishnamachariar, 1970, para. 553 and footnote, p. 535" (McHugh 2012: 94n12).

24 See, for example, the Parijat tree in Kintoor, Uttar Pradesh, which is promoted by the Barabanki District's tourism department as the referenced sacred wish-fulfilling tree but is, botanically speaking, a baobab (https://barabanki.nic.in/tourist-place/parijaat/ [accessed March 19, 2021]).

25 Online Collection, Freer Gallery, Smithsonian Institution, Accession # F1954.6; https://asia.si.edu/object/F1954.6/ (accessed March 19, 2021). *Krishna and the Golden City of Dwarka* from the *Harivamsha* (*Genealogy of Vishnu*), Mughal Dynasty, Reign of Akbar, *c.* 1585.

26 I will use "Mayadevi" or simply "Maya," without diacritics from here on.

27 For example, *Lalitavistara Sūtra* (third century CE); *Buddhacarita* (early second century CE), varying critical editions reflect questions of authorship and manuscript date, see E. H. Johnstone's preface, especially xvi–xviii (1936 repr. new edn. 1984). I have used E. B. Cowell's edition (1894 repr. 1997) for all references and translations.

28 Online Collection, Freer Gallery, Smithsonian Institution, Accession #F1949.9d: https://asia.si.edu/object/F1949.9a-d/ (accessed March 19, 2021). Nativity Scene from Gandhara in *Scenes from the Life of the Buddha*, late second–early third century CE.

29 Four other trees are associated with the Buddha's enlightenment: *aśvattha* or *pippala* (*Ficus religiosa*), *nyagrodha* (*F. benghalensis*), *mucalinda* or possibly *samudraphala* (*Barringtonia acutangula* or *Eugenia acutangula*), *rajayatana* or *āmrāta* (wild mango, *Spondias pinnata* or *S. mangifera*).

30 The sal tree was already sacred through its association with several Jain Tirthankaras: Ajitanatha, Sambavanatha, Mahavira, and sometimes Sumatinatha. See B. C. Bhattacharya (1974); Krishna and Amirthalingam (2014) offers two useful tables of sacred trees and plants before the Buddha: "Sacred Plants of the Rig Veda," 13–14. "Jaina Tirthankaras and Their Trees," 23.

31 Because female figures can be intertwined with many types of trees, they should simply be "bhanjika" figures. However, catalogs tend to call such figures "shalabhanjika," "yakshi," or "yakshini," unless they have been miscatalogued like the so-called "Lakshmi" of Pompeii; see note 32. I have also seen such figures referred to as *vṛkṣaka*, which reflects the generalization of this bent-branch form for all trees. See also Coomaraswamy (1931) on iconography.

32 Online collection, "Lakshmi" (pre-79 CE) held in the Secret Cabinet of the Museo Archeologico Nazionale di Napoli (https://www.museoarcheologiconapoli.it/en/room-and-sections-of-the-exhibition/3490-2/ [accessed March 19, 2021]). See also Weinstein (forthcoming) on the movement of this figurine from India to the Roman world.

33 "Gajamukhane ganapathiye ... *kalpa taru* [my emphasis] neene," popular devotional film song in Kannada by Vijaya Narasimha and M. Ranga Rao, sung by S. Janaki.

34 The term *vṛkṣāyurveda* has been attested as the *Gulma-Vṛkṣāyurveda* in Kautilya's *Arthaśastra* (ch. 24) of disputed date (possibly first century BCE to first century CE; see note 12); Varāhamihira *Bṛhatsaṁhitā* (sixth century CE); *Viṣṇudharmottarapurāṇa* (*c.* seventh century CE); *Agnipurāna* (*c.* seventh to eleventh centuries CE); Sūrapāla, *Vrikshayurveda* (*c.* 1000); Chavundarāya, *Lōkōpakāra* (Kannada, 1025); Someśvara, *Mānasōllasa* (aka. *Abhilaṣitārtha Cintāmaṇi*) (twelfth century CE); *Śārṅgadharapaddhati* (Treatise on Method by Sārngadhara) in a chapter called "Upavanavinoda" (About Pleasure Gardens) (fourteenth century CE); Chakrapāni Mishra, *Vishvavallabha* (*c.* 1577); King Basavarāja of Keladi, *Shivatatvaratnakara* (Kannada, *c.* 1684–1745) [all not in bibliography]. See also Majumdar (1927).

35 Gyula Wojtilla, *History of Kṛṣiśāstra* (2006) has surveyed the known *kṛṣi* literature in several languages (Sanskrit, Tamil, Malayalam, Kannada, Hindi, Gujarati, Marathi, Rajasthani, Bengali, Assamese, Maithili, Persian, Telugu, and Prakrit); the Asian Agri-History Foundation has published two translations as the *Krishi-Parashara* (*c.* 400 BCE) and the *Kashyapiyakrishisukti* (800 CE) [not in bibliography].

36 I will use "Surapala" without diacritics after this point in the body of the text for ease of reading.

37 Conserved in a manuscript at the Bodleian Libraries, Oxford, England, and transcribed and translated by the Asian Agri-History Foundation (1996), Surapala's eleventh-century text, like others of its kind, was compiled from older practices (*vṛkṣāyurveda*). See also Nugteren (2005: 79–83).

38 Several essays within the journal *Pandanus*, a project of the Institute of Indian Studies, Faculty of Arts, Charles University, Prague, Czech Republic, discuss this dichotomy between city and forest (http://iu.ff.cuni.cz/pandanus/publications/ [accessed March 19, 2021]).

For an explanation of how territorial claims in South Asia can relate to caste distinctions between urban and forest dwellers, see, for example, Guha (2013). For ethnographic descriptions of some meanings attributed to trees, forests, and green spaces, see *Journal for the Study of Religions, Nature and Culture*, 4: 2 (2010).

39 After this occurrence, I will abbreviate Surapala's text to *VrksA*.

40 An alternate name for *rambha* is *kadalī* from Krishna and Amirthalinam (2014), "Plant Names in Indian Languages" Table, 283.

41 In this list, the definition of "tree" seems to encompass all woody plants, even smaller herbs like *tulasī*. The text, in fact, defines four types of plants within the discussion on propagation: *vanaspati* (those which bear fruit without visible? flowers), *druma* (those which bear fruit with flowers), *lata* (those which spread with tendrils as creepers), and *gulma* (those which are short but have branches—i.e. bushes). From these verses, we understand that trees (*vṛkṣa*) actually refer to all woody plants. See *VrksA*: 45–7, 45.

42 Here, I accept Meister's proposed correction of *jambu* to mean jackfruit (*Artocarpus heterophyllus*) in early India. See note 10.

43 By planting one *aśvattha* (*Ficus religiosa*), one *picumanda*, one *nyagrodha* (*F. benghalensis*), ten tamarind trees, or the group of three (*kapittha* (*Limonia acidissima*), *bilva* (*Aegle marmelos*), and *āmalakī* (*Phyllanthus emblica*)) or five mango trees, one avoided hell. Latin names listed according to the Pandanus Database.

44 See Sadhale and Nene (2009: 106–11) for some contemporary interpretations of *vṛkṣāyurveda* in India.

45 The description of Daśapura continues from lines 3 to 8 of the inscription, *CII* 3, 1888 (1970): 18, 79–88. "Here the houses have waving flags, (*and*) are full of tender women, (*and*) are very white (*and*) extremely lofty, resembling the peaks of white clouds lit up with forked lightning. And other long buildings on the roofs of the houses, with arbours in them, are beautiful—being like the lofty summits of (the mountain) Kailāsa; being vocal with songs (*like those*) of the Gandharvas; having pictured representations arranged (*in them*); (*and*) being adorned with groves of waving plantain trees. Here, cleaving asunder the earth, there rise up houses which are decorated with successions of storeys; which are like rows of aerial chariots; (*and*) which are as pure as the rays of the full-moon. This *(city)* is beautiful (*through*) being embraced by two charming rivers, with tremulous waves, as if it were the body of (the god) Smara (*embraced*) in secrecy by (*his wives*) Prīti and Rati, possessed of (heaving) breasts."

46 I will use "Kalidasa" without diacritics after this point in the body of the text for ease of reading.

47 After this occurrence, I will abbreviate Kalidasa's text to *MaAg*.

48 This association is slightly different from the previous association of trees to deities, such as the Jain Tirthankaras, because previously the tree and the deified were interconnected and even interchangeable. Now, specific plant products (flowers and leaves as well as plant derivatives like incense) became associated with gods who were believed to take particular pleasure from them.

49 See Ali (2004), *Courtly Culture and Political Life* for a well-researched explanation of courtly culture, style, protocol, and sociability in medieval India.

50 Jain worship practices are excluded here, and contestation about the value of flowers for ritual exists among the traditions of Islam; floral and foliate iconography was used in certain regions and periods.

51 See also Husain (2000) for a discussion of scent in sixteenth-century gardens of the Deccan; see Haq (2001) on humans as caretakers of the environment within Islamic ideology.

52 The word "*dān*" in Hindi is "*dāna*" in Sanskrit; I have used the Sanskrit word throughout this text and therefore converted the word to avoid confusion.

53 I have followed Hazra (1975) to date all subsequently cited *purāṇas*.

54 The larger textual context for *vratas* is summarized in the chapter called "The Concept of *Vrata* in the Hindu Textual Tradition," in Pearson (1996: 45–84; see esp. 61–78).

55 I received a copy of this report from the author, A. N. Yellappa Reddy, an Indian environmentalist and retired member of the Indian Forestry Service; it may not be widely available. The report includes several plant lists in Sanskrit with botanical annotations and an appendix with drawings and details in English about the list of referenced plants. One plant list from this report is reproduced as Figure 6.5, Plant List for "Shree Umāmahēshwara Vrata," with our editorial updates, which include alternate spellings with diacritical marks for plant names as well as corrections to scientific names. Umā-Maheśvara, although worshipped as the masculine Maheśvara, represents the couple, both god and goddess (Umā). Although, as mentioned earlier, it is difficult to link actual plants to texts, these lists were taken from current practice and linked to regional botanical knowledge in local languages about the plants. We should consider the possibility that these botanical names correspond more closely to plant-naming practices in today's Kannada-speaking regions. In that regard, the *kamala* plant (listed as *Nelumbo nucifera*) could have an alternate botanical association, if we take *kamala* to be a Tamil word rather than the generic Sanskrit word for lotus. In Tamil, *kamalā* synonymous with *kuṅ kumam*, *mañcaṉai* (Skt. *kampillaka* or *recanaka*; Hin. *sindūr*) is the Kamala tree (*Mallotus philippensis*). The worship of this goddess (Umā), who is a form of the divine feminine Parvati in a coupled state, might very reasonably have used flowers from this plant: a derivative product is traditionally used by Hindu women across the subcontinent and beyond to mark themselves as married, due to the auspiciousness attributed to Umā and women who hold marital status.

56 For elaboration on the role of fragrance and *alaṁkara*, see remaining articles about the "kama world" in this special issue of the *Journal of Indian Philosophy*, 39: 1 (2011).

57 Vātstyāna's *Kāmasūtra* is most often cited but refers to a wider body of knowledge that was written on multiple occasions. The notion of a "kama world" habilitates these *kāmaśāstra* texts from their overweening association with early Indian sex. On this point, see also Doniger (2016).

58 Often called "areca nut" in Indian references, betel nut is a completely separate plant from "betel leaf," which is used to wrap the nut for chewing in a product conventionally known today as *pān*. In this section, I will use the Sanskrit word *tāmbūla* for the "betel roll" because it figures in the sources when prepared for gifting.

59 *Mānasōllasa* (Refreshment of the Mind) also known as *Abhilāṣitārtha Cintāmaṇi* (the Magical Gem that Fulfills Desires) [not in bibliography].

60 See, for example, Cox et al. (2016) and Science Direct: https://www.sciencedirect.com/topics/neuroscience/betel; https://www.sciencedirect.com/topics/neuroscience/areca-nut (both accessed March 19, 2021).

61 It was a hero stone inscription found in Shimoga District, Sagar Taluk, Karnataka, India.

62 Although I have used the generic term "elite," *nāgaraka* specifically contains within it the notion of someone sophisticated and urbane. It is this city-dweller that participated in the aesthetics of court culture, and future work should nuance our understanding of the kinds of elites involved with various plant practices.

63 Daily courtesies are also attested in eleventh-century decrees naming paid palace staff for this function.

64 The *Anuśāsanaparvan* covers a range of topics, often in the form of dialogues, and is dated by Hiltebeitel (2005) and Fitzgerald (2003) as having been composed and added to the main text of the *Mahābhārata* at some point between the last two centuries BCE and the early few centuries CE.

65 This paragraph is an attempt to clarify Ali (2004: 162–70) who cites Gonda (1975: 271) to develop a theory about the concept of *alaṁkara* as something more than simple ornamentation. Ali incorporates a relational aspect to the practice of self-perfection through the courtesy (*upacāra*) of adornment. This theory as presented by Ali building upon Foucault seems a promising aid for understanding ephemeral plant culture at court, which I explore here.

66 "The Tiruvalangadu Copper Plates of the Sixth Year of Rajendra-Chola I," *SII* 3 (1929: 205, 383–439), cited in Ali (2000: 170–5).

67 To date, the gardens and plants mentioned within inscriptions have not been surveyed systematically in English, although they have in regional languages, see, for example, Swamy (1975). On gardens in inscriptions, see Dasgupta (2016); Hauser and Selvakumar (2021); Wagoner (2012). Other sources for early regional garden cultures are Ali (2003); Ali and Flatt (2012); Satari (2008).

68 Tenth century, "Paschimbhāg Copperplate of Śrīchandra" cited in Sircar (1973: 30) and Dasgupta (2016: 145): includes land for florists (flower sellers) associated with the *maṭha* of Brahman. Four florists worked for the temple precinct supplying flowers for rituals. Each received a half *pāṭaka* of land—the same allotment given to a variety of artisans associated with temple service, such as oilmen, potters, drummers, conch shell blowers, servants, and cobblers, and half of what was allotted to the brahman who built the temple and the temple astrologer/accountant (*gaṇaka*).

69 Twelfth century, Bijapur, Southern Kalachuri inscription cited in *EI* 5 (1898–9: 18, 22–3) and Dasgupta (2016: 146): various lands were donated for a special occasion of the Kālamukha sect to support certain rites. One *mattar* of land was set aside for the flower garden of the god. It was located adjacent on the east and north to gardens of two named individuals and on the west by a garden attached to a Jain temple. This orientation of greenspace suggests an urban zoning practice in which certain areas of the city were designated for gardens owned by different parties; twelfth-century (1111 CE) inscription on the central shrine of the Ulagāḷanda Perumal temple, Kanchipuram published in *ARE* (1921: 39), cited in Heitzman and Rajagopal (2004: 252–3): when describing the boundaries of a land-grant, the inscription calls out the location of "city land used as a flower garden by the garland makers."

70 Twelfth century, Dharwar, Erambarage/Erambapura Inscription, cited in *EI*, 20 (1929–30:12.46–54, 109–22) and Dasgupta (2016: 146): The city was ornamented with temples, lovely women, brahmanas, wealthy vaiśyas as well as sacred groves. In the northeast quarter of this city, a temple of Telligeśvara was built and endowed by fifty local families of Telligas. As part of the endowment, various goods, including rations of betel nuts and betel leaves as well as one *mattar* of garden land, were given from each family.

71 C. 1183, 13th regnal year, Chola Kulottunga I assigned two plots of tax-free land "for raising two flower gardens" along with a quantity of paddy for an annual payment to the temple and "maintenance of two servants for the upkeep of the gardens" at Srirangam, *ARE* (1948–9: B 3). Several similar records for flower gardens are found at Srirangam on the walls of the Ranganathaswamy Temple, *ARE* (1948–9). One cited in *Subject Index to ARE* (1988: 63) registers a sale of land to raise a flower garden in the name of the Pāṇḍya princess Neriyan Mahādēvī and makes provisions for the maintenance of servants employed for the upkeep of the garden.

72 Two Kakatiya-period inscriptions published in Sreenivasachar (1940: 56.161–7) and (1940: 41.17–20), cited in Wagoner (2012: 108): the first discusses the endowment of a triple-shrine temple, a tank, and "beautiful flower gardens and gardens of citrons, cloves, palmyras, and mangoes, so as to give happiness to all people." The second inscription speaks about the good works of Nāmi Reddy who beautified land with large tanks, planted groves for shade, "filling the quarter with the fragrance of flowers." He also set up alms houses and water "sheds" for hungry and thirsty people. These traditions of setting up charitable gardens for shade and food service, not restricted to elites, seem pervasive.

73 I have come across plant lists in the Hire Bevinur Inscription (1190 CE), Indi Taluk, Bijapur District, cited in *SII* 20 (1965 repr. 1988: 175, 218–21) in which a woman named Siridevi performed a *vrata* to plant several trees for a grove. Trees included mango, areca, coconut, date palm, and several others in specific combinations. See also Wagoner (2012: 120–6) for a reconstructed plant list for Amin Khan's garden dated to the late sixteenth century and suggesting continuities for planting practices within the sultanate-period Deccan.

74 The focus of scholarship thus far has been on medieval architectural features in the contemporary cultural landscape and the garden imaginary revealed via medieval texts. My dissertation challenges this convention of separating mental and material worlds for landscape context and cites other efforts attempting the same: "The Experience of Early Designed Landscape in South Asia: Sigiriya, Sri Lanka and Mamallapuram, India," PhD thesis, University of Pennsylvania (forthcoming 2021).

Chapter 7

1 The *Bundahishn*, a religious text written or compiled probably by the late ninth century CE (MacKenzie 1989), hints at the animistic "Mazdean" conception of the vegetal kingdom, especially with respect to tree symbolism. In the Mazdean tradition plants, like other natural phenomena, were believed to have souls; all "Ahura Mazdā-given, clean plants," instilled with divine spirit, were venerable. From the combined seeds of all the plants, a cosmic or archetypal tree, called a "multi-seeded tree," grew in the middle of the mythical sea. This fabulous tree was also known as "the righteous," or "diligent physician," and as "the all-healer tree" ('Alam 1994).

Curiously, these important vegetal species do not seem to be explicitly represented in the visual arts. A gilded silver pitcher in the Hermitage Museum shows two medallions with a *senmurv*, separated by a kind of tree, similar to the motifs on the pediments of Taq-e Bostan (Loukonin and Ivanov 1995: 98–100). The *senmurv* was the mythological hybrid bird, half a lion and half a prey-bird, that could nest on this "multi-seeded tree" and allowed for the seeds to fall in a river, and from there to be carried down to humankind. However, even if we find numerous examples of *senmurv* on silver vessels or textiles, the specific iconography of the interaction of the bird and the tree is never explicitly showed on Sasanian artifacts. We should add here that no Sasanian manuscript has come down to us; silver vessels, together with rock-reliefs and textiles, are among the rare items with a definite Sasanian iconography.

2 Louvre, MAO S 376 and 377.

3 See Grabar and Ettinghausen (1987: 31 and 33).

4 For discussion and images, see Grabar and Ettinghausen (1987: 42) and Degeorge (2010: 146–61, 166–7).

5 For discussion and further images, see Grabar and Ettinghausen (1987: 41, 106) and Degeorge (2010: 186).

6 LACMA, Madina Collection of Islamic Art, M.2002.1.34.

7 Paris, Louvre Museum, Inv. No. 4068; see Grabar and Ettinghausen (1987: 149).

8 See for example a fragment of woven silk with falconers flanking a palm tree, Iran probably tenth century, Washington, DC, The George Washington University Museum and the Textile Museum; reproduced in Grabar and Ettinghausen (1987: fig. 261).

9 An example, with the characteristic double-wing motif, can be seen on a large dish from Khurasan or Transoxiana, dated to the tenth century CE (Louvre MAO 95, gift D. David-Weill, 1950. Reproduced in Bernus-Taylor 1989: 32). Similar fleuron-like decorations can be found on book illuminations from the same period. In twelfth-century Norman Sicily, the palm tree appears as the central pattern on the silk embroidered *Coronation mantle of King Roger II* (Vienna, Kaiserliche Schatzkammer, acc. no. WS XIII 14). In thirteenth-century Arabic illustrated manuscripts, a similar example of "palm tree" appears in the *Kalila wa Dimna* made probably in Syria and dated to *c.* 1200–20 CE (Ettinghausen 1962: 62). A far later reminiscence can also be detected in the extraordinary stone open-work *jali*-screens of the Sayyid Mosque in Ahmedabad, dated 1571.

10 For images and discussion of the Samarra styles, see Grabar and Ettinghausen (1987: 103, figs. 78, 79, 80).

11 The site of Samarra also yielded important ceramic pieces, which fall mainly into two categories: blue and white tin-glazed, and luster painted faience. Both used floral and vegetal patterns. Blue and white tin-glazed bowls and dishes often display a multi-petalled rim and a central floral motif, occasionally framing an inscription in Arabic. Luster bowls can have a double-wing *fleuron* (flower) pattern, sometimes supported by a kind of trunk, thus reminding of the hybrid trees already noted in the Dome of the Rock (Grabar and Ettinghausen 1987: figs. 90–3). The Great Mosque of Kairouan in what is now Tunisia has, on its *qibla*-wall, a remarkable series of polychrome luster tiles coming from Abbasid Iraq dated to the ninth century and imported by the Aghlabid emirs (800–909 CE) to embellish their monuments. The decor of these tiles plays in the double register of geometric and vegetal compositions, familiar with the motifs on the vessels from Samarra (Grabar and Ettinghausen 1987: fig. 98).

12 See Grabar and Ettinghausen (1987: fig. 216, 210–13).

13 See ibid., figs. 264–5.

14 See ibid., 246, fig. 262; and 247, fig. 263; Ettinghausen (1962: 171).

15 See Grabar and Ettinghausen (1987: 261 and 286).

16 Louvre Inv. No. MAO 442 with Bernus-Taylor (2001: 174–5).

17 MS cod. Arab 464. See Vesel et al. (2009: 178).

18 A painted page made at Tabriz in the 1330s, shows this episode, and the tree certainly has human-head finials (Grabar and Blair 1980: 133). A much later version of the *Waq-Waq Tree* is showed in an album-page painting from seventeenth century Golconda, Deccan (Bernus-Taylor 2001: 167–73).

19 Accession number 1927.414. Reproduced in Watson (1985: 71).

20 University Museum, University of Pennsylvania, Philadelphia, NE-P 19. Reproduced in Watson (1985: 92).

21 Istanbul, Topkapi Sarayi Museum Library, Hazine 841.

22 Paris, Bibliothèque nationale de France, MS Or. Arabe 3465. Reproduced in Ettinghausen (1962: 62–3).

23 Paris, Bibliothèque nationale de France, MS Or. Arabe 2964; reproduced in *Arabesques et jardins de Paradis: collections françaises d'art islamique* (Bernus-Taylor 1989: 186–95).

24 We find lotus flowers on the borders and as a central motif on glazed tiles of two types: the so-called low-fired *lajvardina* and metallic luster (Komaroff and Carboni 2002: figs. 97 and 100). The leaves of lotus flowers did not yet appear. Examples on *lâjvardina* tiles display an "unnatural" shrub with lozenge leaves and lotus flowers with six petals

(Porter 1995: fig. 27). At the beginning of the fourteenth century, the outer friezes of monumental *mihrabs* still displayed lotus flowers on their borders (see the frieze kept at Doris Duke Foundation, Shangri-La [Honolulu], signed by Yusuf ibn 'Ali ibn Mohammad ibn Abi Tâher; the tile with the date [710 AH/1310 CE] is kept at the Islamic Art Museum in Cairo: Pope 1971: pl. 725, E). As far as textiles are concerned, the collection kept at Verona's Castello Vecchio, in the tomb of Can Grande I, exhibits several fragments of brocaded satins from the fourteenth century, probably coming from Il-Khanid workshops (Pope 1971: pl. 1005). Several paintings from the so-called *Great Mongol Shahnama* also display textiles adorned with lotus motifs (Grabar and Blair 1980: 135, 137, 161, 165). This theme is also found on metalwork objects, such as a chandelier made of a copper alloy, dated to 1308 CE, with medallions inlaid with silver six-petal lotuses (Pope 1971: pl. 1355; now at Museum of Fine Arts, Boston [55.106; H. 32.5 cm]; reproduced in Komaroff and Carboni 2002: fig. 154).

Some examples of the so-called "Sultanabad" ceramic occasionally show lotus flowers and leaves (Bernus-Taylor 1989: 275). However, this pattern can be best distinguished on underglaze painted tiles produced in Khwarazm. The leaves often adopt a navicular form with four or five finely dented lobes. The earliest example is probably on the double cenotaph in the mausoleum of Seyyed 'Ala al-Din in Khiva (Uzbekistan), *c.* 1305 (Soustiel and Porter 2003: 40). The same kind of pattern appears in an expanded size in the slightly later tiled panels of the tomb of Najm al-Din Kobra at Kohna Urgench (Turkmenistan), dated *c.* 1330 (Soustiel and Porter 2003: 43).

25 The rich decorative repertoire plays with the double register of the *eslimi-khata'i* described earlier. The Timurid albums of paintings and drawings preserved to this day contain abundant examples of sketches and pounced-drawings which were used on several kinds of media (illumination, embroidery, inlaid objects (Lentz and Lowry 1989: 192–8).

Chapter 8

1 This *artemisia* plant has been identified as either field wormwood (*Artemisia campestris*) or feverfew (*Tanacetum parthenium*). Confusion abounds about the modern identification of *Artemisia leptifillos*. On the arguments regarding its identification, see Van Arsdall (2002: 153n121) and Tobyn, Denham, and Whitelegg (2016: 124–35).

2 Stannard (1962) debates whether Homer's "moly" is indeed mandrake. However, Riddle (2010) and other earlier scholars argue that *moly* is indeed mandrake because Dioscorides describes the plant as "Circe's herb" (*De materia medica*, 4.75.1 [ed. Wellmann 1906: 2.233–34; trans. Beck 2005: 280]).

3 For example, Theophrastus, *Enquiry into Plants*, 9.7.8 (ed. and trans. Hort 1916–26: 2.256–9); Dioscorides, *De materia medica*, 4.75 and, on its lethality, 4.75.3 (ed. Wellmann 1906: 2.233–5; trans. Beck 2005: 280–1); Galen, *De simplicium medicamentorum temperamentis et facultatibus*, 5.14 and 5.19 (ed. Kühn 1821–33: 11.751 and 11.766–7); Avicenna, *Liber canonis*, Bk. I, tract IV, cap. 30, ff. 80v–81r (Venice, 1507; ed. Bakhtiar and trans. Gruner 1999–2015: 1.544); *Herbarium Apuleii* CXXXI (eds. Sigerist and Howald 1927: 222–5).

BIBLIOGRAPHY

Abel, Wilhelm. 1980. *Strukturen und Krisen der spätmittelalterlichen Wirtschaft*. Stuttgart: Gustav Fischer.

Abulafia, David. 2012. *The Great Sea: A Human History of the Mediterranean*. London: Penguin Books.

Ackerman, James S. 1985. "Early Renaissance 'Naturalism' and Scientific Illustration." In Allan Ellenius (ed.), *The Natural Sciences and the Arts: Aspects of Interaction from the Renaissance to the Twentieth Century: An International Symposium*. Uppsala/Stockholm: Almqvist and Wiksell.

Adam of Bockenfield, *Glossa super De vegetabilibus et plantis*, R. James Long (ed.). 2013. Leiden, Boston, MA, and Köln: Brill.

Adamson, Melitta Weiss. 2000. *Daz buoch von guoter spise (The Book of Good Food): A Study, Edition, and English Translation of the Oldest German Cookbook*. Sonderband IX. Krems: Medium Aevum Quotidianum.

Adamson, Melitta Weiss (ed.). 2002. *Regional Cuisines of Medieval Europe*. New York and London: Routledge.

Adamson, Melitta Weiss. 2004. *Food in Medieval Times*. Westport, CT: Greenwood Press.

Ahmad ibn Muhammad al-Ghâfiqî (d. c. 1165), Fuat Sezgin (ed.). 1906. *Texts and Studies Collected and Reprinted*. Frankfurt-am-Main: Johann Wolfgang Goethe University.

'Alam, Hushang. 1994. "Derakt." *Encyclopaedia Iranica*, vol. 8, fascicle 3, 316–19. Available online: http://www.iranicaonline.org/articles/derakt (accessed February 22, 2018).

Alan of Lille, *De incarnatione Christi rhythmus perelegans*, Jacques-Paul Migne (ed.). 1855. "De incarnatione Christi rhythmus perelegans." In *Patrologiae cursus completus, Series Latina* 210, cols. 579–80. Paris: Migne.

Albertus Magnus, *De vegetabilibus Libri VII, Historiae naturalis pars XVIII*, Karl Friedrich Wilhelm Jessen and Ernst H. F. Meyer (eds.). 1867. Berlin: Georg Reimer.

Ali, Daud. 2000. "Royal Eulogy as World History: Rethinking Copper-plate Inscriptions in Cola India." In Ronald Inden et al. (eds.), *Querying the Medieval: The History of Practice in South Asia*, 165–229. Oxford: Oxford University Press.

Ali, Daud. 2003. "Gardens in Early Indian Court Life." *Studies in History*, 19(2): 221–52.

Ali, Daud. 2004. *Courtly Culture and Political Life in Early Medieval India*. Cambridge: Cambridge University Press.

Ali, Daud. 2011. "Rethinking the History of the Kāma World in Early India." *Journal of Indian Philosophy*, 39(1): 1–13.

Ali, Daud. 2012. "Botanical Technology and Garden Culture in Someśvara's 'Mānasollāsa'." In Daud Ali and Emma J. Flatt (eds.), *Garden and Landscapes in Pre-colonial India: Histories from the Deccan*, 39–53. London: Routledge (Visual and Media Histories).

Ali, Daud. 2018. "The Betel-Bag Bearer in Medieval South Indian History: A Study from Inscriptions." In Manu Devadevan (ed.), *Clio and her Descendants: Essays for Kesavan Veluthat*, 535–58. Delhi: Primus Books.

Ali, Daud, and Emma J. Flatt. 2012. "Introduction." In Daud Ali and Emma J. Flatt (eds.), *Garden and Landscapes in Pre-colonial India: Histories from the Deccan*, 1–17. London: Routledge (Visual and Media Histories).

Ali, Daud, and Emma J. Flatt (eds.). 2012. *Garden and Landscape Practice in Pre-colonial India: Histories from the Deccan*. London: Routledge (Visual and Media Histories, 2).

Amarakośa (aka *Nāmaliṅgānuśāsanam, with the commentary Amarakośodhāṭana of Bhaṭṭa Kṣīrasvāmin*). 1941. Edited by, with English equivalents, H. D. Sharma and N. G. Sardesai. Poona Oriental Series, no. 43. Poona: Oriental Book Agency.

André, Jacques. 1961. *L'alimentation et la cuisine à Rome*. Paris: Klincksieck.

Angot, Michel. 2007. *Taittiriya-Upanisad avec le commentaire de Samkara*. Paris: Collège de France.

Anthimus, *De observatione ciborum ad Theodoricum regem Francorum epistula*. Edited by Eduard Liechtenhan. 1927. Berlin: B. G. Teubner.

Apicius, *De re coquinaria*. Edited by Christopher Grocock and Sally Grainger. 2006. Totnes: Prospect Books.

Apicius, *De re coquinaria*. Edited by Jacques André. 1974. (Collection des universités de France). Paris: Belles Lettres.

ARE (Annual Report on Indian Epigraphy) Ulagāḷanda Perumal Temple Inscription. 1921. Edited by Archaeological Survey of India (ASI), 39.

ARE (*Annual Report on Indian Epigraphy*) B-3. 1948–9. Edited by N. Lakshminarayan Rao, 125.

ARE Subject Index (*Subject Index to the Annual Reports of Indian Epigraphy [from 1936 to 1973]*). 1988. New Delhi: Archaeological Survey of India (ASI), 63.

Arikha, Noga. 2007. *Passions and Tempers. A History of the Humours*. New York: Harper Collins.

Arthaśāstra (aka *The Kauṭilīya Arthaśāstra*). Edited by R. P. Kangle. 3 vols. 2nd edn. 1972. Bombay: University of Bombay.

Arthaśāstra (see Scharfe 1993; Olivelle 2013).

Asher, Frederick M. 2012. "A Place for the Gods: Gardens of Buddhist and Hindu India." In Stephen Bann (ed.), *Interlacing Words and Things: Bridging the Nature-Culture Opposition in Gardens and Landscapes*, 81–94. Washington, DC. Dumbarton Oaks Research Library and Collection (Dumbarton Oaks Colloquium on the History of Landscape Architecture, 33).

Ashtor, Eliyahu. 1976. *A Social and Economic History of the Near East in the Middle Ages*. London: Collins.

Ashtor, Eliyahu. 1986. *East–West Trade in the Medieval Mediterranean*. Edited by Benjamin Z. Kedar (Collected Studies Series CS245). London: Variorum.

Aśvaghoṣa (see *Buddhacarita*).

Avicenna, *Liber canonis*. Translated by Gerardus Cremonensis. 1507. *Liber canonis Auicenne reuisus [et] ab omni errore me[n]daq[u]e purgatus summaq[ue] cum diligentia Impressus*. Venetiis: Paganinum de Paganinis (reprint: Hildesheim: Olms, 1964). English translation: Laleh Bakhtiar. 1999–2015. *Avicenna (Abū ʿAli āl-Husayn ibn ʿAbd All āh ibn Siñā), The Canon of Medicine (al-Qañuñ fī'l-ṭibb)*, 5 vols. Chicago, IL: Great Books of the Islamic World.

Balducci Pegolotti, Francesco, *La pratica della mercatura*. Edited by Allan Evans. [1936] 1970. (Medieval Academy of America Publication 24). Cambridge, MA: Medieval Academy of America (repr. New York: Kraus).

Ballanfat, Paul 2007. "Arbres." In M. A. Amir-Moezzi (ed.), *Dictionnaire du Coran*, 8–79. Paris: Robert Laffont.

Barker, Randolph. 2011. "The Origin and Spread of Early-Ripening Champa Rice: Its Impact on Song Dynasty China." *Rice*, 4: 184–6.

Barrett, Timothy H. 2001. "Woodblock Dyeing and Printing Technology in China, *c*. 700 A.D.: The Innovations of Ms. Liu, and Other Evidence." *Bulletin of the School of Oriental and African Studies*, 64(2): 240–7.

Barrett, Timothy H. 2008. *The Women who Discovered Printing*. New Haven: Yale University Press.

Bartholomaeus Anglicus, *De proprietatibus rerum, Liber XVII*. Edited by Iolanda Ventura. 2007. Turnhout: Brepols.

Bartholomaeus Mini de Senis (Pseudo-), *Tractatus de herbis (MS London, BL, Egerton 747)*. Edited by Iolanda Ventura. 2009. (Edizione Nazionale La Scuola Medica Salernitana, 5). Firenze: SISMEL-Edizioni del Galluzzo.

Baumann, Felix Andreas. 1974. *Das Erbario Carrarese und die Bildtradition des Tractatus de Herbis. Ein Beitrag zur Geschichte der Pflanzendarstellung im Übergang von Spätmittelalter zur Frührenaissance*. Bern: Benteli.

Bautier, Robert-Henri. 1970. "Les relations économiques des occidentaux avec les pays d'Orient, au Moyen Âge. Points de vue et documents." In Michel Mollat (ed.), *Sociétés et compagnies de commerce en Orient et dans l'Océan Indien: Actes du huitième Colloque international d'histoire maritime (Beyrouth: 5,10 septempre 1966)*. Paris: Sevpen.

Beccaria, Augusto. 1956. *I codici di medicina del periodo presalernitano: secoli IX, X e XI*. Roma: Storia e Letteratura.

Benn, James A. 2015. *Tea in China: A Religious and Cultural History*. Honolulu: University of Hawai'i Press.

Bernus-Taylor, Marthe (ed.). 1989. *Arabesques et jardins de Paradis*. Paris: Musée du Louvre.

Bernus-Taylor, Marthe (ed.). 2001. *L'Etrange et le merveilleux en terres d'Islam*. Paris: Musée du Louvre.

Bertiz, Agnes Acolos. 2003. *Picturing Health: The Garden and Courtiers at Play in the Late Fourteenth-Century Illuminated Tacuinum Sanitatis*. Unpublished PhD thesis. University of Southern California.

Betson, Thomas, John Adams, and Stuart Forbes (eds.). 2014. *The Syon Abbey Herbal: The Last Monastic Herbal in Britain c.* AD *1517*. Surrey: AMCD.

Bhattacharya, B. C. 1974. *The Jaina Iconography*. Delhi: Motilal Banarsidass.

Biewer, Klaus. 1992. *Albertus Magnus "De vegetabilibus" VI. 2: lateinisch-deutsch. Übersetzung und Kommentar* (Quellen und Studien zur Geschichte der Pharmazie 62). Stuttgart: Wissenschaftliche Verlagsgesellschaft.

Bolens, Lucie. 1997. "Sciences humaines et histoire de l'alimentation: conservation des aliments et associations des saveurs culturelles (de l'Andalousie à la Suisse Romande)." In Simonetta Cavaciocchi (ed.), *L'alimentazione e nutrizione secc. XIII–XVIII: Atti della ventottesima settimana di studi, 22–27 aprile 1996*. Grassina, Bagno a Ripoli: Le Monnier.

Book of Eparch. Edited by Johannes Koder. 1991. (Corpus Fontium Historiae Byzantinae, Series Vindobonensis 33). Vienna: Österreichische Akademie der Wissenschaften.

Bottiglieri, Corinna. 2018. "Silvatico, Matteo." *Dizionario biografico degli Italiani*, 92. Available online: www.treccani.it (accessed June 30, 2019).

Boulnois, Lucette. 1963. *La route de la soie*. Paris: Arthaud.

Branfoot, Crispin. 2017. "Making Space for the Sacred: Hindu Art and Material Religion." In Greg Bailey (ed.), *Hinduism in India: The Early Period*, 179–207. New Delhi: SAGE Publications.

Bray, Francesca. 1984. *Science and Civilisation in China*, vol. 6: *Biology and Biological Technology*. Part II: Agriculture. Cambridge: Cambridge University Press.

Bray, Francesca. 1989. "Essence and Utility: The Classification of Crop Plants in China." *Chinese Science*, 9: 1–13.

Bray, Francesca. 1997. *Technology and Gender: Fabrics of Power in Late Imperial China*. Berkeley: University of California Press.

Bṛhatsaṁhitā (aka *Bṛhat Saṁhitā* of Varāhamihira). Translated by M. Ramakrishna Bhat. 1981. Delhi: Motilal Banarsidass.

Buberl, Paul. 1937. *Die byzantinischen Handschriften*. 1. *Der Wiener Dioskurides und die Wiener Genesis* (BeschreibendesVerzeichnis der illuminierten Handschriften in Österreich, Neue Folge, 4). Leipzig: Karl W. Hiersemann.

Buddhacarita (aka *Buddha-Karita or Life of Buddha by Asvaghosa*). Edited by Edward B. Cowell 1894 (repr. 1997). Delhi: Cosmo Publications.

Buddhacarita (aka *Buddhacarita or Acts of the Buddha* by Aśvaghosa). Edited by E. H. Johnstone. 1936 (repr./new enlarged addition 1984). Delhi: Motilal Banarsidass.

Büntgen, Ulf, et al. 2016. "Cooling and societal change during the Late Antique Little Ice Age from 536 to around 660 AD." *Nature Geoscience*, 9: 231–6.

Campbell, A. 1891. *Santal Folk Tales*. Pokhuria: Santal Mission Press.

Cato, *De agricultura*. Edited by William Davis Hooper and Harrison Boyd Ash. 1934. *Cato, Varro. On Agriculture* (Loeb Classical Library 283). Cambridge, MA: Harvard University Press.

Čejka, Jakub. 1999. "Plants in Kāvya Poetry: Problems with Plant-names." *Pandanus '98: Flowers, Nature, Semiotics & Kavya and Sangam*. Edited by Jaroslav Vacek and Blanka Knotková-Čapková, 1: 167–78.

Chaunu, Pierre. 1969. *L'expansion européenne du XIIIe au XVe siècle*. Paris: Presses Universitaires de France.

Chen, Huaiyu. 2009. "A Buddhist Classification of Plants and Animals in Early Tang China." *Journal of Asian History*, 43(1): 31–51.

Chen, Weiji. 1984. *Zhongguo fangzhi kexue jishu shi* (*A History of Textile Production Technology in China*). Beijing: Kexue chubanshe.

CII (*Corpus Inscriptionum Indicarum*) vol. 2, part 2 (aka *Bhārhut Inscriptions*). B13. 1963. Edited by Heinrich Lüders, revised by Ernst Waldschmidt and M. A. Mehendale. Ootacamund: Archaeological Survey of India, 82–3.

CII (*Corpus Inscriptionum Indicarum*) vol. 3 (aka *Inscriptions of the Early Gupta Kings and their Successors*). No. 17, Gangadhar Inscription. 1888 (rev. 1970). Edited by John F. Fleet, revised by A. K. Narain. Calcutta: Superintendent of Government Printing, 77–8.

CII (*Corpus Inscriptionum Indicarum*) vol. 3 (aka *Inscriptions of the Early Gupta Kings and their Successors*). No. 18, Mandasor Inscription. 1888 (rev. 1970). Edited by John F. Fleet, revised by A. K. Narain. Calcutta: Superintendent of Government Printing, 79–88.

Clunas, Craig. 1996. *Fruitful Sites: Garden Culture in Ming Dynasty China*. Durham, NC: Duke University Press.

Codex Badianus. Martin de la Cruz and Juan Badiano. 1552. *Libellus de medicinalibus Indorum herbis quem quidam Indus, Collegii Sanctae Crucis Medicus, composuit, nullis rationibus doctor, sed solis experimentis edoctus Anno Domini Servatoris 1552* ... Formerly Vatican City, Biblioteca Apostolica Vaticana, MS Barberini Latinus 241; since 1990 Mexico, DF, Biblioteca Nacional de Antropologia y Historia.

Cogliati Arano, Luisa. 1976. *The Medieval Health Handbook Tacuinum Sanitatis*. Translated by Oscar Ratti and Adele Westbrook. New York: Braziller.

Collins, Minta, and Sandra Raphael. 2003. *A Medieval Herbal: A Facsimile of British Library Egerton MS 747*. London: British Library.

Collins, Minta. 2000. *Medieval Herbals: The Illustrative Traditions*. Toronto: University of Toronto Press, and London: British Museum.

Columella, *De agricultura*. Edited by Harrison Boyd Ash. 1941–55. *Columella On Agriculture*, 3 vols (Loeb Classical Library 361, 407. 408). Cambridge, MA: Harvard University Press.

Coomaraswamy, Ananda K. 1931. *Yakṣas*. Washington, DC: Freer Gallery, Smithsonian Institution.

Corazzo, Nina. 1996. "The Garden Enclosed (*Hortus Conclusus*) and its Vegetation as Sign." *Semiotics*: 132–43.

Corradini, Piero. 1996. "La Cina." In Valerio Castronovo (ed.), *Storia della economia mondiale*, vol. 1. Roma: Laterza.

Cosmas Indicopleustes, *Topography*. Edited by John Watson McCrindle. 1897/2010. *The Christian Topography of Cosmas, an Egyptian Monk* (Hakluyt Society 98). London: Hakluyt Society (repr. Cambridge: Cambridge University Press).

Cox, Stephen, et al. 2016. "Chapter 78—Oral and Systemic Health Effects of Compulsive Areca Nut Use." In V. R. Preedy (ed.), *Neuropathology of Drug Addictions and Substance Misuse*, 785–93. San Diego: Academic Press. DOI: 10.1016/B978-0-12-800634-4.00078-0.

Crowe, Yolande. 2000. "Jeux de lotus dans la céramique du monde musulman: apparition et adaptation d'un motif chinois." In *Autour du lotus. Actes du colloque tenu au Musée Ariana le 30 octobre 1999*: 3–9, 17–22. Geneva: Musée Ariana & Fondation Amaverunt.

Curley, David L. 2003. "'Voluntary' Relationships and Royal Gifts of Pān in Mughal Bengal." In Stewart Gordon (ed.), *Robes of Honour: Khil'at in Pre-Colonial and Colonial India*, 50–79. New Delhi: Oxford University Press.

d'Andiran, Gérald (ed.). 2010. *Early Medicine: From the Body to the Stars*. Basel: Schwabe.

Daems, Willem F. 1993. *Nomina simplicium medicinarum ex synonymariis Medii Aevi collecta*. Leiden: Brill.

Dainelli, Giotto. 1960. *Missionari e mercadanti rivelatori dell'Asia nel Medioevo*. Torino: Utet.

Dalby, Andrew. 2003. *Flavours of Byzantium*. Totnes: Prospect Books.

Dale, Stephen F. 2009. "Silk Road, Cotton Road or ... Indo-Chinese Trade in Pre-European Times." *Modern Asian Studies*, 43(1): 79–88.

Dallapiccola, Anna L. 2003. *Hindu Myths*. Austin: University of Texas Press.

Dasgupta, Nupur. 2016. "Gardens in Ancient India: Concepts, Practices and Imaginations." *Puravritta: Journal of the Directorate of Archaeology and Museums, Government of West Bengal*, 1: 133–51.

Davidson, Alan (ed.). 2006. *The Oxford Companion to Food*. 2nd edn, edited by Tom Jaine. Oxford and New York: Oxford University Press.

De Karabacek, Iosephus (ed.). 1906. *De codicis Dioscuridei Aniciae Iulianae, nunc Vindobonensis Med. Gr. I historia, forma, scriptura, picturis*. Leiden: A. W. Sijthoff.

De la Cruz, Martín, *Libellus de medicinalibus Indorum herbis. Manuscrito azteca de 1552 según traducción latina de Juan Badiano*. 1991. 2 vols. México: Fondo de culta económica, Instituto mexicano del seguro social.

de Oviedo y Valdés, Gonzalo Fernández. 1526. *De la natural hystoria de las Indias*. Toledo: Remon de Petras.

De Pee, Christian. 2009. "Wards of Words: Textual Geographies and Urban Space in Song-Dynasty Luoyang, 960–1127." *Journal of the Economic and Social History of the Orient*, 52 (1): 85–116.

De Planhol, Xavier 2011. "Kariz i. Terminology." In *Encyclopædia Iranica*, XV/6, 564–5. Available online: http://www.iranicaonline.org/articles/kariz_1(accessed February 22, 2018).

De Sahagún, Bernardino. 1540–77. *Historia general de las cosas de la Nueva España*, 3 vols. Firenze, Biblioteca Medicea Laurenziana, MSS Palatino 218–220 (Facsimile edn.: México, D.F.: Secretaría de Gobernación, 1979).

Decker, Michael. 2009. "Plants and Progress: Rethingking the Islamic Agricultural Revolution." *Journal of World History*, 20(2): 187–206.

Degeorge, Gérard. 2010. *La Grande Mosquée des Omeyyades à Damas*. Paris: Imprimerie Nationale.

Dehejia, Vidya. 2009. *The Body Adorned: Sacred and Profane in Indian Art*. New York: Columbia University Press.

Delatte, Louis (ed.). 1942. *Textes latins et vieux français relatifs aux Cyranides*. Paris: Belles Lettres.

Demange, Françoise (ed.). 2006. *Les Perses Sassanides, Fastes d'un empire oublié (224–642)*, Paris; Findakly.

Dendle, Peter, and Alain Touwaide (eds.). 2008. *Health and Healing from the Medieval Garden*. Woodbridge: Boydell.

Dioscorides, *De materia medica*. Edited by Max Wellmann. 1906–14. *De materia medica libri quinque*, 3 vols. Berlin: Weidmann. English translation: Lily Y. Beck. 2005. *Dioscorides, De Materia Medica* (Altertumswissenschaftliche Texte und Studien 38). Hildesheim: Olms.

Dioscorides. 1970. *Codex Vindobonensis med. gr. 1 der Österreichischen Nationalbibliothek* (Codices selecti phototypice impressi, 12). Graz: Akademische Druck.- u. Verlagsanstalt.

Dols, Michael Walters. 1977. *The Black Death in the Middle East*. Princeton, NJ: Princeton University Press.

Doniger, Wendy. 2016. *Redeeming the Kamasutra*. Oxford: Oxford University Press.

Drège, Jean-Pierre. 1986. "L'analyse fibreuse des papiers et la datation des manuscrits de Dunhuang." *Journal Asiatique*, 274: 403–15.

Duby, Georges. 1962. *L'économie rurale et la vie des campagnes dans l'Occident médiéval: (France, Angleterre, Empire, IXe–XVe siècles): essays de synthèse et perspectives de recherches*. Paris: Aubier.

Duby, Georges. 1973. *Guerriers et paysans, VIIe–XIIe siècles: premier essor de l'économie européenne*. Paris: Gallimard.

Dyer, Christopher. 1996. "Il commercio nell'Europa medievale." In Valerio Castronovo (ed.), *Storia della economia mondiale*, vol. 1. Roma: Laterza.

Eamon, William. 1994. *Science and the Secrets of Nature: Books of Secrets in Medieval and Early Modern Culture*. Princeton: Princeton University Press.

EC (Epigraphia Carnatica) vol. 8, part 2 (aka *Inscriptions in the Shimoga District*). Sagar 86, Hero Stone Inscription. 1904. Edited by B. Lewis Rice. Bangalore: Mysore Government Central Press, 291.

Eck, Diana L. 1985. *Darśan: Seeing the Divine Image in India*. Delhi: Motilal Banarsidass.

Eck, Diana L. 2012. *India: A Sacred Geography*. New York: Harmony Books.

Egan, Ronald. 2006. *The Problem of Beauty: Aesthetic Thought and Pursuits in Northern Song Dynasty China*. Cambridge, MA: Harvard University Asia Center.

Egmond, Florike. 2017. *Eye for Detail: Images of Plants and Animals in Art and Science, 1500–1630*. London: Reaktion.

EI (Epigraphia Indica) vol. 5 (aka *Inscriptions at Managoli*). No. 18, Southern Kalachuri Inscription. 1898–9. Edited by John F. Fleet, 22–3.

EI (*Epigraphia Indica*) vol. 20. No. 12.46–54. 1929–30. Edited by R. S. Panchamukhi, 109–22.

Elkhadem, Hosam. 1990. "Le Taqwīm al-Ṣiḥḥa (*Tacuinum Sanitatis*) d'Ibn Buṭlān: Histoire du texte." *Acta Belgica Historiae Medicinae*, 3: 139–46.

Elvin, Mark. 2004. *The Retreat of the Elephants: An Environmental History of China*. New Haven: Yale University Press.

Enderlein, Volkmar, and Michael Meinecke. 1993. *Mschatta-Fassade*. Berlin: Mann.

Engelhardt, Ute. 2001. "Dietetics in Tang China and the First Extant Works of Materia Dietetica." In *Innovation in Chinese Medicine*, Elisabeth Hsu (ed.), 173–91. Cambridge: Cambridge University Press.

Ettinghausen, Richard. 1962. *Arab Painting*. Geneva: Skira.

Fagan, Brian. 2000. *The Little Ice Age: How Climate Made History 1300–1850*. New York: Basic Books.

Fagan, Brian. 2008. *The Great Warming: Climate Change and the Rise and Fall of Civilizations*. New York: Bloomsbury Press.

Feng, Guangping. 2016. "Tangdai Pingquan shanju zaipei zhiwu kao (Textual Research on Cultivated Plants in Pingquan Villa of Tang Dynasty)." *Zhongguo lishi dili luncong (Journal of Chinese Historical Geography)*, 31(1): 60–72.

Feng, Hanyong. 1957. "Tang Song shidai de zaochuanye (Shipbuilding in the Tang and Song Dynasties)." *Lishi jiaoxue* (*Teaching History*), 10: 10–14.

Feng, Jiren. 2007. "Bracketing Likened to Flowers, Branches and Foliage: Architectural Metaphors and Conceptualization in Tenth to Twelfth-Century China as Reflected in the *Yingzao Fashi*." *T'oung Pao*, 2nd Series, 93(4/5): 369–432.

Findlen, Paula. 1994. *Possessing Nature: Museums, Collecting, and Scientific Culture in Early Modern Italy*. Berkeley: University of California Press.

Fitzgerald, James L. 2003. "Review: The Many Voices of the Mahābhārata." *Journal of the American Oriental Society*, 123(4): 803–18.

Flandrin, Jean-Louis, and Massimo Montanari (eds.). 1999. *Food: A Culinary History from Antiquity to the Present*. Albert Sonnenfeld (English edn.), Clarissa Botsford (trans.). New York: Columbia University Press.

Flatt, Emma J. 2016. "Spices, Smells and Spells: The Use of Olfactory Substances in the Conjuring of Spirits." *South Asian Studies*, 32(1): 3–21. DOI: 10.1080/02666030.2016.1174400.

Fleming, John, and Hugh Honour. 1977. *Dictionary of the Decorative Arts*. New York : Harper & Row.

Fouchécour, Charles-Henri. 1969. *La description de la Nature dans la poésie lyrique persane du XIe siècle*. Paris: Klincksieck.

Fowler-Smith, L. 2018. "Adorning and Adoring: The Sacred Trees of India." *Journal for the Study of Religion, Nature & Culture*, 12(3): 261–84. DOI: 10.1558/jsrnc.33347.

Freeman, Michael. 1977. "Sung." In Kwang-chih Chang (ed.), *Food in Chinese Culture: Anthropological and Human Perspectives*, vol. 1. New Haven: Yale University Press.

Freedman, Paul. 2008/9. *Out of the East: Spices and the Medieval Imagination*. New Haven, CT, and London: Yale University Press.

Freedman, Paul. 2009. *Il gusto delle spezie nel Medioevo*. Italian translation. Bologna: il Mulino.

Fujiansheng Quanzhou haiwai jiaotongshi bowuguan (ed.). 2017. *Quanzhouwan Songdai haichuan fajue yu yanjiu* (*The Discovery and Study on the Sea Boat in the Song Dynasty from the Quanzhou Bay*). Beijing: Haiyang chubanshe.

Fumagalli, Vito. 1978. *Coloni e signori nell'Italia settentrionale: secoli VI–XI*. Bologna: Patron.

Fumagalli, Vito. 1993. *L'alba del Medioevo*. Bologna: il Mulino.

Galen, *De simplicium medicamentorum temperamentis et facultatibus*. Edited by Karl Gottlob Kühn. 1821–33. In *Claudii Galeni Opera Omnia* (Medicorum Graecorum Opera quae exstant 1–20), 22 vols. Lipsiae: C. Cnobloch, vol. 11, 379–892 (repr. Cambridge and New York: Cambridge University Press, 2011).

García González, Alejandro. (2008). "*Hermeneumata medicobotanica vetustiora*. Apuntes para una edición completa de los glosarios médico-botánicos altomedievales (siglos VIII–XI)." *Studi Medievali. Rivista della Fondazione Centro Italiano di Studi sull'Alto Medioevo*, serie II, 49(1): 119–39.
Gasparri, Stefano. 1996. "Tra antichità e medioevo: i modelli sociali ed economici dei popoli barbarici e il loro impatto con il mondo mediterraneo." In Valerio Castronovo (ed.), *Storia della economia mondiale*, vol. 1. Roma: Laterza.
Geoponica. Translated by Andrew Dalby. 2011. *Geoponika. Farm Work. A Modern Translation of The Roman and Byzantine Farming Handbook*. Totnes: Prospect Books.
Geoponica. Edited by Heinrich Beck. 1895. Leipzig: Teubner.
al-Ghāfīqī, Aḥmad ibn Muḥammad, *The Book of Simple Drugs*, Max Meyerhof and George P. Sobhy (eds. and trans.). 1932–40. *The Abridged Version of "The Book of Simple Drugs" of Ahmad Ibn Muhammad Al-Ghafiqi, by Gregorius Abu-l-Farag (Barhebraeus). Edited from the Only Two Known Manuscripts with an English Translation, Commentary and Indices* (The Egyptian University, The Faculty of Medicine, Publications, 4). Cairo: Egyptian University (reprinted as vols 51 and 57 [1996] in the series *Islamic Medicine* of the Publications of the Institute for the History of Arabic-Islamic Science. Frankfurt-am-Main: Johann Wolfgang Goethe University).
Ghirshman, Roman. 1951. *L'Iran des origines à l'Islam*. Paris: Payot.
Ghirshman, Roman. 1963. *Perse. Proto-Iraniens, Mèdes, Achéménides*. Paris: Gallimard.
Gies, Joseph, and Francis Gies. 1974. *Life in a Medieval Castle*. New York: Harper Collins.
Gies, Francis, and Joseph Gies. 1994. *Cathedral, Forge, and Waterwheel: Technology and Invention in the Middle Ages*. New York: Harper Collins.
Givens, Jean A. 2005. *Observation and Image-Making in Gothic Art*. Cambridge: Cambridge University Press.
Givens, Jean A. 2006. "Reading and Writing the Illustrated *Tractatus de herbis*, 1280–1526." In Jean A. Givens, Karen M. Reeds, and Alain Touwaide (eds.), *Visualizing Medieval Medicine and Natural History, 1200–1550* (AVISTA Studies in the History of Medieval Technology, Science and Art 5), 115–46. Aldershot and Burlington, VT: Ashgate.
Givens, Jean A., Karen M. Reeds, and Alain Touwaide (eds.). 2006. *Visualizing Medieval Medicine and Natural History, 1200–1550* (AVISTA Studies in the History of Medieval Technology, Science and Art 5). Aldershot and Burlington, VT: Ashgate.
Glaze, Florence Eliza. 2018. "Salerno's Lombard Prince: Johannes 'Abbas de Curte' as Medical Practitioner." *Early Science and Medicine*, 23: 177–216.
Golas, Peter J. 1980. "Rural China in the Song." *Journal of Asian Studies*, 39(2): 291–325.
Golombek, Lisa. 1993. "The *paysage* as Funerary Imagery in the Timurid Period." *Muqarnas*, 10: 241–52.
Gonda, Jan. 1975. "The Meaning of the Word Alaṃkāra." In J. Gonda (ed.), *Selected Studies*, Vol. II: *Sanskrit Word Studies*. Leiden: Brill.
Goody, Jack. 1993. "Garlands in India: The Marigold and the Jasmine." In *The Culture of Flowers*, 321–46. Cambridge: Cambridge University Press.
Gothein, Marie Luise. 1979. *A History of Garden Art*. New York: Hacher Art Book.
Gourou, Pierre. 1972. *La terre et l'homme en Extrême-Oriente*. Paris: Flammarion.
Grabar, Oleg. 1992. *The Mediation of Ornament*. Princeton, NJ: Princeton University Press.
Grabar, Oleg. 2006. *The Dome of the Rock*. Cambridge, MA: Harvard University Press.
Grabar, Oleg, and Richard Ettinghausen. 1987. *The Art and Architecture of Islam, 650–1250*. New Haven, CT: Yale University Press.

Grabar, Oleg, and Sheila Blair. 1980. *Epic Images and Contemporary History. The Illustrations of the Great Mongol Shanama*. Chicago, IL: University of Chicago Press.
Grafton, Anthony, and Nancy Siraisi (eds.). 1999. *Natural Particulars: Nature and the Disciplines in Renaissance Europe*. Cambridge, MA: MIT Press.
Granoff, Phyllis. 1998. *The Forest of Thieves and the Magic Garden: An Anthology of Medieval Jain Stories*. New Delhi: Penguin Books India.
Grendler, Paul F. 2002. *The Universities of the Italian Renaissance*. Baltimore, MD: Johns Hopkins University Press.
Grieve, Maude. 1931. *A Modern Herbal. The Medicinal, Culinary, Cosmetic and Economic Properties, Cultivation and Folklore of Herbs, Grasses, Fungi, Shrubs and Trees with All Their Modern Scientific Uses*. Edited by C. F. Leyel. London: Tiger Books.
Grimm, Jacob. 1854. *Deutsche Mythologie*, 2 vols. 3rd edn. Göttingen: Dieterich (English trans. James S. Stallybrass [trans.]. 1976. *Teutonic Mythology*, 4 vols. Gloucester, MA: Peter Smith).
Guha, Sumit. 2013. *Beyond Caste: Identity and Power in South Asia, Past and Present*. Leiden: Brill.
Guo, Qinghua. 1998. "*Yingzao Fashi:* Twelfth-Century Chinese Building Manual." *Architectural History*, 41: 1–13.
Gupta, Anil K. 2004. "Origin of Agriculture and Domestication of Plants and Animals Linked to Early Holocene Climate Amelioration." *Current Science*, 87(1): 54–9.
Haq, S. Nomanul. 2001. "Islam and Ecology: Toward Retrieval and Reconstruction." *Daedalus*, 130(4): 141–77.
Harappa.com. n.d.. Available online: https://www.harappa.com/search/site/pipal (accessed March 19, 2021).
Hardy, Frederick G., and Laurence M. V. Totelin. (2016). *Ancient Botany*. London: Routledge.
Hargett, James M. (trans.). 2010. *Treatises of the Supervisor and Guardian of the Cinnamon Sea*, Fan Chengda. Seattle: University of Washington Press.
Harivaṃśa (aka *The Harivaṃśa: Being the Khila or Supplement to the Mahābhārata*). Edited by P. L. Vaidya. 1969–71. Poona: Bhandarkar Oriental Research Institute.
Harivaṃśa (see Kalan, Kesu, and Miskin *c*. 1585).
Harper, Donald. 1986. "Flowers in T'ang Poetry: Pomegranate, Sea Pomegranate, and Mountain Pomegranate." *Journal of the American Oriental Society*, 106(1): 139–53.
Harrison, Freya, et al. 2015. "A 1,000-year-old Antimicrobial Remedy with Antistaphylococcal Activity." *mBio*, 6(4): e01129.
Harrison, R. K. 1956. "The Mandrake and the Ancient World." *The Evangelical Quarterly*, 28(2): 87–92.
Hartlieb, Johannes. 2010. *Kräuterbuch*. Edited by Gerold Hayer and Bernhard Schnell. Wiesbaden: Ludwig Reichert.
Hartwell, Robert M. 1982. "Demographic, Political, and Social Transformations of China, 750–1550." *Harvard Journal of Asiatic Studies*, 42(2): 365–442.
Hauser, Mark W., and V. Selvakumar. 2021. "Gardens of the Coromandel Coast: Landscape Considerations of Commercial Agriculture in Tamil Nadu, South India." *International Journal of Historical Archaeology*, 1–21. DOI:10.1007/S10761-021-00591-X.
Haw, Stephen G. 2017. "Tung Oil and Tong Trees." *Zeitschrift der Deutschen Morgenländischen Gesellschaft*, 167(1): 215–36.
Hazra, R. C. 1975. *Studies in the Purāṇic Records on Hindu Rites and Customs*. Delhi: Motilal Banarsidass.

Heitzman, James, and S. Rajagopal. 2004. "Urban Geography and Land Measurement in the Twelfth Century: The Case of Kanchipuram." *The Indian Economic & Social History Review*, 41(3): 237–68.

Helman-Ważny, Agnieszka. 2016. "More than Meets the Eye: Fibre and Paper Analysis of the Chinese Manuscripts from the Silk Roads." *STAR: Science & Technology of Archaeological Research*, 2(2): 127–40.

Henry of Huntingdon, *Anglicanus ortus, A Verse Herbal of the Twelfth Century*. Winston Black (ed.). 2012. London and Toronto: British Library and Toronto University Press.

Herlihy, David. 1997. *The Black Death and the Transformation of the West*. Cambridge, MA: Harvard University Press.

Hernández de Toledo, Francisco. 1607. *Index medicamentorum*. Mexico: Fernando Balli.

Hernández de Toledo, Francisco. 1615. *Quatro libros de la naturaleza, y virtudes de las plantas, y animales que estan recividos en el uso de Medicina en la Nueva España, y la Methodo, y correccion, y preparacion, que para administrallas se requiere ...* Mexico: Viuda de Diego Lopez Davalos.

Hernández de Toledo, Francisco. 1651. *Rerum medicinarum Novae Hispaniae Thesaurus seu Plantarum animalium mineralium mexicanorum historia ex Francisci Hernandez Novi Orbis Medici Primarii relationibus in ipsa Mexicana Urbe conscriptis a Nardo Antonio Recchio ... collecta ac in ordine digesta ...* Romae: Ex typographeio Vitalis Mascardi.

Herzfeld, Ernst. 1913. "Arabesque." *Encyclopaedia of Islam*, vol. 1, cols. 363–7. Leiden: Brill.

Herzfeld, Ernst. 1923. *Ausgrabungen von Samarra*, Band I, *Der Wandschmuck der Bauten von Samarra und Seine Ornamentik*. Berlin: Dietrich Reimer.

Heyd, Wilhelm von. [1885–6] 1967. *Histoire du commerce du Levant au Moyen-Âge*. Edited by Furcy Raynaud. Leipzig: Harrassowitz (repr. Amsterdam: Hakkert).

Hildegard of Bingen, *Causae et curae*. Edited by Paul Kaiser. 1903. *Causae et curae* (Bibliotheca Scriptorum Graecorum et Romanorum Teubneriana). Leipzig: B. G. Teubner (Edition and English trans. Mary Pawlik, and Patrick Madigan [trans.], Mary Palmquist, and John Kulas [eds]. 1994. *Book of Holistic Healing*. Collegeville, MN: Liturgical Press).

Hildegard of Bingen, *Causae et Curae*. Edited by Laurence Moulinier. 2003. Berlin: Akademie Verlag.

Hildegard of Bingen. 2010. *Physica. Liber subtilitatem diversarum naturarum creaturarum*. Edited by Rainer Hildebrandt and Thomas Gloning. 2010. Berlin and New York: De Gruyter.

Hildegard von Bingen, *Physica*. Edited by Jacques-Paul Migne and Adalberto G. Hamman. 1855. "Physica. Subtilitatum diversarum naturarum creaturarum. Libri novem." In *Patrologiae cursus completus, Series Latina*, 197, cols. 1117–286. Paris: Migne.

Hiltebeitel, Alf. 2005. "Review: On Reading Fitzgerald's Vyāsa." *Journal of the American Oriental Society*, 125(2): 241–61.

Hippocrates, *De locis in homine*. Edited by Elizabeth M. Craik. 1998. *Hippocrates, Places in man. Greek Text and Translation with Introduction and Commentary*. Oxford and New York: Clarendon.

Hobhouse, Penelope. 2004. *The Gardens of Persia*. Carlsbad, CA: Kales.

Hoeniger, Cathleen. 2006. "The Illuminated *Tacuinum sanitatis* Manuscripts from Northern Italy ca. 1380–1400: Sources, Patrons, and the Creation of a New Pictorial Genre." In Jean A. Givens, Karen M. Reeds, and Alain Touwaide (eds.), *Visualizing Medieval Medicine and Natural History, 1200–1550* (AVISTA Studies in the History of Medieval Technology, Science and Art 5), 51–82. Aldershot and Burlington, VT: Ashgate.

Homer, *Iliad*. Edited by Augustus T. Murray. 1924. *The Iliad*, vol. 1 (Books 1–12) (Loeb Classical Library 170). London: W. Heinemann, and New York: Putnam.

Horn, Walter, and Ernest Born. 1979. *The Plan of St. Gall*, 3 vols. Berkeley, CA: University of California Press.

Horowitz, Maryanne Cline. 1998. *Seeds of Virtue and Knowledge*. Princeton, NJ: Princeton University Press.

Hu, Shiu-Ying. 1961. "The Economic Botany of the Paulownias." *Economic Botany*, 15(1): 11–27.

Huang, Chunyan. 2017. *Zaochuanye shiyu xia de Songdai shehui* (Society in Song China: A Perspective of Shipbuilding). Shanghai: Shanghai renmin chubanshe.

Hunt, Tony (ed.). 2017. *An Anglo-Norman Pharmacopoeia (Oxford, Bodleian Library MS Bodley 761)*. Oxford: Anglo-Norman Society.

Husain, Ali Akbar. 2000. *Scent in the Islamic Garden: A Study of Literary Sources in Persian and Urdu*. 2nd edn. Karachi: Oxford University Press.

Ibn al-Baytar, *Kitāb al-Jāmiʿ li-Mufradāt al-Adwiya wa-l-Aghdhiya*. Edited by Ibrahim Ben Mrad. 1990. *Ibn al-Baytar (m. 646 H./1248 J.C.). Commentaire de la "Materia Medica" de Dioscoride*. Tunis: Académie Tunisienne des Sciences des Lettres et des Arts Beït al-Hikma.

Ibn al-Baytar, *Kitāb al-Jāmiʿ li-Mufradāt al-Adwiya wa-l-Aghdhiya*. Edited by Ana María Cabo González. 2002. *Ibn Al-Baytar al-Malaqi (m. 646–1248), Kitab al-Yami li-mufradat al-adwiya wa-l-agdiya, Colección de Medicamentos y Alimentos. Introducción, edición crítica, traducción e índices de las letras sad y dad*. Santander: Mergablum Edición y Comunicación.

Ibn al-Baytar, *Kitāb al-Jāmiʿ li-Mufradāt al-Adwiya wa-l-Aghdhiya*: Lucien Leclerc (trans.). 1877–83. *ibn al Beithar, Traité des simples*, 3 vols. (Notices et extraits des manuscrits de la Bibliothèque nationale et autres bibliothèques 23, 25, 26). Paris: Imprimerie nationale (repr. Paris: Institut du Monde Arabe, 1992, and Frankfurt am Main: Institut für Geschichte der arabisch-islamischen Wissenschaften, Johann Wolfgang Goethe-Universität, 1996).

Ibn Battuta, *The Travels of Ibn Battuta, AD 1325–1354*. Translated by H. A. R. Gibb. 1958. Cambridge: Cambridge University Press.

Ibn Butlan. Edited by Hosam Elkhadem. 1990. *Le 'Taqwim' al-Sihha (Tacuini Sanitatis) d'Ibn Butlan: un traité médical du XIe siècle. Histoire du Texte, Edition Critique, Traduction, Commentaire*. Lovanii: Peeters.

Ibn Sina, *Qanun*. Translated by Laleh Bakhtiar. 2012. *The Canon of Medicine (al-Qanun fi'l-tibb) (The Law of Natural Healing)*, vol. 2: *Natural Pharmaceuticals. Avicenna (Abu ʿAli al-Husayn ibn ʿAbd Allah ibn Sina)*. Chicago, IL: Great Books of the Islamic World.

[Ibn-Qaiyim al-Ǧauzīya], Muḥammad Ibn-Abī-Bakr. [1292–1350]. *Medicine of the Prophet*. Translated by P. Johnstone. 1998. Cambridge: Islamic Texts Society.

Inden, Ronald. 2006. "The Ceremony of the Great Gift." In *Text and Practice: Essays in South Asian History*, 89–101. Oxford: Oxford University Press.

Jacobs, Joseph (ed.). 1892. *Indian Fairy Tales*. New York: Dover Publications, Inc.

Jamison, Stephanie, and Joel Brereton. 2020. *The Rigveda: A Guide*. Oxford: Oxford University Press.

Janaki, S. n.d. *Gajamukhane Ganapathiye*. Available online: https://www.dailymotion.com/video/x1e67y5 (accessed April 7, 2021).

Jones, Owain, and Paul Cloke. 2008. "Non-human Agencies: Trees in Place and Time." In C. Knappett and L. Malafouris (eds.), *Material Agency: Towards a Non-Anthropocentric Approach*, 79–96. Berlin: Springer.

Jones, Peter M. 2006. "Image, Word, and Medicine in the Middle Ages." In Jean A. Givens, Karen M. Reeds, and Alain Touwaide (eds.), *Visualizing Medieval Medicine and Natural History, 1200–1550* (AVISTA Studies in the History of Medieval Technology, Science and Art 5), 1–25. Aldershot and Burlington, VT: Ashgate.

Josephus, Flavius, *Bellum Judaicum*. Edited by Henry St. John Thackeray. 1927. *The Jewish War*, 3 vols (Loeb Classical Library 203, 210, 487). Cambridge, MA: Harvard University Press, and London: W. Heinemann.

Journal for the Study of Religions, Nature and Culture. 2010. 4(2).

Justinian, *Digest of Civil Law*. Edited by Samuel Parsons Scott. 1932. Cincinnati: Central Trust Company.

Kalan, Kesu, and Miskin. *c.* 1585 *Krishna and the Golden City of Dwarka from the Harivamsha (Geneology of Vishnu)*. Available online: https://asia.si.edu/object/F1954.6/ (accessed March 19, 2021).

Kālidāsa (see *Mālavikāgnimitram*).

Kāmasūtra (aka *Kāmasūtram*. Sanskrit text). Edited by Goswami D. Shastri. 1929. Banares: Jaikrishnadas and Haridas Gupta.

Kāmasūtra (aka *Vatsyayana's Kama Sutra: A Complete and Unexpurgated Version*). Translated by Richard F. Burton and F. F. Arbuthnot. 1963. London: Luxor.

Kāmasūtra (see Doniger 2016).

Kanga, Maneck Fardunji. 1988. "Barsom." *Encyclopaedia Iranica*, vol. 3, fascicle 8, 825–7. Available online: http://www.iranicaonline.org/articles/barsom-av (accessed February 22, 2018).

Karnataka Forest Department and A. N. Yellapa Reddy. 1988. *Sacred Plants: A Book on Vratas, Gardens and the Connected Plants Described in our Vedas and Puranas*. Dharwad: Karnataka Forest Department.

Kazhdan, Alexander P. 1996. "Bisanzio." In Valerio Castronovo (ed.), *Storia della economia mondiale*, vol. 1. Roma: Laterza.

[Khusrau], *Nuh Siphr* (aka Khusraw Dīhlavī, Amir. *Nuh Sipihr, 3. India as Seen by Amir Khusrau in 1318 AD*). Translated by R. Nath and F. Gwaliari. 1981. Jaipur: Historical Research Documentation Programme.

Kiilerich, Bente. 2001. "The Image of Anicia Juliana in the *Vienna Dioscorides*: Flattery or Appropriation of Imperial Imagery?" *Symbolae Osloenses*, 76: 169–90.

Al-Kindi, *Aqrābādhīn*. Edited by Martin Levey. 1966. *Al-Kindi, The Medical Formulary or* Aqrābādhīn *of Al-Kindi*. Madison: University of Wisconsin Press.

Koder, Johannes. 1992. "Die Drogisten und ihre Waren im Eparchenbuch (EB 10.1)." *Scripta Classica, Mediaevalia et Archaeologica Sibirica*, 4: 220–36.

Komaroff, Linda, and Stefano Carboni. 2002. *The Legacy of Genghis Khan. Courtly Art and Culture in Western Asia, 1256–1353*. New York: The Metropolitan Museum of Art.

Komaroff, Linda. 2010. "Arabesque." *Encyclopaedia of Islam*, vol. 3. Edited by Kate Fleet, Gudrun Krämer, Denis Matringe, John Nawas, and Everett Rowson. DOI:10.1163/1573-3912_ei3_COM_23685.

König, Eberhard. 2009. "The BnF *Tacuinum* from the Library of Eberhard im Bart." In Alain Touwaide (ed.), *Tacuinum Sanitatis*, 75–146. Barcelona: Moleiro.

Krishna, Nanditha, and M. Amirthalingam. 2014. *Sacred Plants of India*. Gurgaon, Haryana: Penguin Books India.

Kuhn, Dieter. 1988. *Science and Civilisation in China*, vol. 5. Chemistry and Chemical Technology. Part 9. Textile Technology: Spinning and Reeling. Cambridge: Cambridge University Press.

Kühnel, Ernst. 1949. *Die Arabeske: Sinn und Wandlung eines Ornaments*. Wiesbaden: Dietrichsche Verlagsbuchhandlung.

Kumar-Dumas, Divya. 2021. "The Experience of Early Designed Landscape in South Asia: Sigiriya, Sri Lanka and Mamallapuram, India." PhD thesis, University of Pennsylvania.

Kusukawa, Sachiko. 2012. *Picturing the Book of Nature: Image, Text, and Argument in Sixteenth-Century Human Anatomy and Medical Botany*. Chicago, IL: University of Chicago Press.

Kyle, Sarah. 2017. *Medicine and Humanism in Late Medieval Italy: The Carrara Herbal in Padua* (Medicine in the Medieval Mediterranean 7). New York and London: Routledge.

Lagardère, Vincent. 1994. "Canne à sucre et sucriers en el Al-Andalus au Moyen Âge (VIIIᵉ–XVᵉ siècles)." In Expiración García Sanchez (ed.), *Ciencias de la naturaleza en al-Andalus: textos y estudios*, vol. 3: m 337–359. Granada: Consejo Superior de Investigaciones Científicas.

"*Lakshmi*," *(pre-79* CE*) Held in the Secret Cabinet*. n.d. Available online: https://www.museoarcheologiconapoli.it/en/room-and-sections-of-the-exhibition/3490-2/ (accessed March 19, 2021).

Laiou, Angeliki E. 2002. "Exchange and Trade, Seventh-Twelfth Centuries." In Angeliki E. Laiou (ed.), *The Economic History of Byzantium: from the Seventh through the Fifteenth Century*, vol. 2: 697–770. Washington, DC: Dumbarton Oaks Research Library and Collection.

Lalitavistara Sūtra (aka *The Lalitavistara*). Translated by R. Mitra. 1881. Calcutta: Baptist Mission Press.

Lambourn, Elizabeth 2004. "Carving and Communities. Marble Carving for Muslim Patrons at Khambat and around the Indian Ocean Rim, Late Thirteenth-Mid-Fifteenth Centuries." *Ars Orientalis*, 34: 99–133.

Lanfranchi, Giovanni B. 2006. "Prestiti alimentari nelle tradizioni greca e romana." In Massimo Montanari and Françoise Sabban (eds.), *Storia e geografia dell'alimentazione. Risorse, scambi, consumi*, vol. 1. Torino: Utet.

Laurioux, Bruno. 2005. "Les épices dans l'alimentation vegetale." In Bruno Laurioux (ed.), *Une histoire culinaire du Moyen Âge*. Paris: Honoré Champion.

Lentz, Thomas W., and Glenn D. Lowry. 1989. *Timur and the Princely Vision. Persian Art and Culture in the Fifteenth Century*. Los Angeles, CA: Los Angeles County Museum of Art.

Lev, Efraim, and Zohar Amar. 2008. *Practical Materia Medica of the Medieval Eastern Mediterranean According to the Cairo Genizah*. Leiden and Boston, MA: Brill.

Li, Fan. 1984. *Zhongguo zaipei zhiwu fazhan shi* (*A History of the Development of Cultivated Plants in China*). Beijing: Kexue chubanshe.

Li, Hui-lin. 1969. "The Vegetables of Ancient China." *Economic Botany*, 23(3): 253–60.

Li, Hui-lin. 1974a. "The Origin and Use of Cannabis in Eastern Asia Linguistic-Cultural Implications." *Economic Botany*, 28(3): 293–301.

Li, Hui-lin. 1974b. "An Archaeological and Historical Account of Cannabis in China." *Economic Botany*, 28(4): 437–48.

Li, Hui-lin. 1977. "Hallucinogenic Plants in Chinese Herbals." *Botanical Museum Leaflets*, (Harvard University), 25(6): 161–81.

Liu, Changzhi. 1994. "Zhongguo chuantong zhixue de fazhan" (The Development of Chinese Traditional Biology). In *Zhongguo zhixue xuehui* (*Chinese Association for Plant Science*) (ed.), *Zhongguo zhiwu xue shi* (*A History of Plant Science in China*), 35–59. Beijing: Kexue chubanshe.

Liu, Fei-hu, Hua-ren Hu, Gang-hui Du, Gang Deng, and Yang Yang. 2017. "Ethnobotanical Research on Origin, Cultivation, Distribution and Utilization of Hemp (Cannabis sativa L.) in China." *Indian Journal of Traditional Knowledge*, 16(2): 235–42.

Liu, Fei-Hu, Xia Chen, Bo Long, Rui-Yan Shuai, and Chun-Lin Long. 2011. "Historical and Botanical Evidence of Distribution, Cultivation and Utilization of *Linum usitatissimum L.* (flax) in China." *Vegetation History and Archaeobotany*, 20(6): 561–6.

Liu, William Guanglin. 2015. *The Chinese Market Economy, 1000–1500*. Albany: State University of New York Press.

Liu, Xianbo, and Xiong Jun. 2017. *Tangdai jiaju yanjiu* (*A Study on Tang Furniture*). Beijing: Renmin chubanshe.

Liu, Yonghua. 2002. *Zhongguo gudai cheyu maju* (*Carts, Carriages, and Horse Saddles in Traditional China*). Shanghai: Shanghai cishu chubanshe.

Lopez, Robert S. 1966. *La nascita dell'Europa: secoli V–XIV*. Torino: Einaudi.

Lopez, Robert S. 1976. *The Commercial Revolution of the Middle Ages, 950–1350*. Cambridge: Cambridge University Press.

Lopez, Robert S., and Irving W. Raymond. 1955. *Medieval Trade in the Mediterranean World: Illustrative Documents Translated with Introductions and Notes*. New York: Columbia University Press.

Lotus Sutra (see Reeves 2008 and Teiser and Stone 2009).

Loukonine, Vladimir, and Anatoly Ivanov. 1995. *L'art persan*. Bournemouth: Parkstone/Aurora.

Luo, Guihuan. 2002. "Biology in Song China." In Chan Alan Kam Leung, Clancey Gregory K, and Loy Hui-chieh (eds.), *Historical Perspectives On East Asian Science, Technology And Medicine*, 171–81. Singapore: World Scientific.

Luo, Guihuan. 2015. "Biology." In Lu Yongxiang (ed.), *A History of Chinese Science and Technology*, vol. 1, 431–95. Shanghai: Shanghai Jiaotong University Press, Berlin and Heidelberg: Springer.

Mabberley, David J. 2017. *Mabberley's Plant-book: A Portable Dictionary of Plants, their Classification and Uses*. 4th edn. Cambridge: Cambridge University Press. DOI: 10.1017/9781316335581.

MacDougall, Elisabeth (ed.). 1986. *Medieval Gardens* (Dumbarton Oaks Colloquium on the History of Landscape Architecture 9). Washington, DC: Dumbarton Oaks Research Library and Collection.

MacKenzie, D. Neil. 1989. "Bundahishn." *Encyclopaedia Iranica*, vol. 4, fascicle 5, 547–51. Available online: http://www.iranicaonline.org/articles/bundahisn-primal-creation (accessed February 22, 2018).

Mahābhārata (aka *The Mahābhārata for the First Time Critically Edited*). Translated by V. S. Sukthankar. 1933. Poona: Bhandarkar Oriental Research Institute.

Mahābhārata (see Fitzgerald 2003 and Hiltebeitel 2005).

Mair, Victor H., and Erling Hoh. 2009. *The True History of Tea*. New York and London: Thames & Hudson.

Majumdar, Girija P. 1927. *Vanaspati: Plants and Plant-life as in Indian Treatises and Traditions*. Calcutta: University of Calcutta.

Mālavikāgnimitram (aka *Mālavikāgnimitram of* Kālidāsa). Translated by C. R. Devadhar. 1966. Delhi: Motilal Banarsidass.

Martin, Luther, et al. 1988. *Technologies of the Self: A Seminar with Michel Foucault*. Amherst: University of Massachusetts Press.

Matossian, Mary Kilbourne. 1989. *Poisons of the Past: Molds, Epidemics, and History*. New Haven, CT: Yale University Press.

Mayamata (aka *Mayamatam: Treatise of Housing, Architecture and Iconography*) vol. 1. Translated by B. Dagens. 1994 (repr. 2000) Delhi: Indira Gandhi National Centre for the Arts and Motilal Banarsidass.

Mayer, Johannes G., and Konrad Goehl (eds.). 2001. *Höhepunkte der Klostermedizin. Der "Macer floridus" und das Herbarium des Vitus Auslasser*. Holzminden: Reprint Verlag.

Mayer, Johannes G., Konrad Goehl, and Katharina Englert. 2009. *Die Pflanzen der Klostermedizin in Darstellung und Anwendung*. Baden-Baden: Deutscher Wissenschafts-Verlag.

Mazal, Otto. 1981. *Pflanzen, Wurzeln, Säfte, Samen. Antike Heilkunst in Miniaturen des Wiener Dioskurides*. Graz: Akademische Druck- u. Verlagsanstalt.

McCormick, Michael. 2001. *Origins of the European Economy: Communications and Commerce, A.D. 300–900*. Cambridge: Cambridge University Press.

McDermott, Joseph P. 2006. *A Social History of the Chinese Book: Books and Literati Culture in Late Imperial China*. Hong Kong: University of Hong Kong Press.

McHugh, James. 2011. "The Incense Trees of the Land of Emeralds: The Exotic Material Culture of Kāmaśāstra." *Journal of Indian Philosophy*, 39(1): 63–100. DOI: 10.1007/s10781-010-9118-4.

McHugh, James. 2012. *Sandalwood and Carrion: Smell in Indian Religion and Culture*. New York: Oxford University Press.

McLaughlin, Raoul. 2010. *Rome and the Distant East. Trade Routes to Ancient Lands of Arabia, India and China*. London and New York: Continuum/Bloomsbury.

Meister, Michael W. 2007. "Early Architecture and Its Transformations: New Evidence for Vernacular Origins for the Indian Temple." In A. Hardy (ed.), *The Temple in South* Asia. *Volume II of the Proceedings of the 18th Conference of the European Association of South Asian Archaeologists, London, 2005*, 1–19. London: British Association for South Asian Studies London.

Meister, Michael W. 2009. "Exploring Kāfirkot: When is a Rose Apple not a Rose?" *Pakistan Heritage, Research Journal, Department of Archaeology*, 1: 109–28.

Mélikian-Chirvani, A.S. 1984. "Le Shah-nama, la gnose soufie et le pouvoir mongol", *Journal Asiatique*, 272 (3–4): 249–337.

Melis, Giorgio. 1984. "Odorico nella Cina del Trecento: itinerario e realtà sociale secondo la 'Relatio'." In Giorgio Melis (ed.), *Odorico da Pordenone e la Cina: Atti del Convegno storico internazionale: Pordenone, 28–29 maggio 1982*. Pordenone: Concordia sette.

Mendelsohn, Loren D. 2013. "The Tacuinum Sanitatis: A Medieval Health Manual." CUNY Academic Works. Available online: https://academicworks.cuny.edu/cgi/viewcontent.cgi?article=1204&context=cc_pubs (accessed March 18, 2021).

Meyer, Ernst F. H. 1856. *Geschichte der Botanik*, vol. 3. Königsberg: Gebrüder Bornträger.

Meyerhof, Max. 1940–1. "Études de pharmacologie arabe tirées de manuscrits inédits. III. Deux manuscrits illustrés du '*Livre des simples*' d'Aḥmad al-Ġāfiqī." *Bulletin de l'Institut d'Égypte*, 23: 13–29.

Miller, James Innes. 1969. *The Spice Trade of the Roman Empire: 29 B.C. to A.D. 641*. Oxford: Clarendon Press.

Mines, Diane P. 2008. "Waiting for Veḷḷāḷkaṇṭaṉ: Narrative, Movement, and Making Place in a Tamil Village." In M. A. Selby and I. V. Peterson (eds.), *Tamil Geographies: Cultural Constructions of Space and Place in South India*, 199–219. Albany: SUNY Press.

Misra, R. N. 1981. *Yaksha Cult and Iconography*. New Delhi: Munshiram Manoharlal Pvt. Ltd.

Moldenke, Alma. L., and Harold N. Moldenke. 1952. *Plants of the Bible*. Waltham, MA: Chronica Botanica Company.

Montanari, Massimo. 1984. *Campagne medievali. Strutture produttive, rapporti di lavoro, sistemi alimentari*. Torino: Einaudi.

Montanari, Massimo. 1988. *Alimentazione e cultura nel Medioevo*. Roma: Laterza.

Montanari, Massimo. 1990. "Vegetazione e alimentazione." In Centro italiano di studi sull'Alto Medioevo (ed.), *L'ambiente vegetale nell'Alto Medioevo: settimane di studio del Centro italiano di studi sull'Alto Medioevo: 30 marzo–5 aprile 1989*, vol. 1. Spoleto: Centro italiano di studi sull'Alto Medioevo.

Montanari, Massimo. 1996. "L'agricoltura medievale." In Valerio Castronovo (ed.), *Storia della economia mondiale*, vol. 1. Roma: Laterza.

Montanari, Massimo. 2015. *Medieval Tastes: Food, Cooking, and the Table*. Translated by Beth Archer Brombert. New York: Columbia University Press.

Morony, Michael G. 1988. "Bahar-e Kesra." *Encyclopaedia Iranica*, vol. 3, fascicle 5, 479. Available online: http://www.iranicaonline.org/articles/bahar-e-kesra-the-spring-of-kosrow-tabari-fars-e-zamestani-winter-carpet-balami-or-baharestan-spring-gar (accessed February 22, 2018).

Morton, Andrew G. (1981). *History of Botanical Science: An Account of the Development of Botany from Ancient Times to the Present Day*. London: Academic Press.

Mudge, Ken, Jules Janick, Steven Scofield, and Eliezer E. Goldschmidt. 2009. "A History of Grafting." *Horticultural Reviews*, 35: 437–93.

Narayanan, Vasudha. 2001. "Water, Wood, and Wisdom: Ecological Perspectives from the Hindu Traditions." *Daedalus*, 130(4): 179–206.

Nash, Elizabeth G. 1995. *The Hansa*. New York: Barnes and Noble.

Needham, Joseph. 1986. *Science and Civilisation in China*, vol. 6: Biology and Biological Technology. Part I: Botany. With the collaboration of Lu Gwei-Djen and a special contribution by Huang Hsing-tsung. Cambridge: Cambridge University Press.

Nene, Y. L. 2012. "Relevance of Vrikshayurveda in Agriculture Today: A Tele-interview." *Asian Agri-History*, 16(4): 423–9.

Ni'matnamah (aka *The Ni'matnama Manuscript of the Sultans of Mandu: The Sultan's Book of Delights*). Translated by Norah M. Titley. 2004. London: Routledge.

Niederer, Monica (ed.). 2005. *Der St. Galler* Botanicus, *ein frühmittelalterliches Herbar. Kritische Edition, Übersetzung und Kommentar*. Bern: Peter Lang.

Nugteren, Albertina. 2005. *Belief, Bounty, and Beauty: Rituals around Sacred Trees in India*. Leiden: Brill.

Nuh Siphr (see Khusrau).

Nutton, Vivian. 1985. "John Caius and the Eton Galen: Medical Philology in the Renaissance." *Medizenhistorisches Journal*, 20: 227–52.

Nutton, Vivian. 1988. "Prisci Dissectionum Professores: Greek Texts and Renaissance Anatomists." In Anna Carlotta Dionisotti, Anthony Grafton, and Jill Kraye (eds.), *The Uses of Greek and Latin: Historical Essays*, 111–16. London: Warburg Institute.

Nutton, Vivian. 1997. "The Rise of Medical Humanism: Ferrara, 1464–1555." *Renaissance Studies*, 11: 2–19.

Nutton, Vivian. 2004. "Galen of Pergamum." In Hubert Cancik, Helmuth Schneider, and Christine F. Salazar (eds.), *Brill's New Pauly: Encyclopaedia of the Ancient World*, vol. 5, cols. 654–61. Leiden: Brill.

Nutton, Vivian. 2013. *Ancient Medicine*. 2nd edn. (Sciences of Antiquity Series). London and New York: Routledge.

Odoric of Pordenone. *De rebus incognitis*. Edited by Luigi Pontico Virunio. 1513. Pesaro: Girolamo Soncino.

Ogilvie, Brian. 2006. *The Science of Describing: Natural History in Renaissance Europe*. Chicago, IL: University of Chicago Press.

Olivelle, Patrick (trans.). 2013. *King, Governance, and Law in Ancient India: Kauṭilya's Arthaśāstra*. New York: Oxford University Press.
Olson, Glending. 1982. *Literature as Recreation in the Later Middle Ages*. Ithaca, NY: Cornell University Press.
Opsomer, Carmélia. 1989. *Index de la pharmacopée du Ier au Xe siècle*, 2 vols. Hildesheim: Olms.
Orme, Nicholas. 1982. "A Grammatical Miscellany of 1427–1465 from Bristol and Wiltshire." *Traditio*, 38: 301–26.
Orta, Garcia da. 1563. *Coloquios dos Simples e drogas da India*. Goa: Johannes de Endem.
Oslan, Lea. 2003. "Charms and Prayers in Medieval Medical Theory and Practice." *Social History of Medicine*, 16(3): 343–66.
Ovid, *Fasti*. Edited by James G. Frazer. 1931. *Fasti* (Loeb Classical Library 253). Cambridge, MA: Harvard University Press, and London: W. Heinemann.
Pächt, Otto. 1950. "Early Italian Nature Studies and the Early Calendar Landscape." *Journal of the Warburg and Courtauld Institutes*, 13: 13–47.
Pandanus. 1999–2015. Edited by Jaroslav Vacek. Charles University, Faculty of the Arts, Prague: Triton. Available online: http://iu.ff.cuni.cz/pandanus/publications (accessed March 19, 2021).
Pandanus Database of Plants. 1998. Available online: http://iu.ff.cuni.cz/pandanus/database/ (accessed March 19, 2021).
Parijat Tree in Kintoor, UP. n.d. Available online: https://barabanki.nic.in/tourist-place/parijaat/ (accessed March 19, 2021).
Park, Katharine. 2013. "Medical Practice." In David C. Lindberg and Michael H. Shank (eds.), *The Cambridge History of Science*, vol. 2: *Medieval Science*, 611–29. Cambridge: Cambridge University Press.
Park, Katharine, and Lorraine Daston. 2001. *Wonders and the Order of Nature, 1150–1750*. New York: Zone.
Parker, Grant. 2011. *The Making of Roman India*. Cambridge: Cambridge University Press.
Patlagean, Evelyne. 1993. "Byzance et les marchés du grand commerce, vers 830–vers 1030. Entre Pirenne et Polanyi." In Centro italiano di studi dell'Alto Medioevo (ed.), *Mercati e mercanti nell'Alto Medioevo: l'area euroasiatica e l'area mediterranea: settimane di studio del Centro italiano di studi dell'Alto Medioevo: 23–29 aprile 1992*. Spoleto: Centro studi dell'Alto Medioevo.
Pearson, Anne M. 1983. *A Study of the Purāṇic Vratas*. MA thesis, McMaster University, Ontario, Canada.
Pearson, Anne M. 1996. *Because It Gives Me Peace of Mind: Ritual Fasts in the Religious Lives of Hindu Women*. Albany: SUNY Press.
Peduto, Vito A. 2001. "La radice di mandragora nel 'Dioscorides Viennese'." *Minerva Anestesiologica*, 67(10): 751–66.
Pelliot, Paul. 1921. "Notes sur les anciens itinéraires dans l'Orient roman." *Journal Asiatique*, 17(11): 139–45.
Petrus Hispanus, *Thesaurus pauperum*. Edited by Maria Helena da Rocha Pereira. 1973. *Las obras medicas de Pedro Hispano*. Coimbra: Universidade.
Pini, Antonio Ivan. 1990. "Vite e olivo nell'Alto Medioevo." In Centro italiano di studi sull'Alto Medioevo (ed.), *L'ambiente vegetale nell'Alto Medioevo: settimane di studio del Centro italiano di studi sull'Alto Medioevo: 30 marzo–5 aprile 1989*, vol. 1. Spoleto: Centro italiano di studi sull'Alto Medioevo.
Pliny the Elder, *Naturalis Historia*. Edited and translated by Harris Rackham, William H. S. Jones, Alfred C. Andrews, and David E. Eichholz. 1938–63. *Natural History*, 10 vols (Loeb

Classical Library 330, 352–3, 370–1, 392–4, 418–19). Cambridge, MA: Harvard University Press, and London: W. Heinemann.

Pollock, Sheldon. 1996. "The Sanskrit Cosmopolis, 300–1300: Transculturation, Vernacularization, and the Question of Ideology." In Jan E. M. Houben (ed.), *Ideology and Status of Sanskrit: Contributions to the History of the Sanskrit Language*, 197–247. New York: Brill.

Pollock, Sheldon. 2009. *The Language of the Gods in the World of Men*. Berkeley: University of California Press.

Polo, Marco. 1559. *Secondo volume delle Navigationi et viaggi nel quale si contengono l'historia delle cose dei Tartari, et diversi fatti de loro imperatori descritta da M. Marco Polo gentilhuomo venetiano*: Giovanni Battista Ramusio (ed.). 1583. *Navigationi et viaggi*, vol. 2. Venetia: Giunti.

Pope, Arthur Upham (ed.). 1971. *A Survey of Persian Art: from Prehistoric Times to the Present*. London: Oxford University Press.

Porter, Venetia. 1995. *Islamic Tiles*. London: British Museum.

Porter, Yves. 1992. *Peinture et arts du livre. Essai sur la littérature technique indo-persane*. Tehran: Institut Français de Recherche en Iran.

Porter, Yves. 2000. "From the 'Theory of the Two *qalam*-s' to the 'Seven Principles of Painting' Theory, Terminology and Practice of Persian Classical Painting." *Muqarnas*, 17(2000): 109–18.

Porter, Yves. 2004. "Jardins du Paradis et 'tapis-jardins': les métaphores croisées." In Sāzmān-i Mīrāṯ-i Farhangī va Gardishgarī-i Kishvar (ed.). *Tisser le paradis, Tapis-jardins persans*, 24–9. Tehran: Service de Coopération et d'Action Culturelle de l'Ambassade de France en Iran, and Clermont-Ferrand: Ville de Clermont-Ferrand.

Porter, Yves. 2007a. "Beauté." In Mohammad Ali Amir-Moezzi (ed.), *Dictionnaire du Coran*, 125–7. Paris: Laffont.

Porter, Yves. 2007b. "Paradis." In Mohammad Ali Amir-Moezzi (ed.), *Dictionnaire du Coran*, 638–41. Paris: Laffont.

Porter, Yves. 2016. "Lotus Flowers and Leaves, from China and Iran to the Indian Sultanates." In Monique Burési and Eloïse Brac De La Perrière (eds.), *Le Coran de Gwalior. Polysémie d'un manuscrit à peintures* (Orient & Méditerranée 19), 171–89. Paris: de Boccard.

Porter, Yves. 2016b. "Jardins persans." In *Jardins d'Orient, de l'Alhambra au Taj Mahal*, 73–9. Paris: Institut du Monde Arabe.

Porter, Yves, and Gérard Degeorge. 2001. *L'art de la céramique dans l'architecture musulmane*. Paris: Flammarion.

Premerstein, Anton von. 1903. "Anicia Juliana im Wiener Dioskorides-Kodex." *Jahrbuch der Kunsthistorischen Sammlungen des Allerhöchsten Kaiserhauses*, 24: 105–24.

Prodromic Poems. Edited by Dirk Christiaan Hesseling and Hubert Octave Pernot. 1910. *Poèmes Prodromiques en grec vulgaire* (Verhandelingen der Koninklijke Akademie van Wetenschappen te Amsterdam, Afdeeling Letterkunde, Nieuwe Reeks, Deel XI, no. 1). Amsterdam: J. Müller.

Pseudo-Apuleius, *Herbarius*. Edited by Henry Sigerist and Ernst Howald. 1927. *Antonii Musae De herba vettonica liber; Pseudo-Apulei Herbarius; Anonymi De taxone liber; Sexti Placiti Liber medicinae ex animalibus etc* (Corpus Medicorum Latinorum 4), 13–225. Leipzig: Teubner.

Pseudo-Aristote, *Du monde. Positions et dénominations des vents*. Edited by Michel Federspiel and Jean-Pierre Levet. 2019. *Des plantes*. Introduit, traduit et commenté par Michel Federspiel, Paris: Belles Lettres.

Radden, Jennifer. 2000. *The Nature of Melancholy: From Aristotle to Kristeva*. Oxford and New York: Oxford University Press.

Ramanujan, A. K. 2011. *Poems of Love and War: From the Eight Anthologies and the Ten Long Poems of Classical Tamil*. Columbia University Press.

Rawson, Jessica. 1984. *Chinese Ornament: The Lotus and the Dragon*. London: British Museum.

Reeds, Karen M. 1976. "Medical Humanism and Botany." *Annals of Science*, 33: 519–42.

Reeds, Karen M., and Tomomi Kinukawa. 2013. "Medieval Natural History." In David C. Lindberg and Michael H. Shank (eds.), *The Cambridge History of Science*, vol. 2: *Medieval Science*, 569–89. Cambridge: Cambridge University Press.

Reeves, Gene. (trans.). 2008. *The Lotus Sutra: A Contemporary Translation of a Buddhist Classic*. Boston: Wisdom Publications.

Ṛgveda (aka *Ṛgveda-saṁhitā: Sanskrit text, English translation, and notes according to the translation of H. H. Wilson and Bhāṣya of Sāyaṇācārya*.) Translated by H. H. Wilson, R. P. Arya, and K. Jośī. 2001. Delhi: Parimal.

Ṛgveda (see Jamison and Brereton 2020).

Rhodes, J. T., and Clifford Davidson. 1994. "The Garden of Paradise." In Clifford Davidson (ed.), *The Iconography of Heaven*, 69–109. Kalamazoo, MI: Medieval Institute Publications.

Riddle, John M. 1981. "'Pseudo-Dioscorides' *Ex Herbis Femininis* and Early Medieval Medical Botany." *Journal of the History of Biology*, 14: 43–82.

Riddle, John M. 1986. *Dioscorides on Pharmacy and Medicine* (History of Science Series 3). Austin, TX: Texas University Press.

Riddle, John M. 2010. *Goddesses, Elixirs, and Witches: Plants and Sexuality throughout Human History*. New York: Palgrave MacMillan.

Riera-Melis, Antoni. 2002. "Il Mediterraneo, crogiuolo di tradizioni alimentari. Il lascito islamico alla cucina catalana medievale." In Massimo Montanari (ed.), *Il mondo in cucina: storia, identità, scambi*. Roma: Laterza.

Riera Melis, Antoni, and Maria Soler Sala. 2016. "La distribución y los mercados de alimentos en el Mediterráneo occidental (siglos VIII–XII)." In Fondazione Centro italiano di studi sull'Alto Medioevo (ed.), *L'alimentazione nell'Alto Medioevo: pratiche, simboli, ideologie: settimane di studio della Fondazione Centro italiano di studi sull'Alto Medioevo: Spoleto, 9–14 aprile 2015*, vol. 1. Spoleto: Fondazione centro italiano di studi sull'Alto Medioevo.

Robinson, Cynthia. 2006. "Trees of Love, Trees of Knowledge: Toward the Definition of a Cross-Confessional Current in Late Medieval Iberian Spirituality." *Medieval Encounters*, 12(3): 388–435.

Rocher, Ludo. 1986. *The Purāṇas*. Wiesbaden: Otto Harrassowitz Verlag.

Rong, Xinjiang. 2013. *Eighteen Lectures on Dunhuang*. Translated by Imre Galambos. Leiden: Brill.

Rosenberger, Bernard. 1997. "La cucina araba e il suo apporto alla cucina europea." In Jean-Louis Flandrin and Massimo Montanari (eds.), *Storia dell'alimentazione*. Roma: Laterza.

Roth, Nicolas. 2018. "Poppies and Peacocks, Jasmine and Jackfruit: Garden Images and Horticultural Knowledge in the Literatures of Mughal India, 1600–1800." *Journal of South Asian Intellectual History*, 1(1): 48–78.

Ruddiman, William F. 2003. "The Anthropogenic Greenhouse Era Began Thousands of Years Ago." *Climatic Change*, 61(3): 261–93.

Ruggiadi, Martina. 2010. "Marble from the Palace of Mas'ud III in Ghazni." In Pierfrancesco Callieri and Luca Colliva (eds.), *South Asian Archaeology 2007. Proceedings of the 19th Meeting of the European Association of South Asian Archaeology in Ravenna*, Italy, July (BAR International Series 2133), 297–306.

Ruggles, D. Fairchild. 2008. *Islamic Gardens and Landscapes*. Philadelphia: University of Pennsylvania Press.

Sabban, Françoise. 2002. "La cucina cosmopolita dell'imperatore della Cina del XIV secolo. Un nuovo approccio." In Massimo Montanari (ed.), *Il mondo in cucina: storia, identità, scambi*. Roma: Laterza.

Sabban, Françoise. 2006a. "La Cina dei Song." In Françoise Sabban and Massimo Montanari (eds.), *Storia e geografia dell'alimentazione. Risorse, scambi, consumi*, vol. 1, 212–30. Torino: Utet.

Sabban, Françoise. 2006b. "La Cina fra I Mongoli e Ming." In Françoise Sabban and Massimo Montanari (eds.), *Storia e geografia dell'alimentazione. Risorse, scambi, consumi*, vol. 1, 303–19. Torino: Utet.

Sadhale, Nalini, and Y. L. Nene. 2009. "Ancient Indian Traditional and Scientific Thought on Plants: Sir JC Bose and Vrikshayurveda." *Asian Agri–History Journal*, 13(2): 101–12.

Sarton, George, and Lynn Thorndike. 1928. "Tacuinum." *Isis*, 10: 489–93.

Satari, Sri Soejatmi. 2008. "Ancient Gardens and Hindu-Buddhist Architecture in Java." In Elisabeth A. Bacus, Ian C. Glover, and Peter D. Sharrock (eds.), *Interpreting Southeast Asia's Past: Monument, Image and Text. Selected Papers from the 10th International Conference of the European Association of Southeast Asian Archaeologists*, 122–32. Singapore: NUS Press.

Scarborough, John. 2006. "Drugs and Drug Lore in the Time of Theophrastus: Folklore, Magic, Botany, Philosophy and the Rootcutters." *Acta Classica*, 69: 1–29.

Scenes from the Life of the Buddha (late second–early third century CE). Available online: https://asia.si.edu/object/F1949.9a-d/ (accessed March 19, 2021).

Schafer, Edward H. 1963. *The Golden Peaches of Samarkand: A Study of T'ang Exotics*. Berkeley: University of California Press.

Schafer, Edward. 1977. "T'ang." In Kwang-chih Chang (ed.), *Food in Chinese Culture: Anthropological and Historical Perspectives*, vol. 1. New Haven, CT: Yale University Press.

Scharfe, Hartmut. 1993. *Investigations in Kauṭalya's Manual of Political Science*. Wiesbaden: Otto Harrassowitz Verlag.

Schedl, Barbara. 2014. *Der Plan von St. Gallen. Ein Modell europäischer Klosterkultur*. Wien: Böhlau.

Schmitt, Charles B. 1971. "Theophrastus." In Paul O. Kristeller and F. Edward Cranz (eds.), *Catalogus Translationum et Commentariorum*, vol. 2, 239–322. Washington, DC: Catholic University of America Press.

Schmitz, Barbara. 1997. *Islamic and Indian Manuscripts and Paintings in the Pierpont Morgan Library*. New York: The Pierpont Morgan Library.

Schnell, Bernhard, and William Crossgrove (eds.). 2003. *Der deutsche* Macer, *Vulgatfassung. Mit einem Abdruck des lateinischen Macer Floridus* De viribus herbarum. Tübingen: Niemeyer.

Schnell, Bernhard. 2018. *Das "Debrecener Pflanzen- und Tierbuch." Die illustrierte deutsche Ps.-Apuleius Handschrift Debrecen R 459: Faksimile, Edition, Übersetzung und Kommentare*. Wiesbaden: Ludwig Reichert.

ScienceDirect: Areca Nut. n.d. Available online: https://www.sciencedirect.com/topics/neuroscience/areca-nut (accessed March 19, 2021).

ScienceDirect: Betel. n.d. Available online: https://www.sciencedirect.com/topics/neuroscience/betel (accessed March 19, 2021).

Scully. Terence. 1995. *The Art of Cookery in the Middle Ages*. Woodbridge: The Boydell Press.

Segre Rutz, Vera. 2000. "Il *Tacuinum Sanitatis* di Verde Visconti e la miniatura milanese di fine Trecento." *Arte Cristiana*, 88: 375–90.

Segre Rutz, Vera. 2002. "*Historia Plantarum*. Erbe, oro e medicina nei codici medievali." In Vera Segre Rutz, Ennio Lazzarini, Mauro Di Vito, and Silvia Panini (eds.), *Historia plantarum: Ms 459 Biblioteca Casanatense, l'enciclopedia medica dell'imperatore Venceslao*, vol. 2: *Commento*, 11–202. Modena: Franco Cosimo Panini.

Sekar, Uma S. 2012. "Five Landscapes: The 'Tinai' of 'Kuruntogai'." *Journal of Landscape Architecture (India)*, 34: 84–93.

Serapiom, *El libro agregà*. Edited by Gustav Ineichen. 1962–6. *El libro agregà de Serapiom, volgarizzamento di frater Jacobus Philippus de Padua*, 2 vols. Venice and Rome: Istituto per la collaborazione culturale.

Shah, Behula. 2007. "Braj: The Creation of Krishna's Landscape of Power and Pleasure and its Sixteenth-Century Construction through the Pilgrimage of the Groves." In Michel Conan (ed.), *Sacred Gardens and Landscapes: Ritual and Agency*, 153–71. Washington, DC: Dumbarton Oaks Research Library and Collection (Dumbarton Oaks Colloquium on the History of Landscape Architecture, 26).

Shah, U.P. 1987. *Jaina-rūpa-maṇḍana*. Delhi: Abhinav Publications.

Shani, Raya. 1994. "Stucco Decoration in the Gunbad-i 'Alawiyyan at Hamadan." In R. Hillenbrand (ed.), *The Art of the Saljuqs in Iran and Anatolia*, 71–8. Costa Mesa: Mazda.

Shao, Xiaofeng. 2010. *Zhongguo Songdai jiaju: Yanjiu yu tuxiang jicheng* (*Furnitures in Song China: Studies and Collected Images*). Nanjing: Dongnan daxue chubanshe.

Shaw, Miranda. 2015. *Buddhist Goddesses of India*. Princeton, NJ: Princeton University Press.

Shen, Hsueh-man. 2006. *Gilded Splendor: Treasures of China's Liao Empire (907–1125)*. Milano: 5 Continents.

Sheng, Angela. 2013. "Determining the Value of Textiles in the Tang Dynasty: In Memory of Professor Denis Twitchett (1925–2006)." *Journal of the Royal Asiatic Society*, 3rd series, 23 (2): 175–95.

SII (*South Indian Inscriptions*) vol. 3. No. 205, The Tiruvalangadu Copper Plates of the Sixth Year of Rajendra-Chola I. 1929. Edited by H. K. Sastri, 383–439.

SII (*South Indian Inscriptions*) vol. 20 (aka *Bombay Karnataka Inscriptions*). No. 175, Hire Bevinur Inscription. 1965 repr. 1988. New Delhi: Archaeological Survey of India (ASI), 218–21.

Siraisi, Nancy. 1987a. *Avicenna in Renaissance Italy: The Canon and Medical Teaching in Italian Universities after 1500*. Princeton, NJ: Princeton University Press.

Siraisi, Nancy. 1987b. "The Physician's Task: Medical Reputations in Humanist Collective Biographies." In Nancy G. Siraisi and Andrew C. Crombie (eds.), *The Rational Arts of Living* (Smith College Studies in History 50), 105–33. Northampton, MA: Department of History, Smith College.

Siraisi, Nancy. 2007. *History, Medicine, and the Traditions of Renaissance Learning*. Ann Arbor, MI: University of Michigan Press.

Sircar, D. C. (ed.). 1957. *Inscriptions of Aśoka*. Calcutta: Government of India.

Sircar, D. C. (ed.). 1973. *Epigraphic Discoveries in East Pakistan. Sanskrit College Research Series*, 82. Calcutta: Sanskrit College.

Slicher Van Bath, and Bernard Hendrik. 1966. "Le climat et les récoltes en Haut Moyen Âge." In Centro italiano di studi sull'Alto Medioevo (ed.), *Agricoltura e mondo rurale in Occidente nell'Alto Medioevo: settimane di studio del Centro italiano di studi sull'Alto Medioevo, 22–28 aprile 1965*. Spoleto: Centro italiano di studi sull'Alto Medioevo.

Smith, Pamela H. 2006. "Art, Science, and Visual Culture in Early Modern Europe." *Isis*, 97: 83–100.

Soustiel, Jean, and Yves Porter. 2003. *Tombeaux de Paradis. Le Shâh-e Zende de Samarcande et la céramique architecturale d'Asie centrale*. Saint-Rémy-en-l'Eau: Monelle Hayot.

Springer, Lauren, and Rob Proctor. 2000. *Passionate Gardening: Good Advice for Challenging Climates*. Golden, CO: Fulcrum.

Squatriti, Paolo. 2014. "Of Seeds, Seasons, and Seas: Andrew Watson's Medieval Agrarian Revolution Forty Years Later." *Journal of Economic History*, 74 (4): 1205–20.

Sreenivasachar P. (ed.). 1940. *A Corpus of Inscriptions in Telingana Districts of H.E.H. the Nizam's Dominions*, Part 2.

Stannard, Jerry. 1962. "The Plant Called Moly." *Osiris*, 14: 254–307.

Stannard, Jerry. 1999a. *Herbs and Herbalism in the Middle Ages and Renaissance*. Aldershot and Burlington, VT: Ashgate.

Stannard, Jerry. 1999b. *Pristina Medicamenta. Ancient and Medieval Medical Botany*. Aldershot and Burlington, VT: Ashgate.

Stasik, Danuta, and Anna Trynkowsla. 2010. *The City and the Forest in Indian Literature and Art*. Warsaw: Dom Wydawniczy Elipsa.

Stathakopoulos, Dionysios Ch. 2004. *Famine and Pestilence in the Late Roman and Early Byzantine Empire: A Systematic Survey of Subsistence Crises and Epidemics* (Birmingham Byzantine and Ottoman Monographs 9). Aldershot and Burlington, VT: Ashgate.

Stathakopoulos, Dionysios Ch. 2007. "Between the Field and the Plate: How Agricultural Products were Processed into Food." In Leslie Brubaker and Kallirroe Linardou (eds.), *Eat, Drink, and Be Merry (Luke 12:19)—Food and Wine in Byzantium: Papers of the 37th Annual Symposium of Byzantine Studies, in Honour of Professor A. A. M. Bryer*, 27–38 (Society for the Promotion of Byzantine Studies Publications, vol. 13). Aldershot and Burlington, VT: Ashgate Variorum.

Statius, *Achilleid*. Edited by John Henry Mozley. 1928. *Statius II. Thebaid V–XII. Achilleid* (Loeb Classical Library 498), 508–95. London: W. Heinemann, and New York: Putnam.

Steinhardt, Nancy Shatzman. 2004. "The Tang Architectural Icon and the Politics of Chinese Architectural History." *The Art Bulletin*, 86(2): 228–54.

Stoffler, Hans-Dieter (ed.). 1996. *Der Hortulus des Walahfrid Strabo. Aus dem Kräutergarten des Klosters Reichenau*. Sigmaringen: Jan Thorbecke Verlag.

Stoll, Ulrich (ed.). 1992. *Das "Lorscher Arzneibuch". Ein medizinisches Kompendium des 8. Jahrhunderts (Codex bambergensis medicinalis 1). Text Übersetzung und Fachglossar* (Sudhoffs Archiv, Beiheft 28). Stuttgart: Franz Steiner.

Stothers, Richard B. 2000. "Climatic and Demographic Consequences of the Massive Volcanic Eruption of 1258." *Climatic Change*, 45(2): 361–74.

Studholme, Alexander. 2002. *The Origins of Oṃ Maṇipadme Hūṃ: A Study of the Kāraṇḍavyūha Sūtra*. Albany: State University of New York Press.

Subtelny, Maria E. 1993. "A Medieval Persian Agricultural Manual in Context: The Irshād al-Zirā'a a in Late Timurid and Early Safavid Khorasan." *Studia iranica* (Paris), 22(2): 167–217.

Sūrapāla (see *Vṛkṣāyurveda*).

Swamy, B. G. L. 1975. *Shāsanagaḷalli Giḍamaragaḷu* [*Kannada*]. Bangalore: Vishvavidyālaya.

Symeon Seth, *Syntagma de alimentorum facultatibus*. Edited by Bernhard Langkavel. 1868. Leizpig: B. G. Teubner.

Taillieu, Dieter. 2003. "Haoma." *Encyclopaedia Iranica*, vol. 11, fascicle 6, 659–62. Available online: http://www.iranicaonline.org/articles/haoma-i (accessed March 19, 2021).

Tangheroni, Marco. 1996. *Commercio e navigazione nel Medioevo*. Roma: Laterza.

Teiser, Stephen F. and Jacqueline I. Stone (eds.). 2009. *Readings of the Lotus Sutra*. New York: Columbia University Press.

Telelis, Ioannis G. 2005. "Historical-Climatological Information from the Time of the Byzantine Empire (4th–15th Centuries AD)." *History of Meteorology*, 2: 41–50.

Thapar, Romila. 1976. "Dana and Dakshina as Forms of Exchange." *Indica*, 13(1–2): 37–48.
Theophrastus, *Historia plantarum*. Edited by Arthur Hort. 1916–26. *Enquiry into Plants*, 2 vols (Loeb Classical Library 70, 79). Cambridge, MA: Harvard University Press, and London: W. Heinemann.
Thomas, John, and Angela Constantinides-Hero (eds). 2000. *Byzantine Monastic Foundation Documents*, 5 vols. Washington, DC: Dumbarton Oaks Research Library and Collection.
Tobyn, Graeme, Alison Denham, and Midge Whitelegg. 2016. *The Western Herbal Tradition: 2000 Years of Medicinal Plant Knowledge*. London and Philadelphia, PA: Singing Dragon.
Touwaide, Alain. 2006. "Latin Crusaders, Byzantine Herbals." In Jean A. Givens, Karen M. Reeds, and Alain Touwaide (eds.), *Visualizing Medieval Medicine and Natural History, 1200–1550* (AVISTA Studies in the History of Medieval Technology, Science and Art 5), 25–50. Aldershot and Burlington, VT: Ashgate.
Touwaide, Alain (ed.). 2009. *Tacuinum Sanitatis*. Barcelona: Moleiro.
Touwaide, Alain. 2010a. "Botany." In Albrecht Classen (ed.), *Handbook of Medieval Studies. Terms, Methods, Trends*, 3 vols, 145–81. Berlin and New York: Walter de Gruyter.
Touwaide, Alain. 2010b. "Byzantine Science." In Albrecht Classen (ed.), *Handbook of Medieval Studies. Terms, Methods, Trends*, 3 vols, 195–239. Berlin and New York: Walter de Gruyter.
Touwaide, Alain. 2010c. "Pharmacology, Pharmacy." In Albrecht Classen (ed.), *Handbook of Medieval Studies. Terms, Methods, Trends*, 3 vols, 1056–90. Berlin and New York: Walter de Gruyter.
Touwaide, Alain. 2010d. "Transfer of Knowledge and Translation." In Albrecht Classen (ed.), *Handbook of Medieval Studies. Terms, Methods, Trends*, 3 vols, 1368–99. Berlin and New York: Walter de Gruyter.
Touwaide, Alain. 2010e. "Pharmaceutical Literature." In Albrecht Classen (ed.), *Handbook of Medieval Studies. Terms, Methods, Trends*, 3 vols, 1979–2000. Berlin and New York: Walter de Gruyter.
Touwaide, Alain. 2011. "Arabic into Greek: The Rise of an International Lexicon of Medicine in the Medieval Eastern Mediterranean?" In Robert Wisnovsky, Faith Wallis, Jamie C. Fumo, and Carlos Fraenkel (eds.), *Vehicles of Transmission, Translation, and Transformation in Medieval Textual Culture* (Cursor Mundi 4), 195–222. Turnhout: Brepols.
Touwaide, Alain. 2013a. "L'importanza di Dioscoride nel Medioevo, con particolare riferimento al mondo arabo." In *De materia medica. Il Dioscoride di Napoli Ms. Ex-Vindobonense Greco 1 (sec. VI–VII)*, vol. 2. Sansepolcro: Aboca.
Touwaide, Alain. 2013b. *Tractatus de Herbis. Sloane Ms. 4016*. Barcelona: Moleiro.
Touwaide, Alain. 2015. "Medicine." In Albrecht Classen (ed.), *Handbook of Medieval Culture. Fundamental Aspects and Conditions of the European Middle Ages*, 3 vols, 954–98. Berlin and New York: Walter de Gruyter, vol. 2.
Trager, James. 1995. *The Food Chronology: A Food Lover's Compendium of Events and Anecdotes, from Prehistory to the Present*. New York: Henry Holt.
Trombert, Eric. 1997. "Des fleurs rouges en galette: Une plante tinctoriale dans la Chine ancienne: le carthame." *Journal Asiatique*, 285(2): 509–47.
Trombert, Eric. 2004. "Cooking, Dyeing, and Worship: The Uses of Safflower in Medieval China as Reflected in Dunhuang Documents." *Asia Major* 3rd series, 17(1): 59–72.
Tsinghua Daxue jianzhu sheji yanjiuyuan (Institute of Architectural Design at Tsinghua University), and Beijing Qinghua chengshi guihua sheji yanjiuyuan (Institute of Urban Planning and Design at Tsinghua, Beijing). 2011. *Foguangsi dongdadian jianzhu kancha yanjiu (An Examination and Study of the East Hall of Buddha's Light Temple)*. Beijing: Wenwu chubanshe.

Twitchett, Denis Crispin. 1979. *Sui and T'ang China, 589–906* (In Johns J. Fairbank and Denis Crispin Twitchett [eds.], *The Cambridge History of China*, vol. 3, Part i). Cambridge: Cambridge University Press.

Twitchett, Denis. 1983. *Printing and Publishing in Medieval China*. New York: Frederic C. Bell.

Unger, Richard W. 2007. *Beer in the Middle Ages and the Renaissance*. Philadelphia, PA: University of Pennsylvania Press.

Van Arsdall, Anne, and Timothy Graham (eds). 2012. *Herbs and Healers from the Ancient Mediterranean through the Medieval West: Essays in Honor of John M. Riddle* (Medicine in the Medieval Mediterranean). Aldershot and Burlington, VT: Ashgate.

Van Arsdall, Anne. 2009. "Exploring what was understood by 'Mandragora' in Anglo-Saxon England." In Pieter Bierbaumer and Helmut W. Klug (eds.), *Old Names—New Growth. Proceedings of the 2nd Anglo-Plant Name Survey Symposium, Graz, June 6–10, 2007*, 57–74. Frankfurt am Main: Lang.

Van Arsdall, Anne. 2002. *Medieval Herbal Remedies: The Old English Herbarium and Anglo-Saxon Medicine*. New York and London: Routledge.

Varey, Simon. 2000. *The Mexican Treasury. The Writings of Dr. Francisco Hernández*. Translated by Rafael Chabran, Cynthia L. Chamberlin, and Simon Varey. Stanford, CA: Stanford University Press.

Ventura, Iolanda. 2015. "Il *Circa instans* attribuito a Platearius: trasmissione manoscritta, redazioni, criteri di costruzione di un'edizione critica." *Revue d'Histoire des Textes*, 10: 249–362.

Verdon, Jean. 2004. *Travels in the Middle Ages*. Notre Dame, IN: University of Notre Dame Press.

Vesel, Živa, Sergei Tourkin, and Yves Porter. 2009. *Images of Islamic Science* (Bibliothèque Iranienne BI67). Tehran: Institut français de recherche en Iran.

Vinidarius, *Apici Excerpta*. Edited by Jacques André. 1987 (Collection des universités de France). Paris: Belles Lettres.

Voigts, Linda E. 1979. "Anglo-Saxon Plant Remedies and the Anglo-Saxons." *Isis*, 70: 250–68.

Von Glahn, Richard. 2010. "La storia economica della Cina imperiale dal 221 A.C. al 1850 D.C." In Maurizio Scarpari (ed.), *La Cina*, vol. 2, 117–216. Torino: Einaudi.

Von Glahn, Richard. 2016. *The Economic History of China: From Antiquity to the Nineteenth Century*. Cambridge: Cambridge University Press.

Vṛkṣāyurveda (aka *Surapala's Vrikshayurveda: the science of plant life by Surapala*). 1996. Translated by N. Sadhale. Secunderabad: Asian Agri-History Foundation.

Vṛkṣāyurveda (see Sadhale and Nene 2009 and Nene 2012).

Waghorne, Joanne P. 1992. "Dressing the Body of God: South Indian Bronze Sculpture in its Temple Setting." *Asian Art*, 5(3): 9–33.

Wagoner, Phillip B. 2012. "In Amīn Khan's Garden: Charitable Gardens in Qutb Shāhi Andhra." In Daud Ali and Emma J. Flatt (eds.), *Garden and Landscapes in Pre-colonial India: Histories from the Deccan*, 98–126. London: Routledge (Visual and Media Histories).

Walahfrid Strabo, *Hortulus*. Edited by Raef Payne. 1966. *Hortulus*. Pittsburgh, PA: Hunt Botanical Library.

Wallis, Faith (ed). 2010. *Medieval Medicine: A Reader*. Toronto: University of Toronto Press.

Walters, Terrence W. 1989. "Historical Overview on Domesticated Plants in China with Special Emphasis on the Cucurbitaceae." *Economic Botany*, 43(3): 297–313.

Wang, Binghua, and Helen Wang. 2013. "A Study of the Tang Dynasty Tax Textiles (Yongdiao Bu) from Turfan." *Journal of the Royal Asiatic Society* 3rd series, 23(2): 263–80.

Wang, Hongjun. 1987. "Tangdai de chaye shengchan" (The Tea Production in the Tang Dynasty). *Qilu xuekan* (*Journal of Qi Lu*), 6: 14–21.

Watson, Andrew. 1974. “The Arab Agricultural Revolution and Its Diffusion, 700–1100.” *Journal of Economic History*, 34: 8–35.
Watson, Andrew. 1981. “A Medieval Garden Revolution.” In Abraham Udovitch (ed.), *The Islamic Middle East, 700–1900*, 29–58. Princeton, NJ: Darwin Press.
Watson, Andrew. [1983] 2008. *Agricultural Innovation in the Early Islamic World: The Diffusion of Crops and Farming Techniques, 700–1100*. (Cambridge Studies in Islamic Civilization). Cambridge: Cambridge University Press.
Watson, Andrew M. 1998. *Innovaciones en la agricultura en los primeros tiempos del mundo islamico: difusión de los distintos cultivos y tecnicas agricolas del año 700 al 1000*. Ana Martina Vela (trans.). Granada: Editorial Universidad de Granada.
Watson, Oliver 1985. *Persian Lustre Wares*. London: Faber & Faber.
Weinstein, Laura Renee. Forthcoming. “The Indian Figurine from Pompeii as an Emblem of East-West Trade in the Early Roman Imperial Era'.” In Serena Autiero and Matthew A. Cobb (eds.), *Globalization and Transculturality from Antiquity to the Pre-Modern World*. London: Routledge.
Weisheipl, James A. (ed.). 1980. *Albertus Magnus and Sciences: Commemorative Essays 1980* (Studies and Texts). Toronto: Pontifical Institute of Mediaeval. Studies.
Williams, Steven J. 2003. *The Secret of Secrets: The Scholarly Career of a Pseudo-Aristotelian Text in the Latin Middle Ages*. Ann Arbor: University of Michigan Press.
Wojtilla, Gyula. 2006. *History of Kṛṣiśāstra: A History of Indian Literature on Traditional Agriculture*. Wiesbaden: Harrassowitz Verlag.
Yang, Xiaoshan. 2004a. “Li Gefei’s ‘Luoyang mingyuan ji’ (A Record of the Celebrated Gardens of Luoyang) Text and Context.” *Monumenta Serica*, 52: 221–55.
Yang, Xiaoshan. 2004b. “Li Deyu’s Pingquan Villa: Forming an Emblem from the Tang and to the Song.” *Asia Major* 3rd series, 17(2): 45–88.
Young, Gary K. 2001. *Rome’s Eastern Trade: International Commerce and Imperial Policy 31 BC–AD 305*. London: Routledge.
Yule, H., and A. C. Burnell. 1903 (rev. 2013). *Hobson-Jobson: The Definitive Glossary of British India*. Oxford: Oxford University Press.
Zangheri, Luigi, Brunella Lorenzi, and Nausikaa M. Rahmati. 2006. *Il giardino islamico*. Firenze: Olschki.
Zarcone, Thierry. 2005. “The Myth of Mandrake, the ‘Plant Human’.” *Diogenes*, 207: 115–29.
Zarra, Giuseppe (ed.). 2018. *Il volgarizzamento pisano del* Thesaurus pauperum. *Edizione critica, commento linguistico e glossario*. Berlin and Boston, MA: De Gruyter.
Zeng, Xiongsheng. 2015. “Agriculture.” In Lu Yongxiang (ed.), *A History of Chinese Science and Technology*, vol. 1, 351–429. Shanghai: Shanghai Jiaotong University Press, Berlin and Heidelberg: Springer.
Zhang, Ling. 2016. *The River, The Plain, and the State: An Environmental Drama in Northern Song China*. Cambridge: Cambridge University Press.
Ziegler, Joseph. 1998. *Medicine and Religion c. 1300*. Oxford: Clarendon.
Zohary, Michael. 1982. *Plants of the Bible*. Cambridge: Cambridge University Press.
Zurndorfer, Harriet T. 2011. “Cotton Textile Manufacture and Marketing in Late Imperial China and the ‘Great Divergence’.” *Journal of the Economic and Social History of the Orient*, 54(5): 701–38.

CONTRIBUTORS

Melitta Weiss Adamson, Professor of German and Comparative Literature at the University of Western Ontario, Canada, is a specialist in medieval cookery and preventive medicine. Her many publications include the books *Medieval Dietetics* (1995), *Food in the Middle Ages* (1995), *Daz buoch von guoter spise* (2000), *Regional Cuisines of Medieval Europe* (2002), *Food in Medieval Times* (2004), and with Francine Segan the two-volume encyclopedia *Entertaining from Ancient Rome to the Super Bowl* (2008).

Huaiyu Chen is an associate professor in the School of Historical, Philosophical and Religious Studies at Arizona State University, USA. He has numerous publications on Buddhism, religions on the Silk Road, and medieval Chinese cultural and social history. He has held fellowships from the Institute for Advanced Study in Princeton, Clare Hall of Cambridge University, and the Max Planck Institute for the History of Science in Berlin.

Tony Hunt is a Fellow of the British Academy. An Emeritus Fellow of St. Peter's College, Oxford, UK, he is the President of the Anglo-Norman Text Society. He has published a series of medico-botanical texts and collections, in Anglo-Norman and English, starting with *Plant Names of Medieval England* (1989) and *Popular Medicine in Thirteenth-Century England* (1990), followed by (with Michael Benskin) *Three Receptaria from Medieval England* (2001), and *An Old French Herbal* (2008).

Divya Kumar-Dumas is a lecturer in Art History, Theory, and Criticism at the Maryland Institute College of Art (MICA), USA. Her PhD in South Asia Studies from the University of Pennsylvania (2021) reconsiders sites in medieval South Asia as designed landscapes. Research areas include art, architecture, and archaeology, landscape and cultural history of pre-modern Asia, and relationships between South Asia and the world. She is a regional editor for the digital humanities project, *Gardens of the Roman Empire*, and a Visiting Research Scholar at ISAW, NYU.

Sarah R. Kyle is Professor and Chair of the Department of Art and Visual Culture at Iowa State University, USA. Her research focuses on illustrated botanical manuscripts as sites of convergence for pan-Mediterranean medical traditions, humanist enterprises, and artistic currents, particularly in the courts of northern Italy and in Venice during the fourteenth and fifteenth centuries. Her publications include *Medicine and Humanism in Late Medieval Italy: The Carrara Herbal in Padua* (2017).

Johannes Gottfried Mayer (d. 2019) devoted his scientific life to the technical prose of the Middle Ages and the history of medicinal plants in Europe. He taught medicinal ethics in Würzburg (JMU) and history of pharmacy in Erlangen (FAU), both in Germany, and co-founded a research group on monastic medicine in 1999. Convinced of the importance of developing awareness about the relevance of medicinal plants and their history among general audiences, he held multiple public lectures, repeatedly appeared in the media, and wrote numerous popular books on monastic and herbal medicine, among them *Das große Buch der Klosterheilkunde* (2013), *Meine Reise in die Welt der Gewürze* (2011), *Das geheime Heilwissen der Klosterfrauen* (2008), and *Handbuch der Klosterheilkunde* (2002).

Yves Porter works on the arts and techniques of the Iranian world and Islamized India. A specialist of Oriental languages (Persian, Turkish, Arabic, Pashtu, Hindi), he earned a PhD (Paris III, 1988) and a Habilitation à diriger des recherches (Paris IV, 2000). He has been teaching Islamic Art at Aix Marseille Université, France since 1993, where he is a member of the Laboratoire d'Archéologie Méditerranéenne Médiévale et Moderne. In October 2018, he joined the Institut Universitaire de France.

Federica Rotelli earned a Law degree (Bologna) and a PhD in Bioeconomics (Verona, 2007). After a study of the transformation of medicine and medical debates provoked by the introduction of foreign plants into European medicine in the sixteenth to eighteenth centuries, she is currently preparing a monograph on the transformation of Mediterranean environment through the introduction of Asian and American plants from the Middle Ages to the seventeenth century.

Alain Touwaide is a specialist in the history of botany, medicine, and medicinal plants in the Mediterranean world from antiquity to the Renaissance and beyond. He is particularly interested in the exchanges of plants, practices, and knowledge between the different populations around the Mediterranean, and has published abundantly on this topic, with a particular focus on Greece and Byzantium. Trained as a philologist, he interprets historical texts at the light of ethnobotany/ethnopharmacology.

Iolanda Ventura is an associate professor of Medieval Latin at the University of Bologna, Italy. She researches the history of medieval pharmacotherapy and the reception and manuscript dissemination of related texts. In 2007, she published a critical edition of Bartholomew the Englishman's *De proprietatibus rerum*, XVII (on plants) (2007) and, in 2009, of the *Tractatus de herbis*. She is currently preparing a critical edition of the Salernitan collection *Circa instans* and a book on pharmacy in a thirteenth–fourteenth-century university context.

INDEX

Note: Page locators in *italic* refer to figures or tables.